The Americ

The American Atom

A Documentary History of Nuclear Policies

from the Discovery of Fission to the Present

Second Edition

PHILIP L. CANTELON,
RICHARD G. HEWLETT,
AND ROBERT C. WILLIAMS
Editors

upp

UNIVERSITY OF PENNSYLVANIA PRESS
Philadelphia

Copyright © 1984 by the University of Pennsylvania Press
Second edition 1991

Library of Congress Cataloging in Publication Data

The American atom : a documentary history of nuclear policies from the discovery of fission to the present / Philip L. Cantelon, Richard G. Hewlett, and Robert C. Williams, editors—2nd ed.
     p.   cm.
    Includes index.
    ISBN 0-8122-3096-5 (cloth).—ISBN 0-8122-1354-8 (paper)
    1. United States—Military policy.   2. Nuclear weapons—United States—History—Sources. 3. Nuclear energy—United States—History—Sources.
4. Nuclear nonproliferation—History—Sources.   5. Deterrence (Strategy)—History—20th century—Sources.   I. Cantelon, Philip L. (Philip Louis). 1940–
II. Hewlett, Richard G.   III. Williams, Robert Chadwell, 1938–
    UA23.A597   1991
    355.02'17'0973—dc20   91-31676
                                       CIP

Printed in the United States of America

What concerns me is really not the technical problem. I am not sure the miserable thing will work, nor that it can be gotten to a target except by oxcart. It seems likely to me even further to worsen the unbalance of our war plans. What does worry me is that this thing appears to have caught the imagination, both of the Congressional and military people, as the answer to the problem posed by the Russians' advance. It would be folly to oppose the exploration of this weapon. We have always known it had to be done; and it does have to be done, though it appears to be singularly proof against any form of experimental approach. But that we become committed to it as the way to save the country and the peace appears to me full of dangers.

                    J. Robert Oppenheimer to James Conant
                    October 12, 1949

In 1950 our research group became part of a special institute. For the next eighteen years I found myself caught up in the rotation of a special world of military designers and inventors, special institutes, committees and learned councils, pilot plants and proving grounds. Every day I saw the huge material, intellectual, and nervous resources of thousands of people being poured into creating the means of total destruction, a force potentially capable of annihilating all human civilization.

                    Andrei Sakharov
                    1974

Contents

Contents

Contents

Contents

Introduction

Historical understanding is essential to an informed public that must live in a world of nuclear weapons and nuclear reactors because of decisions made years or even decades ago. The purpose of this volume is to tell the story of America's nuclear policy through its primary documents. Our hope is that a documentary collection will contribute historical evidence to a public concern with nuclear policy that is often more rhetorical than substantive.

Since 1979, when the SALT II treaty was withdrawn from Senate consideration and the Three Mile Island nuclear accident drew international attention, a public debate on nuclear policy without precedent since the mid-1950s has developed. A generation "born nuclear" and raised in a world of tens of thousands of nuclear warheads and hundreds of reactors is asking questions that had been the secret concern of only a few experts since 1945: should we proceed with the further development of nuclear weapons? Of nuclear reactors? Is the benefit worth the cost and the risk? Does deterrence work? Is arms control beneficial? Should America use nuclear weapons first? Are national technical means of verification reliable? Can you trust the Russians? Can a nuclear war be limited? Can it be won?

The early decisions to build nuclear and thermonuclear weapons were usually made in secret by nuclear and military specialists. Yet since 1979 decisions to install Cruise and Pershing II missiles in Europe, to build the MX missile, and to develop antimissile defense systems all have come under immediate concerned public scrutiny. Nuclear issues have assumed mythic proportions: nuclear war is the Apocalypse and Armageddon; nuclear power is a Faustian bargain; missiles are named after the gods Nike, Zeus, and Titan. What has often been missing is a knowledge of the facts.

After nearly fifty years the development of nuclear policy is certainly history, but its primary sources are scattered and often difficult to obtain, especially for the layman or student. All history involves selecting the significant documents in order to make sense of the past. Here we have focused on neither the technical calculations of nuclear war and its effects on familiar cities nor the endless proposals for a disarmed world. Rather, we have gathered together those documents that will best show how our present policies have evolved in a world that is neither engaged in a nuclear war nor disarmed, but living in the twilight zone of arms control and regulated nuclear power.

In selecting these historical documents we have applied three general tests: first, does the document help tell the story of the development of American nuclear policy in a nontechnical way; second, is the source primary rather than secondary, written by an actor in the drama rather than by a member of the audience; third, does the document provide coverage of the major chapters in the story?

In showing how the vision of a nuclear world preceded the reality by several decades, we eschewed the physics of Einstein and Bohr for the science fiction of H. G. Wells. We also included the correspondence of Leo Szilard, the dynamic Hungarian refugee who engineered the famous Einstein letter to Roosevelt and helped move America from neutrality to defense. Since the original contributions to the theory of the atom bomb were British, not American, we provided some of the early speculations of British nuclear physicists between 1940 and 1941. We also sought to show the reader how the nuclear age began in the minds of men living in a Europe at war.

The Manhattan Project was America's $2 billion secret project to build an atomic bomb. Many documents associated with the project have come to light only in recent years. In Section II we have used the letters of J. Robert Oppenheimer and the recently declassified minutes of policy committees to tell the story of how the bomb was designed and built and how the decision was made to drop the first uranium and plutonium devices on the Japanese cities of Hiroshima and Nagasaki in 1945.

How did a weapon of war become the key to a peacetime industry? In considering atomic energy after World War II, we focused in Section III on the legislative enabling acts that established the Atomic Energy Commission, the short-lived dream of international control of nuclear weapons under the Baruch Plan, and the "atoms for peace" program of President Dwight D. Eisenhower. By 1954 the highly classified work on nuclear weapons paralleled a new development of nuclear energy and power reactors. Knowledge was shared with both private industry and other countries. The fruits of this program are considered in the later section on nuclear power.

Until 1949 America was apparently secure in its monopoly of nuclear weapons. But the detonation of the first Soviet nuclear device, code named

"Joe I," in August 1949 precipitated a frenzied race to construct the most awesome nuclear weapons of our time, based on the thermonuclear or fusion principle. In Section IV on the hydrogen bomb, we have tried to show how the Soviet threat impelled America toward a crash program to build the "super" weapon, opposed by J. Robert Oppenheimer and the Atomic Energy Commission's General Advisory Committee but approved by President Harry S. Truman in 1950. We have included the recently declassified minutes of the General Advisory Committee meeting of October 30, 1949, and Hans Bethe's account of the race for the H-bomb.

Why do scientists build nuclear weapons? In Section V on the Oppenheimer hearing of 1954, we examine a case study of science and conscience versus national security; a courtroom drama that drew the attention of both the public and concerned scientists, many of them Manhattan Project veterans. The withdrawal of Oppenheimer's security clearance after a decade of service to his country—based not on his supposed security risk but on his moral opposition to the H-bomb—seemed to many a dangerous insult to the entire scientific community. In selecting testimony from "In the Matter of J. Robert Oppenheimer," we have tried to show how the Oppenheimer case brought to light the entire history of the Manhattan Project and dramatized the question of scientific conscience and loyalty to the state.

Nuclear testing was an essential part of the research and development of nuclear weapons, from the flats of Nevada to the sands of Enewetak, in the 1950s. Testing presented health hazards to a concerned public and raised fears of a nuclear war between the Soviet Union and the United States. For a time citizens sought security in a civil defense program of fallout shelters, but the dangers of worldwide nuclear testing continued. In 1963 a limited ban on atmospheric testing was effected by America and the Soviet Union, the first of a series of limited arms-control measures that were bilateral and outside the international framework of the United Nations. In Section VI we have selected government documents, statements reflecting public opinion, and treaties to indicate the level of concern over testing and its diminution after 1963.

What is deterrence? Does it work? How did a doctrine of "mutual assured destruction" come into being? In Section VII on the evolution of the central doctrine of the nuclear age—that the only rational use of nuclear weapons is to deter an aggressor by threatening retaliation if he strikes first—we have provided selections from RAND strategic thinkers Bernard Brodie and Herman Kahn along with statements from various administrations, from John Foster Dulles on "massive retaliation" in 1954 to President Ronald Reagan's definition of deterrence in 1982. Deterrence is often more an accepted assumption than a stated doctrine, but these selections help give the reader an understanding of how deterrence has been articulated over three decades as the central rationale for American nuclear weapons policy. We have included Secretary of Defense Caspar Weinberger's outline of this history of U.S. deterrence policy and concluded the section with what many consider an extension of that policy—if also mixed with a good deal of technological fantasy—in Reagan's support of "Star Wars" or the Strategic Defense Initiative (SDI).

Like deterrence, arms control is a policy developed over several decades and lies in a kind of no-man's-land between the paradise of disarmament and the hell of nuclear war. Both deterrence and arms control depend on the careful management of new and old weapons systems and careful communication with the Soviet Union. Can arms control work? In Section VIII we have tried to show how a series of limited and bilateral agreements between the United States and the Soviet Union following the Cuban missile crisis of 1962 attempted to manage communications and to control force levels of the two nuclear superpowers.

We show how the dream of international disarmament in the 1950s gave way to the more limited task of arms control through the hotline communications system between Moscow and Washington, the limiting of antiballistic missile systems in SALT I (1972), and the signing by President Jimmy Carter and Premier Leonid Brezhnev of the SALT II agreement in Vienna in 1979. Limited bilateral measures adopted in the common interest of the two superpowers continue to operate and form the basis of the ongoing Strategic Arms Reduction Talks (START) in Geneva in 1983. In our final selection William P. Clark of the National Security Council attempts to explain this historical process of arms control to the National Conference of Catholic Bishops.

Does nuclear power have a future? Experts and the public may disagree, but it clearly has a past. Beginning with the revision of the Atomic Energy Act in 1954, government and private industry became partners in a major effort to develop nuclear reactors as a source of electrical power in America. For a time in the 1960s the "great bandwagon market" that produced dozens of new light water reactors across the country convinced many that nuclear power could provide electricity that was "too cheap to meter." Gradually, however, the nuclear dream dissolved, in part because of increasing public concern with radiological dangers and increasing government regulation in the interest of public health and safety. The documents on nuclear power in Section IX show how a growing public concern about nuclear energy and the environment precipitated a series of new legal decisions and regulations that helped slow the growth of nuclear power and made it increasingly expensive. By the time of the Three Mile Island accident of 1979, orders for new nuclear plants had all but disappeared, and the public seemed more concerned with the risks than the benefits of nuclear power. One selection is not from the highly publicized Kemeny or Rogovin reports on the accident, but from a nuclear industry account written within a week of the accident. Like so many of the documents collected here, it provides a sense of immediate historical presence, rather than a qualified subsequent analysis. A more historical perspective is offered by the former head of the Oak Ridge National Laboratory and early nuclear power pioneer Alvin M. Weinberg. Since Weinberg's piece appeared, the Chernobyl disaster in the Soviet Union, especially as described in Grigori Medvedev's *The Truth About Chernobyl* (1991), has emphasized the global impact a nuclear accident of significant proportions can have.

The American atom is no longer a secret project but a historical legacy that we must try to understand and with which we must live. Perhaps no other issue of our times is as heavily weighted with history as is the nuclear issue. Without a clear understanding of the nature, purpose, and limits of arms control and regulated nuclear power we cannot understand the world of nuclear weapons and nuclear energy in which we live. We hope that this volume offers a starting point for understanding how we all came to live in the nuclear age, and for distinguishing between the human choices we have inherited and the ones we must still make.

We are greatly indebted to Washington University in St. Louis and to the Alfred P. Sloan Foundation for initially supporting this project, which is based on the experience of classroom teaching, as well as historical research. Many of these materials were developed in connection with a course on the history of atomic energy given at Yale University and a later course entitled "Nuclear Energy and Contemporary History" offered in 1981–1983 at Washington University. We appreciate the assistance of Anne Foster, Dr. Ruth R. Harris, James Lide, Dr. William Burgess, James Gilchrist, and Thomas Burke of History Associates Incorporated; J. Samuel Walker of the Nuclear Regulatory Commission; Franklin C. Miller of the Department of Defense; Roger M. Anders of the Department of Energy; Joseph P. Harahan of the On-Site Inspection Agency; and Wilbert B. Mahoney of the National Archives and Records Administration in suggesting and locating appropriate documents for this edition. Professor Jack M. Holl of Kansas State University has been a long time friend and intellectual parent of this volume. Professor Martin Sherwin of Tufts University also made a number of valuable suggestions. The result, of course, is our responsibility and not theirs, but we are grateful for their assistance.

We would also like to acknowledge the assistance of Antoinette Zemel, who typed the bulk of the initial manuscript, Judy Bressler-Saad of History

Associates Incorporated, who typed much of the revisions, and Malcolm L. Call, who originally encouraged the project. Venita Lake was a devoted and effective reader, editor, critic, and typist. We are also indebted to our wives, Eileen McGuckian, Marilyn Hewlett, and Ann Williams, for their support and patience.

Philip L. Cantelon
Richard G. Hewlett
Robert C. Williams
Rockville, Maryland

The Nuclear Age: Background and Visions

*T*he American atom had its origins in Europe at the turn of the century in *the minds of physicists. Between 1900 and 1914 a veritable revolution oc-curred in our view of the material universe and its laws. The discovery of radioactivity in the late 1890s was an alchemist's dream; one chemical ele-ment could decay into another in a process of transmutation that gave off new sources of energy. A very small amount of matter, by Albert Einstein's 1905 calculations, could be transformed into a very large amount of energy proportionate to the square of the speed of light, a constant. X-rays saw through matter, which could be understood as waves rather than particles. The atom itself, that smallest and most discrete material object, appeared to be made up of still smaller particles, a nucleus surrounded by electrons.*

The technological wonders of World War I—tanks, airplanes, flame-throwers, and submarines—demonstrated the power of science at war. But the discoveries of the 1920s and 1930s extended the revolution in physics still further. Small particles at high speeds could be located and measured only with a finite degree of uncertainty. Energy appeared to be released in discrete amounts as "light quanta," behaving sometimes as waves and at other times as particles. In 1932 experiments verified that the nucleus was composed of positively charged protons and uncharged neutrons, continuing the breakdown of the material world.

Even before World War I, when drinkers consumed radium cocktails that glowed in the dark in Paris and New York nightclubs, a few visionaries an-ticipated the coming of the nuclear age. In 1914 the British science-fiction writer and prophet of future technologies H. G. Wells wrote a book entitled The World Set Free, *in which he envisaged by the 1950s a world of "atomic bombs" for war and nuclear reactors for peace.*

In December 1938 two German physicists, Otto Hahn and Fritz Strassmann, working in a Berlin laboratory, discovered that the heavy element uranium upon bombardment by neutrons had split, or fissioned, into separate, lighter elements with a consequent release of a large amount of energy. When it was later determined that the reaction also produced high energy neutrons, it seemed possible that under proper control the additional neutrons might continue to split other atoms, producing a continuous process of chain reaction. Thus in 1939 (and possibly in 1934 in Italy) the atom had been split, but not in America.

Refugee scientists fleeing Hitler's Germany and Stalin's Russia soon brought the revolution in physics to America. In early 1939 the Hungarian physicist Leo Szilard, a reader of H. G. Wells's prophetic book, learned of the Hahn-Strassmann fission experiment and became concerned. Szilard, then a refugee living in New York, correctly anticipated that a new world of atomic weapons would be "headed for grief," especially if such weapons came into the hands of Adolf Hitler's Nazi Germany.

In 1939, with the support of influential patrons such as the banker Lewis Strauss, Szilard, and another Hungarian refugee, Eugene Wigner, persuaded the world-famous refugee physicist Albert Einstein, living in Princeton, New Jersey, to send a letter to President Franklin Delano Roosevelt warning him of the dangers of nuclear weapons and of the possibility that Hitler would soon acquire the valuable Joachimsthal uranium mines in Czechoslovakia. The letter was conveyed to Roosevelt shortly after World War II began with the German Blitzkrieg against Poland on September 1, 1939.

Despite a proliferation of committees on "the uranium question," little was accomplished in the United States until after Pearl Harbor. The initial key theoretical work on the possibility of building an atomic bomb was performed in England in 1940 and 1941 at Birmingham by two refugee German physicists, Rudolf Peierls and Otto Frisch. In the spring of 1940 they produced a working paper for the British government, then fighting for its life against aerial attacks by the Luftwaffe, on the theory of an atomic bomb. In it they concluded that a relatively small amount of uranium could constitute a "critical mass," which, if properly brought together at sufficient speed, could yield an explosion equivalent to many thousands of tons of TNT. The conclusions of Frisch and Peierls were reinforced by a second study carried out in 1941 by a secret British government working group known as the "MAUD Committee."

For further information on the early history of nuclear energy, see Ronald Clark, The Greatest Power on Earth: The International Race for Nuclear Supremacy from Earliest Theory to Three Mile Island (New York: Harper & Row, 1980); Martin D. Kamen, Radiant Science, Dark Politics: A Memoir of the Nuclear Age *(Berkeley: University of California Press, 1985); Rudolph Peierls,* Bird of Passage *(Princeton, N.J.: Princeton University Press, 1985). The papers and correspondence of Leo Szilard have been collected by his widow, Gertrude Weiss Szilard, and Spencer R. Weart as* Leo Szilard: His Version of the Facts; Selected Recollections and Correspondence, *vol. 2 (Cambridge, Mass. and London: MIT Press, 1980). The British story is told best in Margaret Gowing,* Britain and Atomic Energy 1939–1945 *(London: Macmillan, 1964). For a popular account see Richard Rhodes,* The Making of the Atomic Bomb *(New York: Simon & Schuster, 1986).*

1. Nuclear Energy: H. G. Wells's Vision, 1914

It was in 1953 that the first Holsten-Roberts engine brought induced radio-activity into the sphere of industrial production, and its first general use was to replace the steam-engine in electrical generating stations. Hard upon the appearance of this came the Dass-Tata engine—the invention of two among the brilliant galaxy of Bengali inventors the modernisation of Indian thought was producing at this time—which was used chiefly for automobiles, aeroplanes, water-planes and such-like mobile purposes. The American Kemp engine, differing widely in principle but equally practicable, and the Drupp-Erlanger came hard upon the heels of this, and by the autumn of 1954 a gigantic replacement of industrial methods and machinery was in progress all about the habitable globe. Small wonder was this when the cost even of these earliest and clumsiest of atomic engines is compared with that of the power they superseded. Allowing for lubrication, the Dass-Tata engine, once it was started, cost a penny to run thirty-seven miles, and added only nine and a quarter pounds to the weight of the carriage it drove. It made the heavy alchohol-driven automobile of the time ridiculous in appearance as well as preposterously costly. For many years the price of coal and every form of liquid fuel had been clambering to levels that made even the revival of the draft-horse seem a practicable possibility, and now,

This selection first appeared in H. G. Wells, *The World Set Free* (New York: Dutton, 1914), pp. 51–53, 114–19, 152–53. Used with permission of the Executors of the Estate of H. G. Wells.

with the abrupt relaxation of this stringency, the change in appearance of the traffic upon the world's roads was instantaneous. In three years the frightful armoured monsters that had hooted and smoked and thundered about the world for four awful decades were swept away to the dealers in old metal, and the highways thronged with light and clean and shimmering shapes of silvered steel. At the same time a new impetus was given to aviation by the relatively enormous power for weight of the atomic engine; it was at last possible to add Redmayne's ingenious helicopter ascent and descent engine to the vertical propeller that had hitherto been the sole driving force of the aeroplane without overweighing the machine, and men found themselves possessed of an instrument of flight that could hover or ascend or descend vertically and gently as well as rush wildly through the air. The last dread of flying vanished. As the journalists of the time phrased it, this was the epoch of the Leap into Air. The new atomic aeroplane became indeed a mania; everyone of means was frantic to possess a thing so controllable, so secure, and so free from the dust and danger of the rod, and in France alone in the year 1943 thirty thousand of these new aeroplanes were manufactured and licensed and soared humming softly into the sky.

Never before in the history of warfare had there been a continuing explosive; indeed, up to the middle of the twentieth century the only explosives known were combustibles whose explosiveness was due entirely to their instantaneousness; and these atomic bombs which science burst upon the world that night were strange even to the men who used them. Those used by the Allies were lumps of pure Carolinum, painted on the outside with unoxidised cydonator inducive enclosed hermetically in a case of membranium. A little celluloid stud between the handles by which the bomb was lifted was arranged so as to be easily torn off and admit air to the inducive, which at once became active and set up radio-activity in the outer layer of the Carolinum sphere. This liberated fresh inducive, and so in a few minutes the whole bomb was blazing continual explosion. The Central European bombs were the same, except that they were larger and had a more complicated arrangement for animating the inducive.

Always before in the development of warfare the shells and rockets fired had been but momentarily explosive, they had gone off in an instant once and for all, and if there was nothing living or valuable within reach of the concussion and the flying fragments, then they were spent and over. But Carolinum, which belonged to the β-Group of Hyslop's so-called; "suspended degenerator" elements, once its degenerative process had been induced, continued a furious radiation of energy, and nothing could arrest it. Of all Hyslop's artificial elements, Carolinum was the most heavily stored with energy and the most dangerous to make and handle. To this day it remains the most potent degenerator known. What the earlier twentieth-century chemists called its half period was seventeen days; that is to say, it poured out half of the huge store of energy in its great molecules in the space of seventeen days, the next seventeen days' emission was a half of that first period's outpouring, and so on. As with all radio-active substances, this Carolinum, though every seventeen days its power is halved, though is never entirely exhausted, and to this day the battle-fields and bomb-fields of that frantic time in human history are sprinkled with radiant matter and so centres of inconvenient rays. . . .

What happened then when the celluloid stud was opened was that the inducive oxydised and became active. Then the surface of the Carolinum began to degenerate. This degeneration passed only slowly into the substance of the bomb. A moment or so after its explosion began it was still mainly an inert sphere exploding superficially, a big, inanimate nucleus 6wrapped in flame and thunder. Those that were thrown from aeroplanes fell in this state; they reached the ground still mainly solid and, melting soil and rock in their progress, bored into the earth. There, as more and more of the Carolinum became active, the bomb spread itself out into a monstrous cavern of fiery energy at the base of what became very speedily a miniature active volcano. The Carolinum, unable to disperse freely, drove into and mixed up with a boiling confusion of molten soil and superheated steam, and so remained, spinning furiously and maintaining an eruption that lasted for years or months or weeks according to the size of the bomb employed and the chances of its dispersal. Once launched, the bomb was absolutely unapproachable and uncontrollable until its forces were nearly exhausted, and from the crater that burst open above it, puffs of heavy incandescent vapour and fragments of viciously punitive rock and mud, saturated with Carolinum, and each a centre of scorching and blistering energy, were flung high and far.

Such was the crowning triumph of military science, the ultimate explosive, that was to give the "decisive touch" to war. . . .

A recent historical writer has described the world of that time as one that "believed in established words and was invincibly blind to the obvious in things." Certainly it seems now that nothing could have been more obvious to the people of the early twentieth century than the rapidity with which war was becoming impossible. And as certainly they did not see it. They did not see it until the atomic bombs burst in their fumbling hands. Yet the broad facts must have glared upon any intelligent mind. All through the nineteenth and twentieth centuries the amount of energy that men were able to command was continually increasing. Applied to warfare that meant that the power to inflict a blow, the power to destroy, was continually increasing. There was no increase whatever in the ability to escape. Every sort of passive defence, armour, fortifications, and so forth, was being outmastered by this tremendous increase on the destructive side. Destruction was becoming so facile that any little body of malcontents could use it; it was revolutionising the problems of police and internal rule. Before the last war began it was a matter of common knowledge that a man could carry about in a handbag an amount of latent energy sufficient to wreck half a city. These facts were before the minds of everybody; the children in the streets knew them. And yet the world still, as the Americans used to phrase it, "fooled around" with the paraphernalia and pretensions of war.

It is only by realising this profound, this fantastic divorce between the scientific and intellectual movement on the one hand and the world of the lawyer-politician on the other that the men of a later time can hope to understand this preposterous state of affairs. Social organisation was still in the barbaric stage. There were already great numbers of actively intelligent men and much private and commercial civilisation, but the community as a whole was aimless, untrained, and unorganised to the pitch of imbecility. Collective civilisation, the "Modern State," was still in the womb of the future. . . .

For the whole world was flaring then into a monstrous phase of destruction. Power after power about the armed globe sought to anticipate attack by aggression. They went to war in a delirium of panic, in order to use their bombs first. China and Japan had assailed Russia and destroyed Moscow, the United States had attacked Japan, India was in anarchistic revolt with Delhi a pit of fire spouting death and flame; the redoubtable King of the Balkans was mobilising. It must have seemed plain at last to everyone in those days that the world was slipping headlong to anarchy. By the spring of 1959 from nearly two hundred centres, and every week added to their number, roared the unquenchable crimson conflagrations of the atomic bombs, the flimsy fabric of the world's credit had vanished, industry was completely disorganised, and every city, every thickly populated area was starving or trembled on the verge of starvation. Most of the capital cities of the world were burning; millions of people had already perished, and over great areas government was at an end. Humanity has been compared by one contemporary writer to a sleeper who handles matches in his sleep and wakes to find himself in flames.

For many months it was an open question whether there was to be found throughout all the race the will and intelligence to face these new conditions and make even an attempt to arrest the downfall of the social order. For a time the war spirit defeated every effort to rally the forces of preservation and construction. Leblanc seemed to be protesting against earthquakes and as likely to find a spirit of reason in the crater of Etna. Even though the

shattered official Governments now clamored for peace, bands of irreconcil-
ables and invincible patriots, usurpers, adventurers and political despera-
does were everywhere in possession of the simple apparatus for the
disengagement of atomic energy and the initiation of new centres of destruction. The stuff exercised an irresistible fascination upon a certain type of
mind. Why should anyone give in while he can still destroy his enemies?
Surrender? While there is still a chance of blowing them to dust? The power
of destruction which had once been the ultimate privilege of government
was now the only power left in the world—and it was everywhere.

2. Leo Szilard and the Discovery of Fission

A. Szilard to Sir Hugo Hirst

6, Halliwick Road
London, N.10.
17th March, 1934

Dear Sir Hugo,

As you are on holiday you might find pleasure in reading a few pages out
of a book by H. G. Wells which I am sending you. I am certain you will find
the first three paragraphs of Chapter The First (The New Source of Energy,
page 42) interesting and amusing, whereas the other parts of the book are
rather boring. It is remarkable that Wells should have written those pages
in 1914.

Of course, all this is moonshine, but I have reason to believe that in so
far as the industrial applications of the present discoveries in physics are
concerned, the forecast of the writers may prove to be more accurate than
the forecast of the scientists. The physicists have conclusive arguments as to
why we cannot create at present new sources of energy for industrial purposes; I am not so sure whether they do not miss the point.

It is perhaps possible to be more definite some time after your return, and in the meantime I hope you will in any case enjoy glancing through these few pages.

With best wishes for a pleasant stay,

Yours very truly,
Leo Szilard

B. Szilard to Lewis Strauss

Hotel King's Crown
Opposite Columbia University
420 West 116th Street
New York City
January 25th, 1939

Mr. Lewis L. Strauss
% Kuhn, Loeb & Co.
52 William Street
New York City

Dear Mr. Strauss:

I feel that I ought to let you know of a very sensational new development in nuclear physics. In a paper in the *Naturwissenschaften* Hahn reports that he finds when bombarding uranium with neutrons the uranium breaking up into two halves giving elements of about half the atomic weight of uranium. This is entirely unexpected and exciting news for the average physicist. The Department of Physics at Princeton, where I spent the last few days, was like a stirred-up ant heap.

Apart from the purely scientific interest there may be another aspect of this discovery, which so far does not seem to have caught the attention of those to whom I spoke. First of all it is obvious that the energy released in this new reaction must be very much higher than in all previously known cases. It may be 200 million (electron-) volts instead of the usual 3–10 million volts. This in itself might make it possible to produce power by means of nuclear energy, but I do not think that this possibility is very exciting, for if the energy output is only two or three times the energy input, the cost of investment would probably be too high to make the process worthwhile.

Unfortunately, most of the energy is released in the form of heat and not in the form of radioactivity.

I see, however, in connection with this new discovery potential possibilities in another direction. These might lead to a large-scale production of energy and radioactive elements, unfortunately also perhaps to atomic bombs. This new discovery revives all the hopes and fears in this respect which I had in 1934 and 1935, and which I have as good as abandoned in the course of the last two years. At present I am running a high temperature and am therefore confined to my four walls, but perhaps I can tell you more about these new developments some other time. Meanwhile you may look out for a paper in "Nature" by Frisch and Meitner which will soon appear and which might give you some information about this new discovery.

With best wishes,
Yours sincerely,
Leo Szilard

---

3. Albert Einstein to Franklin D. Roosevelt, August 2, 1939

Albert Einstein
Old Grove Road
Peconic, Long Island
August 2nd, 1939

F. D. Roosevelt
President of the United States
White House
Washington, D.C.

Sir:

Some recent work by E. Fermi and L. Szilard, which has been communicated to me in manuscript, leads me to expect that the element uranium may be turned into a new and important source of energy in the immediate future. Certain aspects of the situation which has arisen seem to call for watchfulness and, if necessary, quick action on the part of the Administration. I believe therefore that it is my duty to bring to your attention the following facts and recommendation.

In the course of the last four months it has been made probable through the work of Joliot in France as well as Fermi and Szilard in America—that it may become possible to set up a nuclear chain reaction in a large mass of uranium, by which vast amounts of power and large quantities of new radium-like elements would be generated. Now it appears almost certain that this could be achieved in the immediate future.

This new phenomenon would also lead to the construction of bombs, and it is conceivable—though much less certain—that extremely powerful bombs of a new type may thus be constructed. A single bomb of this type, carried by boat and exploded in a port, might very well destroy the whole port together with some of the surrounding territory. However, such bombs might very well prove to be too heavy for transportation by air.

The United States has only very poor ores of uranium in moderate quantities. There is some good ore in Canada and the former Czechoslovakia, while the most important source of uranium is Belgian Congo.

In view of this situation you may think it desirable to have some permanent contact maintained between the Administration and the group of physicists working on chain reactions in America. One possible way of achieving this might be for you to entrust with this task a person who has your confidence and who could perhaps serve in an unofficial capacity. His task might comprise the following:

a) to approach Government Departments, keep them informed of the further development, and put forward recommendations for Government action, giving particular attention to the problem of securing a supply of uranium ore for the United States.

b) to speed up the experimental work, which is at present being carried on within the limits of the budgets of University laboratories, by providing funds, if such funds be required, through his contacts with private persons who are willing to make contributions for this cause, and perhaps also by obtaining the co-operation of industrial laboratories which have the necessary equipment.

I understand that Germany has actually stopped the sale of uranium from the Czechoslovakian mines which she has taken over. That she should have taken such early action might perhaps be understood on the ground that the son of the German Under-Secretary of State, von Wiezsacker, is attached to the Kaiser-Wilhelm-Institut in Berlin where some of the American work on uranium is now being repeated.

Yours very truly,
/s/ Albert Einstein

4. The Frisch-Peierls Memorandum, 1940

On the Construction of a "Super-Bomb"; Based on a
Nuclear Chain Reaction in Uranium

The possible construction of "super-bombs" based on a nuclear chain reaction in uranium has been discussed a great deal and arguments have been brought forward which seemed to exclude this possibility. We wish here to point out and discuss a possibility which seems to have been overlooked in these earlier discussions.

Uranium consists essentially of two isotopes, ^{238}U (99.3%) and ^{235}U (0.7%). If a uranium nucleus is hit by a neutron, three processes are possible: (1) scattering, whereby the neutron changes direction and, if its energy is above about 0.1 MeV, loses energy; (2) capture, when the neutron is taken up by the nucleus; and (3) fission, i.e. the nucleus breaks up into two nuclei of comparable size, with the liberation of an energy of about 200 MeV.

The possibility of a chain reaction is given by the fact that neutrons are emitted in the fission and that the number of these neutrons per fission is greater than 1. The most probable value for this figure seems to be 2.3, from two independent determinations.

However, it has been shown that even in a large block of ordinary uranium no chain reaction would take place since too many neutrons would be slowed down by inelastic scattering into the energy region where they are strongly absorbed by ^{238}U.

Several people have tried to make chain reactions possible by mixing the

This selection appears in Margaret Gowing, *Britain and Atomic Energy 1939–1945* (New York: St. Martin's Press, 1964), pp. 389–93. Reprinted by permission of St. Martin's Press and the authors.

uranium with water, which reduces the energy of the neutrons still further and thereby increases their efficiency again. It seems fairly certain, however that even then it is impossible to sustain a chain reaction.

In any case, no arrangement containing hydrogen and based on the action of slow neutrons could act as an effective super-bomb, because the reaction would be too slow. The time required to slow down a neutron is about 10^{-5} sec and the average time loss before a neutron hits a uranium nucleus is even 10^{-4} sec. In the reaction, the number of neutrons would increase exponentially, like $e^{t/\tau}$ where τ would be at least 10^{-4} sec. When the temperature reaches several thousand degrees the container of the bomb will break and within 10^{-4} sec the uranium would have expanded sufficiently to let the neutrons escape and so to stop the reaction. The energy liberated would, therefore, be only a few times the energy required to break the container, i.e. of the same order of magnitude as with ordinary high explosives.

Bohr has put forward strong arguments for the suggestion that the fission observed with slow neutrons is to be ascribed to the rare isotope ^{235}U, and that this isotope has, on the whole, a much greater fission probability than the common isotope ^{238}U. Effective methods for the separation of isotopes have been developed recently, of which the method of thermal diffusion is simple enough to permit separation on a fairly large scale.

This permits, in principle, the use of nearly pure ^{235}U in such a bomb, a possibility which apparently has not so far been seriously considered. We have discussed this possibility and come to the conclusion that a moderate amount of ^{235}U would indeed constitute an extremely efficient explosive.

The behaviour of ^{235}U under bombardment with fast neutrons is not known experimentally, but from rather simple theoretical arguments it can be concluded that almost every collision produces fission and that neutrons of any energy are effective. Therefore it is not necessary to add hydrogen, and the reaction, depending on the action of fast neutrons, develops with very great rapidity so that a considerable part of the total energy is liberated before the reaction gets stopped on account of the expansion of the material.

The critical radius γ_0—i.e. the radius of a sphere in which the surplus of neutrons created by the fission is just equal to the loss of neutrons by escape through the surface—is, for a material with a given composition, in a fixed ratio to the mean free path of neutrons, and this in turn is inversely proportional to the density. It therefore pays to bring the material into the densest possible form, i.e. the metallic state, probably sintered or hammered. If we assume for ^{235}U, no appreciable scattering, and 2.3 neutrons emitted per fission, then the critical radius is found to be 0.8 times the mean free path. In the metallic state (density 15), and assuming a fission cross-section of 10^{-23} cm^2, the mean free path would be 2.6 cm and γ_0 would be 2.1 cm, corresponding to a mass of 600 grams. A sphere of metallic [235]U of a radius greater than γ_0 would be explosive, and one might think of about 1 kg as a suitable size for the bomb.

The speed of the reaction is easy to estimate. The neutrons emitted in the fission have velocities of about 10^{-9} cm/sec and they have to travel 2.6 cm before hitting a uranium nucleus. For a sphere well above the critical size the loss through neutron escape would be small, so we may assume that each neutron after a life of 2.6×10^{-9} sec, produces fission, giving birth to two neutrons. In the expression of e^t/τ for the increase of neutron density with time, it would be about 4×10^{-9} sec, very much shorter than in the case of a chain reaction depending on slow neutrons.

If the reaction proceeds until most of the uranium is used up, temperatures of the order of 10^{10} degrees and pressue of about 10^{13} atmospheres are produced. It is difficult to predict accurately the behaviour of matter under these extreme conditions, and the mathematical difficulties of the problem are considerable. By a rough calculation we get the following expression for the energy liberated before the mass expands so much that the reaction is interrupted:

$$E \;=\; 0 \cdot 2M(\gamma^2/\tau^2)(\sqrt{(\gamma/\gamma_0)} - 1)$$

(M, total mass of uranium; γ, radius of sphere; γ_0, critical radius; τ, time required for neutron density to multiply by a factor e). For a sphere of radius 4.2 cm ($\gamma_0 = 2.1$ cm), M $= 4700$ grams, $\tau = 4 \times 10^{-9}$ sec, we find E $= 4 \times 10^{20}$ ergs, which is about one-tenth of the total fission energy. For a radius of about 8 cm (M $= 32$ kg) the whole fission energy is liberated, according to formula (1). For small radii the efficiency falls off even faster than indicated by formula (1) because τ goes up as γ approaches γ_0. The energy liberated by a 5 kg bomb would be equivalent to that of several thousand tons of dynamite, while that of a 1 kg bomb, though about 500 times less, would still be formidable.

It is necessary that such a sphere should be made in two (or more) parts which are brought together first when the explosion is wanted. Once assembled, the bomb would explode within a second or less, since one neutron is sufficient to start the reaction and there are several neutrons passing through the bomb in every second, from the cosmic radiation. (Neutrons originating from the action of uranium alpha rays on light-element impurities would be negligible provided the uranium is reasonably pure.) A sphere with a radius of less than about 3 cm could be made up in two hemispheres, which are pulled together by springs and kept separated by a suitable structure which is removed at the desired moment. A larger sphere would have to be composed of more than two parts, if the parts, taken separately, are to be stable.

It is important that the assembling of the parts should be done as rapidly as possible, in order to minimise the chance of a reaction getting started at a moment when the critical conditions have only just been reached. If this happened, the reaction rate would be much slower and the energy liberation would be considerably reduced; it would, however, always be sufficient to destroy the bomb.

It may be well to emphasize that a sphere only slightly below the critical size is entirely safe and harmless. By experimenting with spheres of gradually increasing size and measuring the number of neutrons emerging from them under a known neutron bombardment, one could accurately determine the critical size, without any danger of a premature explosion.

For the separation of the ^{235}U, the method of thermal diffusion, developed by Clusius and others, seems to be the only one which can cope with the large amounts required. A gaseous uranium compound, for example uranium hexafluoride, is placed between two vertical surfaces which are kept at a different temperature. The light isotope tends to get more concentrated near the hot surface, where it is carried upwards by the convection current. Exchange with the current moving downwards along the cold surface produces a fractionating effect, and after some time a state of equilibrium is reached when the gas near the upper end contains markedly more of the light isotope than near the lower end.

For example, a system of two concentric tubes, of 2 mm separation and 3 cm diameter, 150 cm long, would produce a difference of about 40% in the concentration of the rare isotope between its ends, and about 1 gram per day could be drawn from the upper end without unduly upsetting the equilibrium.

In order to produce large amounts of highly concentrated ^{235}U, a great number of these separating units will have to be used, being arranged in parallel as well as in series. For a daily production of 100 grams of ^{235}U of 90% purity, we estimate that about 100,000 of these tubes would be required. This seems a large number, but it would undoubtedly be possible to design some kind of a system which would have the same effective area in a more compact and less expensive form.

In addition to the destructive effect of the explosion itself, the whole material of the bomb would be transformed into a highly radioactive stage. The energy radiated by these active substances will amount to about 20% of the energy liberated in the explosion, and the radiations would be fatal to living beings even a long time after the explosion.

The fission of uranium results in the formation of a great number of active bodies with periods between, roughly speaking, a second and a year. The resulting radiation is found to decay in such a way that the intensity is about inversely proportional to the time. Even one day after the explosion the radiation will correspond to a power expenditure of the order of 1,000 kW, or to the radiation of a hundred tons of radium.

Any estimates of the effects of this radiation on human beings must be rather uncertain because it is difficult to tell what will happen to the radio-active material after the explosion. Most of it will probably be blown into the air and carried away by the wind. This cloud of radioactive material will kill everybody within a strip estimated to be several miles long. If it rained the danger would be even worse because active material would be carried down to the ground and stick to it, and persons entering the contaminated area would be subjected to dangerous radiations even after days. If 1% of the active material sticks to the debris in the vicinity of the explosion and if the debris is spread over an area of, say, a square mile, any person entering this area would be in serious danger, even several days after the explosion.

In the estimates, the lethal dose of penetrating radiation was assumed to be 1,000 Roentgen; consultation of a medical specialist on X-ray treatment and perhaps further biological research may enable one to fix the danger limit more accurately. The main source of uncertainty is our lack of knowl-edge as to the behaviour of materials in such a super-explosion, and an ex-pert on high explosives may be able to clarify some of these problems.

Effective protection is hardly possible. Houses would offer protection only at the margins of the danger zone. Deep cellars or tunnels may be comparatively safe from the effects of radiation, provided air can be supplied from an uncontaminated area (some of the active substances would be noble gases which are not stopped by ordinary filters).

The irradiation is not felt until hours later when it may be too late. There-fore it would be very important to have an organisation which determines the exact extent of the danger area, by means of ionisation measurements, so that people can be warned from entering it.

O. R. Frisch
R. Peierls

(The University, Birmingham)

5. The MAUD Report, 1941

Report by MAUD Committee on the Use of Uranium for a Bomb

PART I

1. General Statement

Work to investigate the possibilities of utilising the atomic energy of uranium for military purposes has been in progress since 1939, and a stage has now been reached when it seems desirable to report progress.

We should like to emphasise at the beginning of this report that we entered the project with more scepticism than belief, though we felt it was a matter which had to be investigated. As we proceeded we became more and more convinced that release of atomic energy on a large scale is possible and that conditions can be chosen which would make it a very powerful weapon of war. We have now reached the conclusion that it will be possible to make an effective uranium bomb which, containing some 25 lb of active material, would be equivalent as regards destructive effect to 1,800 tons of T.N.T. and would also release large quantities of radioactive substance, which would make places near to where the bomb exploded dangerous to human life for a long period. The bomb would be composed of an active constituent (referred to in what follows as ^{235}U) present to the extent of about a part in 140 in ordinary Uranium. Owing to the very small difference in properties (other than explosive) between this substance and the rest of the Uranium, its extraction is a matter of great difficulty and a plant to produce 2¼ lb (1 kg) per day (or 3 bombs per month) is estimated to cost approximately £5,000,000 pounds, of which sum a considerable proportion would be spent on engineering, requiring labour of the same highly skilled character as is needed for making turbines.

In spite of this very large expenditure we consider that the destructive effect, both material and moral, is so great that every effort should be made to produce bombs of this kind. As regards the time required, Imperial Chemical Industries after consultation with Dr. Guy of Metropolitan-Vickers, estimate that the material for the first bomb could be ready by the end of 1943. This of course assumes that no major difficulty of an entirely

This text appears in Margaret Gowing, *Britain and Atomic Energy 1939–1945* (New York: St. Martin's Press, 1964), pp. 394–98. Reprinted by permission of St. Martin's Press.

unforeseen character arises. Dr. Ferguson of Woolwich estimates that the time required to work out the method of producing high velocities required for fusing (see paragraph 3) is 1–2 months. As this could be done concurrently with the production of the material no further delay is to be anticipated on this score. Even if the war should end before the bombs are ready the effort would not be wasted, except in the unlikely event of complete disarmament, since no nation would care to risk being caught without a weapon of such decisive possibilities.

We know that Germany has taken a great deal of trouble to secure supplies of the substance known as heavy water. In the earlier stages we thought that this substance might be of great importance for our work. It appears in fact that is usefulness in the release of atomic energy is limited to processes which are not likely to be of immediate war value, but the Germans may by now have realised this, and it may be mentioned that the lines on which we are now working are such as would be likely to suggest themselves to any capable physicist.

By far the largest supplies of Uranium are in Canada and the Belgian Congo, and since it has been actively looked for because of the radium which accompanies it, it is unlikely that any considerable quantities exist which are unknown except possibly in unexplored regions.

2. Principle Involved

This type of bomb is possible because of the enormous store of energy resident in atoms and because of the special properties of the active constituent of uranium. The explosion is very different in its mechanism from the ordinary chemical explosion, for it can occur only if the quantity of ^{235}U is greater than a certain critical amount. Quantities of the material less than the critical amount are quite stable. Such quantities are therefore perfectly safe and this is a point which we wish to emphasise. On the other hand, if the amount of material exceeds the critical value it is unstable and a reaction will develop and multiply itself with enormous rapidity, resulting in an explosion of unprecedented violence. Thus all that is necessary to detonate the bomb is to bring together two pieces of the active material each less than the critical size but which when in contact form a mass exceeding it.

3. Method of Fusing

In order to achieve the greatest efficiency in an explosion of this type, it is necessary to bring the two halves together at high velocity and it is proposed to do this by firing them together with charges of ordinary explosive in a form of double gun.

The weight of this gun will of course greatly exceed the weight of the bomb itself, but should not be more than 1 ton, and it would certainly be within the carrying capacity of a modern bomber. It is suggested that the bomb (contained in the gun) should be dropped by parachute and the gun should be fired by means of a percussion device when it hits the ground. The time of drop can be made long enough to allow the aeroplane to escape from the danger zone, and as this is very large, great accuracy of aim is not required.

4. Probable Effect

The best estimate of the kind of damage likely to be produced by the explosion of 1,800 tons of T.N.T. is afforded by the great explosion at Halifax N.S. in 1917. The following account is from the *History of Explosives*. "The ship contained 450,000 lb of T.N.T., 122,960 lb of guncotton, and 4,661,794 lb of picric acid wet and dry, making a total of 5,234,754 lb. The zone of the explosion extended for about 3/4 mile in every direction and in this zone the destruction was almost complete. Severe structural damage extended generally for a radius of 1-1/8 to 1-1/4 miles, and in one direction up to 1-3/4 miles from the origin. Missiles were projected to 3–4 miles, window glass broken up to 10 miles generally, and in one instance up to 61 miles."

In considering this description it is to be remembered that part of the explosives cargo was situated below water level and part above.

5. Preparation of Material and Cost

We have considered in great detail the possible methods of extracting the ^{235}U from ordinary uranium and have made a number of experiments. The scheme which we recommend is described in Part II of this report and in greater detail in Appendix IV. It involves essentially the gaseous diffusion of a compound of uranium through gauzes of very fine mesh.

In the estimates of size and cost which accompany this report, we have only assumed types of gauze which are at present in existence. It is probable that a comparatively small amount of development would enable gauzes of smaller mesh to be made and this would allow the construction of a somewhat smaller and consequently cheaper separation plant for the same output.

Although the cost per lb of this explosive is so great it compares very favourably with ordinary explosives when reckoned in terms of energy released and damage done. It is, in fact considerably cheaper, but the points which we regard as of overwhelming importance are the concentrated destruction which it would produce, the large moral effect, and the saving in air effort the use of this substance would allow, as compared with bombing with ordinary explosives.

6. Discussion

One outstanding difficulty of the scheme is that the main principle cannot be tested on a small scale. Even to produce a bomb of the minimum critical size would involve a great expenditure of time and money. We are however convinced that the principle is correct, and whilst there is still some uncertainty as to the critical size it is most unlikely that the best estimate we can make is so far in error as to invalidate the general conclusions. We feel that the present evidence is sufficient to justify the scheme being strongly pressed.

As regards the manufacture of the ^{235}U we have gone nearly as far as we can on a laboratory scale. The principle of the method is certain, and the application does not appear unduly difficult as a piece of chemical engineering. The need to work on a larger scale is now very apparent and we are beginning to have difficulty in finding the necessary scientific personnel. Further, if the weapon is to be available in say two years from now, it is necessary to start plans for the erection of a factory, though no really large expenditure will be needed till the 20-stage model has been tested. It is also important to begin training men who can ultimately act as supervisors of the manufacture. There are a number of auxiliary pieces of apparatus to be developed, such as those for measuring the concentration of the ^{235}U. In addition, work on a fairly large scale is needed to develop the chemical side for the production in bulk of uranium hexafluoride, the gaseous compound we propose to use.

It will be seen from the foregoing that a stage in the work has now been reached at which it is important that a decision should be made as to whether the work is to be continued on the increasing scale which would be necessary if we are to hope for it as an effective weapon for this war. Any considerable delay now would retard by an equivalent amount the date by which the weapon could come into effect.

7. Action in U.S.

We are informed that while the Americans are working on the uranium problem the bulk of their effort has been directed to the production of energy, as discussed in our report on uranium as a source of power, rather than to the production of a bomb. We are in fact co-operating with the United States to the extent of exchanging information, and they have undertaken one or two pieces of laboratory work for us. We feel that it is important and desirable that development work should proceed on both sides of the Atlantic irrespective of where it may be finally decided to locate the plant for separating the ^{235}U, and for this purpose it seems desirable that certain members of the committee should visit the United States. We are informed that such a visit would be welcomed by the members of the United States committees which are dealing with this matter.

8. Conclusions and Recommendations

(i) The committee considers that the scheme for a uranium bomb is practicable and likely to lead to decisive results in the war.

(ii) It recommends that this work be continued on the highest priority and on the increasing scale necessary to obtain the weapon in the shortest possible time.

(iii) That the present collaboration with America should be continued and extended especially in the region of experimental work.

II

The Manhattan Project

*B*y 1942 scientists in both England and the United States were convinced that a nuclear weapon based on the principle of fission could be constructed within a period of three to four years. Yet the complex process of separating fissionable U235 from natural uranium or producing man-made plutonium, also fissionable, was expensive and unproven.

On December 2, 1942, in a racquet court beneath the West Stands of Stagg Field at the University of Chicago, a team of scientists led by Nobel Prize physicist Enrico Fermi succeeded in achieving the first controlled, self-sustaining nuclear chain reaction, using a "pile" of graphite and uranium blocks. James Conant, project chief, received the news of the successful experiment in a coded message: "The Italian navigator has landed in the New World." Within months ground was broken for three supersecret atomic cities: Hanford, Washington, where plutonium production would occur; Oak Ridge, Tennessee, where uranium separation plants were under construction; and Los Alamos, New Mexico, where bomb design, experimental testing, and assembly would be centered. Thus, the "Manhattan Engineer District" established under General Leslie Groves and the Army Corps of Engineers in the summer of 1942 was not in Manhattan but in the scattered university research laboratories and new atomic towns across the country.

In 1943 a team of scientists was formed at Los Alamos under the direction of J. Robert Oppenheimer, a brilliant physicist educated at Harvard and Göttingen who had taught students at Berkeley and the California Institute of Technology in the 1930s. Oppenheimer was able to persuade General Groves that the most auspicious approach to building an atomic bomb was to have civilian scientists working in an isolated military environment at Los Alamos, where free discussion could still occur. Groves wanted compartmentalization of all knowledge and information so that each individual would know only what he or she needed to know. Oppie, as he was called, wanted an open exchange of ideas. In the end Oppenheimer had his way, and Los

Alamos became a legendary community of brilliant "crackpots," as Groves labeled them, working feverishly to create a nuclear weapon before Adolf Hitler might do the same. Although all the production plants were built in the United States, the British continued to provide key personnel for the project. Winston Churchill, the British prime minister, used this point in negotiations with Roosevelt during the war to gain Britain a foothold in post-war development.

In fact, Hitler had sacrificed his nuclear physicists to the caldron of the Eastern Front, where Stalin's Red Army held firm, and to the budgetary demands of German rocketry. Dependent on a single heavy water plant in Norway, and believing that graphite would not serve to moderate the flow of neutrons (thus making fission more probable), German scientists by 1944 had made no significant progress toward either a reactor or a bomb. By autumn 1944 German failure was known to American military and political leaders, and the focus shifted to Japan.

It was not until early 1945 that the first plutonium and uranium began to emerge from Hanford and Oak Ridge in usable quantities. In the meantime Los Alamos scientists had discovered that a gun-type uranium bomb was quite feasible, but that plutonium was so fissionable that it could predetonate unless a critical mass were achieved almost instantly. The new method, "imploding" a plutonium sphere by exploding chemical "lenses" around its surface, needed to be tested. The Hiroshima bomb needed no test; the Nagasaki bomb was first exploded in the sands of Alamogordo, New Mexico, in July 1945.

American military leaders now assumed that if and when an atomic bomb was developed it would be used first against Japan. By 1945 conventional mass firebombing had reduced most Japanese cities to rubble; the "Meeting-house" raid over Tokyo in March killed more than 100,000 people. In this light it was not difficult for American military and scientific committees to decide to use the first nuclear weapons in order to bring the bloody fighting in the Pacific to an end. Scientists, however, were not so certain about the military use of the atomic bomb. A debate erupted among those who had developed the weapon in mid-1945 before the implosion device, "Trinity," was tested at Alamagordo, New Mexico, in July. Dr. Glenn T. Seaborg's letter to Ernest O. Lawrence, which is published for the first time in this edition, is an opening discussion of this debate. The concerns of the scientists in Chicago follow.

When President Roosevelt died in April 1945, the new president, Harry S. Truman, had no knowledge of the Manhattan Project to build an atomic bomb at a cost approaching $2 billion. He had suddenly inherited a massive technological race against Germany, Japan, and the Soviet Union conducted *almost entirely in secret. Truman delayed his trip to Potsdam long enough for the successful plutonium bomb test and then told Stalin in oblique terms about the existence of the new weapon. Only later did Truman learn that Stalin's noncommittal response veiled an intimate knowledge of the Manhattan Project gained through Soviet spies operating in the United States.*

The decision to drop the atomic bomb will continue to be the subject of a debate without end. Given the Japanese attack on Pearl Harbor on December 7, 1941, the time and money invested in the project, and the desire to end the war, the bureaucratic momentum toward dropping the bomb was enormous. Japanese surrender did not appear to be forthcoming, and an American invasion estimated to cost half a million American lives was imminent. After some debate among scientists about the morality of using the bomb, as opposed to demonstrating its power to enemy observers, Truman made the decision to use it. On August 6, 1945, the Japanese city of Hiroshima was destroyed by an atomic bomb dropped from the "Enola Gay," a B-29 operating out of the island base of Tinian. On August 9 the city of Nagasaki, a secondary target, was likewise destroyed by a plutonium bomb. The Japanese surrendered within a few more days, and Truman was a hero for having made use of "the greatest thing in history" to bring a terrible war to an end.

The official history of the Manhattan Project is by Richard G. Hewlett and Oscar E. Anderson, Jr., The New World, 1939/1946, vol. 1 of A History of the United States Atomic Energy Commission *(University Park: Pennsylvania State University Press, 1962). Two valuable memoirs are Arthur Holly Compton,* Atomic Quest: A Personal Narrative *(Oxford: Oxford University Press, 1956); and Leslie R. Groves,* Now It Can Be Told: The Story of the Manhattan Project *(New York: Harper and Row, 1962). The abridged version of the declassified multi-volume project history is available in Anthony Cave Brown and Charles B. MacDonald, eds.,* The Secret History of the Atomic Bomb *(New York: Delta, 1977). On the Los Alamos environment, see James W. Kunetka,* City of Fire: Los Alamos and the Birth of the Atomic Age 1943–1945 *(Englewood Cliffs, N.J.: Prentice-Hall, 1978). The best study of the diplomacy of the atomic bomb is Martin Sherwin,* A World Destroyed: The Atomic Bomb and the Grand Alliance *(New York: Random House, 1977). On Oppenheimer, see Peter Goodchild,* J. Robert Oppenheimer: Shatterer of Worlds *(Boston: Houghton Mifflin, 1981). A superb history of the development of radiation protection standards is Barton C. Hacker,* The Dragon's Tail: Radiation Safety in the Manhattan Project, 1942–1945 *(Berkeley: University of California Press, 1987). On the cultural impact of the bomb, see Paul Boyer,* By the Bomb's Early Light: American Thought and Culture at the Dawn of the Atomic Age *(New York: Pantheon, 1985). On the first detonation, see Ferenc Morton Szasz,* The Day the Sun Rose Twice: The Story of the Trinity Nuclear Explosion, July 16, 1946 *(Albuquerque: University of New Mexico Press, 1984).*

6. Letters of J. Robert Oppenheimer, 1942–45

A. To James B. Conant

Berkeley
November 30, 1942

Dear Dr. Conant:

Your letter reached me with some delay since I returned to Berkeley only a day or so ago. I should like to answer first your P.S. You are quite right that the purities listed in Groves' compulsory memo are a little misleading. The reason for this is that Groves defined a satisfactory bomb as one that had a 50 per cent chance of exceeding a 1,000-ton TNT equivalent. The absolute requirements are figured on this basis. It is, of course, my opinion that we should be wanton to strive for such a low goal, but I believe that some good was in fact done by indicating at that time that the purity requirements are not fantastic. The desirable requirements are equally undefined since the purer the material (up to a purity of about 100-fold that given) the less must we be worried about getting the maximum speed for the firing mechanism of the detonator; and this will make for simplicity and reliability in operation. In the Washington memo all impurities were listed on the assumption that not more than five elements would reach the tabulated values. I have, in the meantime, given a much more careful account of what the actual situation is to the committee. I met with them one day in Chicago, came out with them on the train, and have spent two days with them here in Berkeley, and we have had ample opportunity to discuss the purity question and many other

These letters appear in Alice Kimball Smith and Charles Weiner, eds., *Robert Oppenheimer: Letters and Recollections* (Cambridge, Mass.: Harvard University Press, 1980), pp. 240–46, 256, 262–63, 267–68, 270–72, 282, 286–87, 293, 294.

aspects of our problem. The information which I have given them now is contained in a Chicago report on the feasibility of the 49 project and is as follows: If the concentrations by weight are as given in the accompanying table, then the chance of pre-detonation is 5 per cent, if only one element is present in the listed amount. If n elements are present in the listed amount the chance of pre-detonation is 5n per cent. The chance of a pre-detonation in which the energy release is less than 10,000 tons TNT equivalent is 0.5n per cent. In this range the effects of impurities are additive, and from the actual concentrations and figures listed one can figure out the probability of any given energy release. In any case, unless the firing mechanism fails completely the energy release will be more than sufficient to destroy the material and to make its recovery impossible. The figures given in the table are in part based on experimental values. In the case of O and C the figures represent highly conservative estimates based on the assumption that those isotopes which are dangerous will be as dangerous as the worst element, namely, Be.

Element Concentration by Weight

Be	10^{-7}	F	5×10^{-6}
Li	5×10^{-7}	Na	2×10^{-5}
B	2×10^{-7}	Mg	10^{-4}
C	2×10^{-5}	Al	2×10^{-5}
O	10^{-4}	Si	5×10^{-4}
		P	10^{-4}

(Some purity requirements on elements between P and Fe,
none beyond Cu)

The only essential changes since the Washington memo are that we have sufficiently examined the experiments somewhat; and that we have studied the case of N carefully enough to be positive that there are no purity requirements on that element. The committee was of the opinion that the purity requirements as they now stand could, with a very high probability, be met. In fact, the Chicago uranium is good enough except for C and O, and they have made no effort at all to solve this problem. If it were necessary it would be possible to work with depleted C and O and so considerably relax the chemical conditions on these elements. In fact, the committee was of the opinion that the major extraction processes which have to be handled automatically and the removal of traces of active material, coupled with the necessity for working in lots of less than 100 grams or of introducing suitable neutron absorbers as "safers," would present greater technical difficulties than the purification. Nevertheless, in our last discussions they seemed convinced that the helium-cooled graphite pile was a good bet.

Now to the second point, the main subject of your letter, where I feel myself on less secure ground. It is, of course, natural that the men we are after will leave a big hole. I may though, in this connection, remind you that when McMillan himself left the Radiation Laboratory for San Diego there were the same dire predictions of disastrous disruption. Nevertheless, the Radiation Laboratory has not only survived but has, as you know, flourished and expanded. In view of this and of the very large number of men of the first rank who are now working on that project, I am inclined not to take too seriously the absolute no's with which we shall be greeted. I believe that it is important to emphasize that we should in any case be willing to let these men have time enough in their old positions to try to minimize the disruption of their leaving. I also agree that a fundamental clarification on this personnel problem, which can hardly be complete without Dr. Bush's participation, will be necessary. The job we have to do will not be possible without personnel substantially greater than that which we now have available, and I should only be misleading you and all others concerned with the S-1 project if I were to promise to get the work done without this help.

The suggestion of Eckhardt as a substitute for Kurie is a welcome one and we shall arrange to talk with him on our next trip east.[1] There are, however, two reasons more substantial than prejudice why the limitation to men who are known to us is sound: 1) that the technical details of this work will in large part have to do with atomic physics so that any man whose experience has been in another field will necessarily be of more limited usefulness; Kurie, for instance, would have had as one of his responsibilities the installation and servicing of the cyclotron. The second reason is that in a tight isolated group such as we are now planning, some warmth and trust in personal relations is an indispensable prerequisite, and we are, of course, able to insure this only in the case of men whom we have known in the past. You will have had from me a note on possible alternatives to Kurie. If none of our suggestions seem practicable we shall see whether Dr. Eckhardt could fill the bill.

With good wishes,

very sincerely yours,
Robert Oppenheimer

1. Probably E. A. Eckhardt, a physicist who was vice-president of Gulf Research and Development Company in 1942.

B. Memorandum to Leslie R. Groves

Los Alamos
April 30, 1943

In accordance with our discussion of last week, I have given some thought to the question of a story about the Los Alamos Project which, if disseminated in the proper way, might serve somewhat to reduce the curiosity of the local population, and at least to delay the dissemination of the truth.

We propose that it be let known that the Los Alamos Project is working on a new type of rocket and that the detail be added that this is a largely electrical device. We feel that the story will have a certain credibility; that the loud noises which we will soon be making here will fit in with the subject; and that the fact, unfortunately not kept completely secret, that we are installing a good deal of electrical equipment, and the further fact that we have a large group of civilian specialists would fit in quite well. We further believe that the remoteness of the site for such a development and the secrecy which has surrounded the project would both be appropriate, and that the circumstance that a good deal of work is in fact being done on rockets, together with the appeal of the word, makes this story one which is both exciting and credible.

This question has been discussed with the governing board of the laboratory who approve it and who further recommend that the technical staff of the laboratory be specially warned neither to contradict nor to support a story of this kind if they should run into it.

J. R. Oppenheimer

C. Leslie R. Groves to J. Robert Oppenheimer

July 29, 1943

Dear Dr. Oppenheimer:

In view of the nature of the work on which you are engaged, the knowledge of it which is possessed by you and the dependence which rests upon you for its successful accomplishment, it seems necessary to ask you to take certain special precautions with respect to your personal safety.

It is requested that:

(a) You refrain from flying in airplanes of any description; the time saved is not worth the risk. (If emergency demands their use my prior consent should be requested.)

(b) You refrain from driving an automobile for any appreciable distance (above a few miles) and from being without suitable protection on any lonely road, such as the road from Los Alamos to Santa Fe. On such trips you should be accompanied by a competent, able bodied, armed guard. There is no objection to the guard serving as chauffeur.

(c) Your cars be driven with due regard to safety and that in driving about town a guard of some kind should be used, particularly during hours of darkness. The cost of such guard is a proper charge against the United States.

I realize that these precautions may be personally burdensome and that they may appear to you to be unduly restrictive but I am asking you to bear with them until our work is successfully completed.

Sincerely,
L. R. Groves
Brigadier General, C.E.

D. To Leslie R. Groves

Los Alamos
November 2, 1943

Dear General Groves:

After you gave me the list during your last visit of the men whom we may expect from the United Kingdom, it occurred to me that it might be wise before they arrive here to give them new names. This refers especially to Niels Bohr. I am thinking of the fact that mail will be addressed to them, that they may on occasion originate or receive long-distance calls, that they will be making some local purchases, and that for all these routine matters it would be preferable if such well known names were not put in circulation.

It has, in fact, troubled us some that we are forced to place calls for Dr. Conant, Fermi, Lawrence, etc. This does not happen very often, but in view of the fact that we try not to use these names over the telephone, the placing of the calls themselves seems to us rather unwise. I doubt whether at this late date it would be practicable to assign new names to those who have been associated with the project in the past. In the case of Bohr and Chadwick I think it would be advisable to do so before they get here.

Sincerely yours,
J. R. Oppenheimer

E. To Leslie R. Groves

(Los Alamos)
October 6, 1944

Dear General Groves:

I am glad to transmit the enclosed report of Captain Parsons, with the general intent and spirit of which I am in full sympathy. There are a few points on which my evaluation differs somewhat from that expressed in the report and it seems appropriate to mention them at this time.

1. I believe that Captain Parsons somewhat misjudges the temper of the responsible members of the laboratory. It is true that there are a few people here whose interests are exclusively "scientific" in the sense that they will abandon any problem that appears to be soluble. I believe that these men are now in appropriate positions in the organization. For the most part the men actually responsible for the prosecution of the work have proven records of carrying developments through the scientific and into the engineering stage. For the most part these men regard their work here not as a scientific adventure, but as a responsible mission which will have failed if it is let drop at the laboratory phase. I therefore do not expect to have to take heroic measures to insure something which I know to be the common desire of the overwhelming majority of our personnel.

2. I agree completely with all the comments of Captain Parsons' memorandum on the fallacy of regarding a controlled test as the culmination of the work of this laboratory. The laboratory is operating under a directive to produce weapons; this directive has been and will be rigorously adhered to. The only reason why we contemplate making a test, and why I have in the past advocated this, is because with the present time scales and the present radical assembly design this appears to be a necessary step in the production of a weapon. I do not wish to prejudge the issue: it is possible that information available to us within the next months may make such a test unnecessary. I believe, however, that the probability of this is extremely small.

3. The developmental program of the laboratory, whether or not it has been prosecuted with intelligence and responsibility, is still far behind the minimal requirements set by our directive. This fact, which rests on no

perfectionist ideals for long-range development, means that there must inevitably be some duplication of effort and personnel if the various phases of our program,—scientific, engineering and military,—are to be carried out without too great mutual interference. It is for this reason that I should like to stress Captain Parsons' remark that a very great strengthening in engineering is required. The organizational experience which the last year has given us is no substitute for competent engineers.

Sincerely yours,

J. R. Oppenheimer

F. To the Secretary of War [1]

August 17, 1945

Dear Mr. Secretary:

The Interim Committee has asked us to report in some detail on the scope and program of future work in the field of atomic energy. One important phase of this work is the development of weapons; and since this is the problem which has dominated our war time activities, it is natural that in this field our ideas should be most definite and clear, and that we should be most confident of answering adequately the questions put to us by the committee. In examining these questions we have, however, come on certain quite general conclusions, whose implications for national policy would seem to be both more immediate and more profound than those of the detailed technical recommendations to be submitted. We, therefore, think it appropriate to present them to you at this time.

1. We are convinced that weapons quantitatively and qualitatively far more effective than now available will result from further work on these problems. This conviction is motivated not alone by analogy with past developments, but by specific projects to improve and multiply the existing weapons, and by the quite favorable technical prospects of the realization of the super bomb.

2. We have been unable to devise or propose effective military countermeasures for atomic weapons. Although we realize that future work may reveal possibilities at present obscure to us, it is our firm opinion that no military countermeasures will be found which will be adequately effective in preventing the delivery of atomic weapons.

The detailed technical report in preparation will document these conclusions, but hardly alter them.

1. Henry Stimson

3. We are not only unable to outline a program that would assure to this nation for the next decades hegemony in the field of atomic weapons; we are equally unable to insure that such hegemony, if achieved, could protect us from the most terrible destruction.

4. The development, in the years to come, of more effective atomic weapons, would appear to be a most natural element in any national policy of maintaining our military forces at great strength; nevertheless we have grave doubts that this further development can contribute essentially or permanently to the prevention of war. We believe that the safety of this nation—as opposed to its ability to inflict damage on an enemy power—cannot lie wholly or even primarily in its scientific or technical prowess. It can be based only on making future wars impossible. It is our unanimous and urgent recommendation to you that, despite the present incomplete exploitation of technical possibilities in this field, all steps be taken, all necessary international arrangements be made, to this one end.

5. We should be most happy to have you bring these views to the attention of other members of the Government, or of the American people, should you wish to do so.

Very sincerely,
J. R. Oppenheimer
For the Panel

7. The Quebec Agreement, August 19, 1943

The Citadel, Quebec.

Articles of Agreement Governing Collaboration Between The Authorities of the U.S.A. and the U.K. in the Matter of Tube Alloys[1]

Whereas it is vital to our common safety in the present War to bring the Tube Alloys project to fruition at the earliest moment; and whereas this may

This text appears in Foreign Relations of the United States, *The First Quebec Conference* (Washington, D.C.: U.S. Government Printing Office, 1970), pp. 1117–19; reproduced in Margaret Gowing, *Britain and Atomic Energy 1939–1945* (New York: St. Martin's Press, 1964), pp. 439–40.

1. These Articles of Agreement are typed on four pages of stationery on each of which

be more speedily achieved if all available British and American brains and resources are pooled; and whereas owing to war conditions it would be an improvident use of war resources to duplicate plants on a large scale on both sides of the Atlantic and therefore a far greater expense has fallen upon the United States;

It is agreed between us

First, that we will never use this agency against each other.

Secondly, that we will not use it against third parties without each other's consent.

Thirdly, that we will not either of us communicate any information about Tube Alloys to third parties except by mutual consent.

Fourthly, that in view of the heavy burden of production falling upon the United States as the result of a wise division of war effort, the British Government recognize that any post-war advantages of an industrial or commercial character shall be dealt with as between the United States and Great Britain on terms to be specified by the President of the United States to the Prime Minister of Great Britain. The Prime Minister expressly disclaims any interest in these industrial and commercial aspects beyond what may be considered by the President of the United States to be fair and just and in harmony with the economic welfare of the world.

And Fifthly, that the following arrangements shall be made to ensure full and effective collaboration between the two countries in bringing the project to fruition:

---

appears the letterhead "The Citadel Quebec." For a photocopy of the British signed original of these Articles, see "Articles of Agreement governing collaboration between the authorities of the U.S.A. and U.K. in the matter of Tube Alloys" (Cmd. 9123; London: H. M. Stationery Office, 1954). The text of the articles printed by the Department of State in 1954 as Treaties and Other International Acts Series No. 2993 was prepared from a photocopy of the British signed original. The United States signed original used as the source text for the document printed here is identical with the British signed original except that (a) the three United States members of the Combined Policy Committee appear in typed form in the United States original whereas they are in Roosevelt's handwriting in the British original; (b) the form of the date at the end of the document (in Roosevelt's handwriting on both copies) reads "August 19 1943" in the British original and "Aug. 19th 1943" in the United States original (in which also the first digit of 19th appears to have been written over a figure 2, presumably because Roosevelt began to write 20th and then corrected it to 19th); and (c) there is a period in the United States original after the surname of C. D. Howe.

The source text of the Articles of Agreement is attached to a memorandum from Churchill's Principal Private Secretary to Roosevelt's Naval Aide, dated at Quebec, August 19, 1943, which reads as follows: "Admiral Wilson Brown. I attach, for retention, one of the two copies of the Articles of Agreement relating to Tube Alloys, signed by the President and Mr. Churchill today. J. M. Martin."

(a) There shall be set up in Washington a Combined Policy Committee composed of:

The Secretary of War.	(United States)
Dr. Vannevar Bush.	(United States)
Dr. James B. Conant.	(United States)
Field-Marshal Sir John Dill, G.C.B., C.M.G., D.S.O.	(United Kingdom)
Colonel the Right Hon. J. J. Llewellin, C.B.E., M.C., M.P.	(United Kingdom)
The Honourable C. D. Howe.	(Canada)

The functions of this Committee, subject to the control of the respective Governments, will be:

(1) To agree from time to time upon the programme of work to be carried out in the two countries.

(2) To keep all sections of the project under constant review.

(3) To allocate materials, apparatus and plant, in limited supply, in accordance with the requirements of the programme agreed by the Committee.

(4) To settle any questions which may arise on the interpretation or application of this Agreement.

(b) There shall be complete interchange of information and ideas on all sections of the project between members of the Policy Committee and their immediate technical advisers.

(c) In the field of scientific research and development there shall be full and effective interchange of information and ideas between those in the two countries engaged in the same sections of the field.

(d) In the field of design, construction and operation of large-scale plants, interchange of information and ideas shall be regulated by such ad hoc arrangements as may, in each section of the field, appear to be necessary or desirable if the project is to be brought to fruition at the earliest moment. Such ad hoc arrangements shall be subject to the approval of the Policy Committee.

Approved[2]

Aug. 19th 1943[2]

Franklin D. Roosevelt
Winston S. Churchill

2. In Roosevelt's handwriting in the source text.

8. Anglo-American Declaration of Trust, June 13, 1944

This Agreement and Declaration of Trust is made the thirteenth day of June One thousand nine hundred and fourty four by Franklin Delano Roosevelt on behalf of the Government of the United States of America, and by Winston Leonard Spencer Churchill on behalf of the Government of the United Kingdom of Great Britain and Northern Ireland. The said Governments are hereinafter referred to as "the Two Governments";

Whereas an agreement (hereinafter called the Quebec Agreement) was entered into on the nineteenth day of August One thousand nine hundred and forty three by and between the President of the United States and the Prime Minister of the United Kingdom; and

Whereas it is an object vital to the common interests of those concerned in the successful prosecution of the present war to insure the acquisition at the earliest practicable moment of an adequate supply of uranium and thorium ores; and

Whereas it is the intention of the Two Governments to control to the fullest extent practicable the supplies of uranium and thorium ores within the boundaries of such areas as come under their respective jurisdictions; and

Whereas the Government of the United Kingdom of Great Britain and Northern Ireland intends to approach the Governments of the Dominions and the Governments of India and of Burma for the purpose of securing that such Governments shall bring under control deposits of the uranium and thorium ores within their respective territories; and

Whereas it has been decided to establish a joint organisation for the purpose of gaining control of the uranium and thorium supplies in certain areas outside the control of the Two Governments and of the Governments of the Dominions and of India and of Burma;

Now it is Hereby Agreed and Declared as Follows:

1. (1) There shall be established in the City of Washington, District of Columbia, a Trust to be known as "The Combined Development Trust".
(2) The Trust shall be composed of and administered by six persons who shall be appointed, and be subject to removal, by the Combined Policy Committee established by the Quebec Agreement.

This text appears in Margaret Gowing, *Britain and Atomic Energy 1939–1945* (New York: St. Martin's Press, 1964), pp. 441–42.

2. The Trust shall use its best endeavours to gain control of and develop the production of the uranium and thorium supplies situate in certain areas other than the the the areas under the jurisdiction of the Two Governments and of the governments of the Dominions and of India and of Burma and for that purpose shall take such steps as it may in the common interest think fit to:

a Explore and survey sources of uranium and thorium supplies.

b Develop the production of uranium and thorium by the acquisition of mines and ore deposits, mining concessions or otherwise.

c Provide with equipment any mines or mining works for the production of uranium and thorium.

d Survey and improve the methods of production of uranium and thorium.

e Acquire and undertake the treatment and disposal of uranium and thorium and uranium and thorium materials.

f Provide storage and other facilities.

g Undertake any functions or operations which conduce to the effective carrying out of the purpose of the Trust in the common interest.

3. (1) The Trust shall carry out its functions under the direction and guidance of the Combined Policy Committee, and as its agent, and all uranium and thorium ores and supplies and other property acquired by the Trust shall be held by it in trust for the Two Governments jointly, and disposed of or otherwise dealt with in accordance with the direction of the Combined Policy Committee.

(2) The Trust shall submit such reports of its activities as may be required from time to time by the Combined Policy Committee.

4. For the purpose of carrying out its functions, the Trust shall utilize whenever and wherever practicable the established agencies of any of the Two Governments, and may employ and pay such other agents and employees as it considers expedient, and may delegate to any agents or employees all or any of its functions.

5. The Trust may acquire and hold any property in the name of nominees.

6. All funds properly required by the Trust for the performance of its functions shall be provided as to one-half by the Government of the United States of America and the other half by the Government of the United Kingdom of Great Britain and Northern Ireland.

7. In the event of the Combined Policy Committee ceasing to exist, the functions of the Committee under the Trust shall be performed by such other body or person as may be designated by the President for the time being of the United States of America and the Prime Minister for

the time being of the United Kingdom of Great Britain and Northern Ireland.

8. The signatories of this Agreement and Declaration of Trust will, as soon as practicable after the conclusion of hostilities, recommend to their respective Governments the extension and revision of this war-time emergency agreement to cover post war conditions and its formalization by treaty or other proper method. This Agreement and Declaration of Trust shall continue in full force and effect until such extension or revision.

(Signed) Franklin D. Roosevelt
On Behalf of the Government of the United States of
America

(Signed) Winston S. Churchill
On Behalf of the Government of the United Kingdom of
Great Britain and Northern Ireland

---

9. Roosevelt-Churchill Hyde Park Aide-Memoire, September 19, 1944

---

1. The suggestion that the world should be informed regarding tube alloys, with a view to an international agreement regarding its control and use, is not accepted. The matter should continue to be regarded as of the utmost secrecy; but when a "bomb" is finally available, it might perhaps, after mature consideration, be used against the Japanese, who should be warned that this bombardment will be repeated until they surrender.

2. Full collaboration between the United States and the British Government in developing tube alloys for military and commercial purposes should continue after the defeat of Japan unless and until terminated by joint agreement.

3. Enquiries should be made regarding the activities of Professor Bohr and steps taken to ensure that he is responsible for no leakage of information particularly to the Russians.

This text appears in Foreign Relations of the United States, 1944, vol. 2 (Washington, D.C.: U.S. Government Printing Office, 1967), pp. 1026–28.

10. Military Policy Committee Minutes, May 5, 1943

The point of use of the first bomb was discussed and the general view appeared to be that its best point of use would be on a Japanese fleet concentration in the Harbor of Truk. General Styer suggested Tokio [sic], but it was pointed out that the bomb should be used where, if it failed to go off, it would land in water of sufficient depth to prevent easy salvage. The Japanese were selected as they would not be so apt to secure knowledge from it as would the Germans.

11. Stimson and the Interim Committee, May 1945

War Department
Washington

1 May 1945

Memorandum for the Secretary of War
Subject: Interim Committee on S-1.

Last week you presented to President Truman a fairly complete memorandum on the S-1 project, outlining its genesis, its present state of development and in general its availability for military usage. Your presentation was accompanied by a brief memo which you prepared relative to the broader political and international implications of the problem and the need for post war controls, both national and international. You had in mind the advisability of setting up a committee of particular qualifications for recommending action to the executive and legislative branches of the government when secrecy is no longer fully required. The committee would also be

(Selection 10). The document from which this text is excerpted is found in Manhattan Engineer District-Top Secret, folder 23A, Record Group 77, National Archives, Washington, D.C.

(Selection 11). This document is found in Manhattan Engineer District-Top Secret, Harrison-Bundy files, folder 69, National Archives, Washington, D.C.

expected to recommend actions to be taken by the War Department in anticipation of the post war problems.

In view of the possibly short time available before actual military use and the relaxation of secrecy, it seems to me,—and as you know both Dr. Conant and Dr. Bush agree—that it is becoming more and more important to organize such a committee as promptly as possible. This committee should, I think, be a relatively small committee which should be prepared to serve temporarily or until Congress might appoint a permanent Post War commission to supervise, regulate and control the use of the product.

Certain things, however, must be done now before use if we are to avoid the risk of grave repercussions on the public in general and on Congress in particular. For instance, the committee will need to prepare appropriate announcements to be available for issue (a) by the President and (b) by the War Department as soon as the first bomb is used. These announcements or later publicity would presumably give some of the history of the project, its importance from a military standpoint, its scientific background, and some of its dangers. Most importantly as soon as possible after use some assurance must be given of the steps to be taken to provide the essential controls over post-war use and development, both at home and abroad. With that in mind it will be necessary as soon as possible after use to make recommendations for the necessary Congressional legislation covering patents, use, controls, etc.

All of these and many other factors will have to be studied by the committee with the understanding that all recommendations must be for your own approval and for submission to the President for his approval.

It seems clear that some machinery is essential now to provide the way for continuous and effective controls and to insure or provide for the necessary and persistent research and development of the possibilities of atomic energy in which the United States now leads the way. If properly controlled by the peace loving nations of the world this energy should insure the peace of the world for generations. If misused it may lead to the complete destruction of civilization.

In the circumstances I suggest that a committee of six or seven be set up at once to study and report on the whole problem of temporary war controls and publicity, and to survey and make recommendation on post-war research, development and controls, and the legislation necessary to effectuate them.

The members of this committee should be appointed by you as Secretary of War subject to the approval of the President. When appointed the committee will need promptly to organize appropriate panels to aid in its work—panels of specially qualified scientists, Army and Navy personnel, Congressional advisers, legislative draftsmen and others.

George L. Harrison

12. Interim Committee Minutes, May 31, 1945

10:00 A.M. to 1:15 P.M.—2:25 P.M. to 4:15 P.M.

Present:

Members of the Committee
Secretary Henry L. Stimson, Chairman
Hon. Ralph A. Bard
Dr. Vannevar Bush
Hon. James F. Byrnes
Hon. William L. Clayton
Dr. Karl T. Compton
Dr. James B. Conant
Mr. George L. Harrison

Invited Scientists
Dr. J. Robert Oppenheimer
Dr. Enrico Fermi
Dr. Arthur H. Compton
Dr. E. O. Lawrence

By Invitation
General George C. Marshall
Major Gen. Leslie R. Groves
Mr. Harvey H. Bundy
Mr. Arthur Page

This document is found in Manhattan Engineer District-Top Secret, Harrison-Bundy files, folder 100, National Archives, Washington, D.C.

I. Opening Statement of the Chairman:

Secretary Stimson explained that the Interim Committee had been appointed by him, with the approval of the President, to make recommendations on temporary war-time controls, public announcement, legislation and post-war organization. The Secretary gave high praise to the brilliant and effective assistance rendered to the project by the scientists present for their great contributions to the work and their willingness to advise on the many complex problems that the Interim Committee had to face. He expressed the hope that the scientists would feel completely free to express their views on any phase of the subject.

The Committee had been termed an "Interim Committee" because it was expected that when the project became more widely known a permanent organization established by Congressional action or by treaty arrangements would be necessary.

The Secretary explained that General Marshall shared responsibility with him for making recommendations to the President on this project with particular reference to its military aspects; therefore, it was considered highly desirable that General Marshall be present at this meeting to secure at first hand the views of the scientists.

The Secretary expressed the view, a view shared by General Marshall, that this project should not be considered simply in terms of military weapons, but as a new relationship of man to the universe. This discovery might be compared to the discoveries of the Copernican theory and of the laws of gravity, but far more important than these in its effect on the lives of men. While the advances in the field to date had been fostered by the needs of war, it was important to realize that the implications of the project went far beyond the needs of the present war. It must be controlled if possible to make it an assurance of future peace rather than a menace to civilization.

The Secretary suggested that he hoped to have the following questions discussed during the course of the meeting:

1. Future military weapons.
2. Future international competition.
3. Future research.
4. Future controls.
5. Future developments, particularly non-military.

II. Stages of Development:

As a technical background for the discussions, Dr. A. H. Compton explained the various stages of development. The first stage involved the separation of uranium 235. The second stage involved the user of "breeder" piles to produce enriched materials from which plutonium or new types of uranium could be obtained. The first stage was being used to produce material for the present bomb while the second stage would produce atomic bombs with a tremendous increase in explosive power over those now in production. Production of enriched materials was now on the order of pounds or hundreds of pounds and it was contemplated that the scale of operations could be expanded sufficiently to produce many tons. While bombs produced from the products of the second stage had not yet been proven in actual operation, such bombs were considered a scientific certainty. It was estimated that from January 1946 it would take one and one-half years to prove [sic] this second stage in view of certain technical and metallurgical difficulties; that it would take three years to get plutonium in volume, and that it would take perhaps six years for any competitor to catch up with us.

Dr. Fermi estimated that approximately twenty pounds of enriched material would be needed to carry on research in current engineering problems and that a supply of one-half to one ton would be needed for research on the second stage.

In response to the Secretary's question, Dr. A. H. Compton stated that the second stage was dependent upon vigorous exploitation of the first stage and would in no way vitiate the expenditure already made on the present plant.

Dr. Conant mentioned a so-called "third stage" of development in which the products of the "second stage" would be used simply as a detonator for heavy water. He asked Dr. Oppenheimer for an estimate of the time factor involved in developing this phase. Dr. Oppenheimer stated that this was a far more difficult development than the previous stages and estimated that a minimum of three years would be required to reach production. He pointed out that heavy water (hydrogen) was much cheaper to produce than the other materials and could eventually be obtained in far greater quantity.

Dr. Oppenheimer reviewed the scale of explosive force involved in these several stages. One bomb produced in the first stage was estimated to have the explosive force of 2,000–20,000 tons of TNT. The actual blast effect would be accurately measured when the test was made. In the second stage the explosive force was estimated to be equal to 50,000–100,000 tons of TNT. It was considered possible that a bomb developed from the third stage might produce an explosive force equal to 10,000,000–100,000,000 tons of TNT. . . .

VI. Russia:

In considering the problem of controls and international collaboration the question of paramount concern was the attitude of Russia. *Dr. Oppenheimer* pointed out that Russia had always been very friendly to science and suggested that we might open up this subject with them in a tentative fashion and in the most general terms without giving them any details of our productive effort. He thought we might say that a great national effort had been put into this project and express a hope for cooperation with them in this field. He felt strongly that we should not prejudge the Russian attitude in this matter.

At this point *General Marshall* discussed at some length the story of charges and counter-charges that have been typical of our relations with the Russians, pointing out that most of these allegations have proven unfounded. The seemingly uncooperative attitude of Russia in military matters stemmed from the necessity of maintaining security. He said that he had accepted this reason for their attitude in his dealings with the Russians and had acted accordingly. As to the post-war situation and in matters other than purely military, he felt that he was in no position to express a view. With regard to this field he was inclined to favor the building up of a combination among like-minded powers, thereby forcing Russia to fall in line by the very force of this coalition. General Marshall was certain that we need have no fear that the Russians, if they had knowledge of our project, would disclose this information to the Japanese. He raised the question whether it might be desirable to invite two prominent Russian scientists to witness the test.

Mr. Byrnes expressed a fear that if information were given to the Russians, even in general terms, Stalin would ask to be brought into the partnership. He felt this to be particularly likely in view of our commitments and pledges of cooperation with the British. In this connection *Dr. Bush* pointed out that even the British do not have any of our blue prints on plants. *Mr. Byrnes* expressed the view, *which was generally agreed to by all present*, that the most desirable program would be to push ahead as fast as possible in production and research to make certain that we stay ahead and at the same time make every effort to better our political relations with Russia. . . .

VIII. Effect of the Bombing on the Japanese and Their Will to Fight:

It was pointed out that one atomic bomb on an arsenal would not be much different from the effect caused by any Air Corps strike of present dimensions. However, *Dr. Oppenheimer* stated that the visual effect of an atomic bombing would be tremendous. It would be accompanied by a brilliant luminescence which would rise to a height of 10,000 to 20,000 feet. The neutron effect of the explosion would be dangerous to life for a radius of at least two-thirds of a mile.

After much discussion concerning various types of targets and the effects to be produced, *the Secretary expressed the conclusion, on which there was general agreement, that we could not give the Japanese any warning; that we could not concentrate on a civilian area; but that we should seek to make a profound psychological impression on as many of the inhabitants as possible. At the suggestion of Dr. Conant the Secretary agreed that the most desirable target would be a vital war plant employing a large number of workers and closely surrounded by workers' houses.*

There was some discussion of the desirability of attempting several strikes at the same time. *Dr. Oppenheimer's* judgment was that several strikes would be feasible. *General Groves*, however, expressed doubt about this proposal and pointed out the following objections: (1) We would lose the advantage of gaining additional knowledge concerning the weapon at each successive bombing; (2) such a program would require a rush job on the part of those assembling the bombs and might, therefore, be ineffective; (3) the effect would not be sufficiently distinct from our regular Air Force bombing program.

13. Glenn T. Seaborg to Ernest O. Lawrence, June 13, 1945[1]

June 13, 1945

Mr. Ernest O. Lawrence
Radiation Laboratory
University of California
Berkeley, California

Dear Ernest:

I am writing to give you my opinions and suggestions on the question of the course to be taken for nuclear weapons in the immediate future, and also on the question of the post war future for neucleonics. My purpose is to express myself briefly here upon such political and social questions as the release of information, use of the weapon in the present war and post war control of the weapon, rather than upon the actual research program which would be difficult to cover in a short communication. As you know, it is difficult to express unqualified opinions on such political and social questions as these on the basis of information available to us, and therefore I have the feeling that my views could change if there is important information, especially in respect to the present war, which is not at our disposal. I do want to say, however, that these present opinions are shared almost unanimously by the people associated with me in my section of the Chemistry Division here. For the purpose of brevity I shall list our conclusions with little or no discussion of the basis for the development of these conclusions. These opinions of course are based on the assumption that the development of a nuclear weapon of great destructive capacity is now essentially an accomplished fact.

I believe that the basic facts concerning the successful release of nuclear energy and its immense destructive possibilities should be made public and impressed upon public opinion in this country and all over the world very soon. There should be essentially two stages in the release of this information, disclosure to the general public of the results which have been obtained, and the publication of these results through more or less regular

This letter is found in the Seaborg papers, Lawrence Berkeley Laboratory. Reprinted by permission of Glenn T. Seaborg.

1. Glenn T. Seaborg was a student of Lawrence and one of the discoverers of plutonium, for which he received the Nobel Prize in 1951. Ernest O. Lawrence was a leading physicist and director of the Radiation Laboratory at Berkeley.

scientific channels. The first of these, disclosure to the public, should come soon and probably need go no further than to describe the destructive possibilities of the self-sustaining chain reaction with the heavy isotopes, with some non-technical description of the achievements in the manufacturing of such fissionable material. The method to be employed for this release should be chosen only after much careful study; perhaps a stepwise release, studying the effect at each step, should be used. The publication in regular scientific channels, which is not an urgent matter, should come later but should then go at least so far as to cover the entire scientific basis of the accomplishments. This would include such items as the existence of and the important nuclear properties of the heavy isotopes, the fundamental information about the nuclear chain reactions, the basic information concerned with the methods for the separation of the uranium isotopes, the fundamental information about chain-reacting structures used in the manufacture of the heavy synthetic isotopes, and the fundamental chemical properties of the new synthetic elements. Perhaps it would be all right to withhold indefinitely some of the information with respect to the actual detailed designs of the major manufacturing installations. This might not be construed as too unnatural a procedure in that the maintenance of secrecy in regard to ordnance information and in respect to many industrial operations has an established precedent in this country.

With respect to use in the present war we suggest the following. Our country would probably lose some of the confidence of our Allies and deteriorate our moral position with respect to the outlawing of future use of the weapon if we were to use it directly upon Japan without warning. It seems certain that the moral position of our country would be greatly strengthened if the first demonstration of this weapon were made upon some uninhabited island in the presence of the invited representatives of all the leading countries of the world, including Japan. Following such a successful demonstration Japan would be given an ultimatum to surrender and if this ultimatum was not accepted, the question of then using the weapon would be decided by the United States together with the United Nations; the sanction of other leading nations of the world would be important. The question of international control of this weapon, touched upon in the next paragraph, should of course be vigorously pursued immediately after the demonstration.

The question of the post war control of nucleonics is a most difficult one. The above-described disposition of the weapon in the present war amounts essentially to subordinating its use now toward the broader goal of insuring control over it in the longer post war future. One method of post war control lies in the complete outlawing of nucleonics research throughout the world; I believe that this method, which amounts to advocating the suppression of science, is too unnatural for it to succeed. We would favor, rather, if it could possibly be made consistent with our national security and with world security, free research in nucleonics throughout the world with complete exchange of all the basic information and some degree of control through an international organization. Probably the best method of control lies in the control of the raw materials, although this is admittedly difficult. Completely free research in nucleonics, unfortunately, makes it possible for any country to accumulate a stockpile of fissionable material. It is the opinion of some that probably the only method of maintaining control under such conditions would involve world-wide pooling to form a stockpile of fissionable material to be used by the international organization for policing purposes; I do not feel qualified to express an opinion on this complicated possibility. As suggested by Szilard, perhaps control could be effected, at least in the case of some of the fissionable material, by denaturization, i.e. by mixing it with suitable isotopes to spoil its use for explosive purposes without interfering too much with its use for research purposes such as power pile developments.

With respect to the organization of post war research in nucleonics in our country, I believe that the establishment, with government aid, of about four large research laboratories at four of the major universities is a good idea. These laboratories should form a sort of a foundation for the country's research program and should include men who are able and willing to advise outlying laboratories as to research program. The outlying laboratories might consist of Government laboratories working on the more practical aspects of the field, and also regular university and industrial laboratories supported by Government contracts or grants-in-aid. This government-aided research would be concerned with the application of nucleonics to military and defense purposes and with such other applications as are recognized as governmental purposes. There should be no reason to restrict the development of nucleonics along other lines and industry should be free to work on its application to such fields as power piles, the manufacture of radioactive isotopes and other fields which they may wish to develop.

Sincerely yours,
G. T. Seaborg

14. Science Panel Recommendations on the Immediate Use of Nuclear Weapons, June 16, 1945

TOP SECRET
RECOMMENDATIONS ON THE IMMEDIATE USE OF
NUCLEAR WEAPONS

June 16, 1945

You have asked us to comment on the initial use of the new weapon. This use, in our opinion, should be such as to promote a satisfactory adjustment of our international relations. At the same time, we recognize our obligation to our nation to use the weapons to help save American lives in the Japanese war.

(1) To accomplish these ends we recommend that before the weapons are used not only Britain, but also Russia, France, and China be advised that we have made considerable progress in our work on atomic weapons, that these may be ready to use during the present war, and that we would welcome suggestions as to how we can cooperate in making this development contribute to improved international relations.

(2) The opinions of our scientific colleagues on the initial use of these weapons are not unanimous: they range from the proposal of a purely technical demonstration to that of the military application best designed to induce surrender. Those who advocate a purely technical demonstration would wish to outlaw the use of atomic weapons, and have feared that if we use the weapons now our position in future negotiations will be prejudiced. Others emphasize the opportunity of saving American lives by immediate military use, and believe that such use will improve the international prospects, in that they are more concerned with the prevention of war than with the elimination of this specific weapon. We find ourselves closer to these latter views; we can propose no technical demonstration likely to bring an end to the war; we see no acceptable alternative to direct military use.

(3) With regard to these general aspects of the use of atomic energy, it is clear that we, as scientific men, have no proprietary rights. It is true that we are among the few citizens who have had occasion to give thoughtful consideration to these problems during the past few years. We have, however, no

This text is found in Manhattan Engineer District-Top Secret, Harrison-Bundy files, folder 76, National Archives, Washington, D.C.

claim to special competence in solving the political, social, and military problems which are presented by the advent of atomic power.

A. H. Compton
E. O. Lawrence
J. R. Oppenheimer
E. Fermi
[signed] J. R. Oppenheimer
For the Panel

---

15. Stimson and the Scientists' Concerns, June 26, 1945

WAR DEPARTMENT
WASHINGTON

26 June 1945
TOP SECRET

Memorandum for the Secretary of War:

Many of the scientists who have been working on S-1 have expressed considerable concern about the future dangers of the development of atomic power. Some are fearful that no safe system of international control can be established. They, therefore, envisage the possibility of an armament race that may threaten civilization.

One group of scientists, working in the Chicago Laboratories, urges that we should not make use of the bomb, so nearly completed, against any enemy country at this time. They feel that to do so might sacrifice our whole moral position and thus make it more difficult for us to be the leaders in proposing or enforcing any system of international control designed to make this tremendous force an influence towards the maintenance of world peace rather than an uncontrollable weapon of war.

This anonymous statement of the Chicago scientists was submitted for comment to the Panel of Scientists appointed by the Interim Committee.

This text is found in Manhattan Engineer District-Top Secret, Harrison-Bundy files, folder 77, National Archives, Washington, D.C.

Their answer was that they saw no acceptable alternative to direct military use since they believe that such use would be an obvious means of saving American lives and shortening the war.

It is interesting that practically all of the scientists, including those on the panel, feel great concern for the future if atomic power is not controlled through some effective international mechanism. Accordingly, most of them believe that one of the effective steps in establishing such a control is the assurance that, after this war is over, there shall be a free interchange of scientific opinion throughout the world supplemented, if possible, by some system of inspection. This they admit is a problem of the future. In the meantime, however, they feel that we must, even before actual use, briefly advise the Russians of our progress.

This matter of notice to the Russians was made a subject of thorough discussion at the last meeting of the Interim Committee on June 21. It was unanimously agreed that in view of the importance of securing an effective future control, and in view of the fact that most of the story, other than production secrets, will become known in————[sic] in any event, there would be considerable advantage, if a suitable opportunity arises at the "Big Three" meeting, in having the President advise the Russians simply that we are working intensely on this weapon and that, if we succeed as we think we will, we plan to use it against the enemy. Such a statement might well be supplemented by the statement that in the future, after the war, we would expect to discuss the matter further with a view to insuring that this means of warfare will become a substantial aid in preserving the peace of the world rather than a weapon of terror and destruction.

It was felt by the Committee that if the Russians should ask for more details now rather than later or if they should raise questions as to time-tables, methods of production, etc., they should be told that we are not yet ready to discuss the subject beyond the simple statement suggested above. Our purpose is merely to let them know that we did not wish to proceed with actual use without giving them prior information that we intend to do so. Not to give them this prior information at the time of the "Big Three" Conference and within a few weeks thereafter to use the weapon and to make fairly complete statements to the world about its history and development, might well make it impossible ever to enlist Russian cooperation in the set-up of future international controls over this new power.

It was agreed by the Committee that in view of the provisions of the Quebec Agreement it would be desirable to discuss this whole aspect of the question with the Prime Minister in advance of the "Big Three" Conference.

[signed] George L. Harrison

16. O. R. Frisch, Eyewitness Account of "Trinity" Test, July 1945

I watched the explosion from a point said to be about 20 (or 25) miles away and about north of it, together with the members of the co-ordinating council. Fearing to be dazzled and to be burned by ultraviolet rays, I stood with my back to the gadget, and behind the radio truck. I looked at the hills, which were visible in the first faint light of dawn (0530 M.W. Time). Suddenly and without any sound, the hills were bathed in brilliant light, as if somebody had turned the sun on with a switch. It is hard to say whether the light was less or more brilliant than full sunlight, since my eyes were pretty well dark adapted. The hills appeared kind of flat and colourless like a scenery seen by the light of a photographic flash, indicating presumably that the retina was stimulated beyond the point where intensity discrimination is adequate. The light appeared to remain constant for about one or two seconds (probably for the same reason) and then began to diminish rapidly.

After that I turned round and tried to look at the light source but found it still too bright to keep my eyes on it. A few short glances gave me the impression of a small very brilliant core much smaller in appearance than the sun, surrounded by decreasing and reddening brightness with no definite boundary, but not greater than the sun. After some seconds I could keep my eye on the thing and it now looked like a pretty perfect red ball, about as big as the sun, and connected to the ground by a short grey stem. The ball rose slowly, lengthening its stem and getting gradually darker and slightly larger. A structure of darker and lighter irregularities became visible, making the ball look somewhat like a raspberry. Then its motion slowed down and it flattened out, but still remained connected to the ground by its stem, looking more than ever like the trunk of an elephant. Then a hump grew out of its top surface and a second mushroom grew out of the top of the first one, slowly penetrating the highest cloud layers. As the red glow died out it became apparent that the whole structure, in particular the top mushroom, was surrounded by a purplish blue glow. A minute or so later the whole top mushroom appeared to glow feebly in this colour, but this was no longer easy to see, in the increasing light of dawn.

A very striking phenomenon was the sudden appearance of a white patch on the underside of the cloud layer just above the explosion; the patch

This text appears in Foreign Relations of the United States, *Potsdam*, vol. 2 (Washington, D.C.: U.S. Government Printing Office, 1960), p. 1371.

spread very rapidly, like a pool of spilt milk, and a second or two later, a similar patch appeared and spread on another cloud layer higher up. They marked no doubt the impact of the blast wave on the cloud layers. They appeared, I believe, before the red ball had started to flatten out.

When I thought it was soon time for the blast to arrive, I sat on the ground, still facing the explosion, and put my fingers in my ears. Despite that, the report was quite respectable and was followed by a long rumbling, not quite like thunder but more regular, like huge noisy waggons running around in the hills.

---

17. General Groves's Report on "Trinity," July 18, 1945

TOP SECRET   WASHINGTON, 18 July 1945.

Memorandum for the Secretary of War[1]

Subject: The Test.

1. This is not a concise, formal military report but an attempt to recite what I would have told you if you had been here on my return from New Mexico.

This text appears in Foreign Relations of the United States, *Potsdam*, vol. 2 (Washington, D.C.: U.S. Government Printing Office, 1960), pp. 1361–68.

1. [Henry] Stimson's diary entry for July 21 contains the following information relating to this document:

At eleven thirty-five General Groves' special report was received by special courier. It was an immensely powerful document, clearly and well written and with supporting documents of the highest importance. It gave a pretty full and eloquent report of the tremendous success of the test and revealed far greater destructive power than we expected in S-1. . . .

At three o'clock I found that Marshall had returned from the Joint Chiefs of Staff, and to save time I hurried to his house and had him read Groves' report and conferred with him about it.

I then went to the "Little White House" and saw President Truman. I asked him to call in Secretary Byrnes and then I read the report in its entirety and we then discussed it. They were immensely pleased. The President was tremendously pepped up by it and spoke to me of it again and again when I saw him. He said it gave him an entirely new feeling of confidence and he thanked me for having come to the Conference and being present to help him in this way.

2. At 0530,[2] 16 July 1945, in a remote section of the Alamogordo Air Base, New Mexico, the first full scale test was made of the implosion of the atomic fission bomb. For the first time in history there was a nuclear explosion. And what an explosion! . . . The bomb was not dropped from an airplane but was exploded on a platform on top of a 100-foot high steel tower.

3. The test was successful beyond the most optimistic expectations of anyone. Based on the data which it has been possible to work up to date, I estimate the energy generated to be in excess of the equivalent of 15,000 to 20,000 tons of TNT; and this is a conservative estimate. Data based on measurements which we have not yet been able to reconcile would make the energy release several times the conservative figure. There were tremendous blast effects. For a brief period there was a lighting effect within a radius of 20 miles equal to several suns in midday; a huge ball of fire was formed which lasted for several seconds. This ball mushroomed and rose to a height of over ten thousand feet before it dimmed. The light from the explosion was seen clearly at Albuquerque, Santa Fe, Silver City, El Paso and other points generally to about 180 miles away. The sound was heard to the same distance in a few instances but generally to about 100 miles. Only a few windows were broken although one was some 125 miles away. A massive cloud was formed which surged and billowed upward with tremendous power, reaching the substratosphere at an elevation of 41,000 feet, 36,000 feet above the ground in about five minutes, breaking without interruption through a temperature inversion at 17,000 feet which most of the scientists thought would stop it. Two supplementary explosions occurred in the cloud shortly after the main explosion. The cloud contained several thousand tons of dust picked up from the ground and a considerable amount of iron in the gaseous form. Our present thought is that this iron ignited when it mixed with the oxygen in the air to cause these supplementary explosions. Huge concentrations of highly radioactive materials resulted from the fission and were contained in this cloud.

4. A crater from which all vegetation had vanished, with a diameter of

---

Stimson showed Groves's report to [Gen. H. H.] Arnold on July 22.

Truman later stated that, following receipt of news that the Alamogordo test had been successful, he had called together [Sec. James F.] Byrnes, Stimson, [Adm. William D.] Leahy, [Gen. George] Marshall, Arnold, [Gen. Dwight] Eisenhower, and [Adm. Ernest J.] King and had asked them for their opinions as to whether the bomb should be used, and the consensus had been that it should . . . Truman apparently also received at this meeting an oral estimate of the casualties to be expected in the assault of Japan if the new weapon were not used.

2. That is, 5:30 A.M. All times in this memorandum are expressed in military style, i.e., from 0001 hours (12:01 A.M.) to 2400 hours (midnight).

1200 feet and a slight slope toward the center, was formed. In the center was a shallow bowl 130 feet in diameter and 6 feet in depth. The material within the crater was deeply pulverized dirt. The material within the outer circle is greenish and can be distinctly seen from as much as 5 miles away. The steel from the tower was evaporated. 1500 feet away there was a four-inch iron pipe 16 feet high set in concrete and strongly guyed. It disappeared completely.

5. One-half mile from the explosion there was a massive steel test cylinder weighing 220 tons. The base of the cylinder was solidly encased in concrete. Surrounding the cylinder was a strong steel tower 70 feet high, firmly anchored to concrete foundations. This tower is comparable to a steel building bay that would be found in typical 15 or 20 story skyscraper or in warehouse construction. Forty tons of steel were used to fabricate the tower which was 70 feet high, the height of a six story building. The cross bracing was much stronger than normally used in ordinary steel construction. The absence of the solid walls of a building gave the blast a much less effective surface to push against. The blast tore the tower from its foundations, twisted it, ripped it apart and left it flat on the ground. The effects on the tower indicate that, at that distance, unshielded permanent steel and masonry buildings would have been destroyed. I no longer consider the Pentagon a safe shelter from such a bomb. Enclosed are a sketch showing the tower before the explosion and a telephotograph showing what it looked like afterwards.[3] None of us had expected it to be damaged.

6. The cloud traveled to a great height first in the form of a ball, then mushroomed, then changed into a long trailing chimney-shaped column and finally was sent in several directions by the variable winds at the different elevations. It deposited its dust and radioactive materials over a wide area. It was followed and monitored by medical doctors and scientists with instruments to check its radioactive effects. While here and there the activity on the ground was fairly high, at no place did it reach a concentration which required evacuation of the population. Radioactive material in small quantities was located as much as 120 miles away. The measurements are being continued in order to have adequate data with which to protect the Government's interests in case of future claims. For a few hours I was none too comfortable about the situation.

7. For distances as much as 200 miles away, observers were stationed to check on blast effects, property damage, radioactivity and reactions of the population. While complete reports have not yet been received, I now know that no persons were injured nor was there any real property damage outside

3. Neither reproduced.

our Government area. As soon as all the voluminous data can be checked and correlated, full technical studies will be possible.

8. Our long range weather predictions had indicated that we could expect weather favorable for our tests beginning on the morning of the 17th and continuing for four days. This was almost a certainty if we were to believe our long range forecasters. The prediction for the morning of the 16th was not so certain but there was about an 80% chance of the conditions being suitable. During the night there were thunder storms with lightning flashes all over the area. The test had been originally set for 0400 hours and all the night through, because of the bad weather, there were urgings from many of the scientists to postpone the test. Such a delay might well have had crippling results due to mechanical difficulties in our complicated test set-up. Fortunately, we disregarded the urgings. We held firm and waited the night through hoping for suitable weather. We had to delay an hour and a half, to 0530, before we could fire. This was 30 minutes before sunrise.

9. Because of bad weather, our two B-29 observation airplanes were unable to take off as scheduled from Kirtland Field at Albuquerque and when they finally did get off, they found it impossible to get over the target because of the heavy clouds and the thunder storms. Certain desired observations could not be made and while the people in the airplanes saw the explosion from a distance, they were not as close as they will be in action. We still have no reason to anticipate the loss of our plane in an actual operation although we cannot guarantee safety.

10. Just before 1100 the news stories from all over the state started to flow into the Albuquerque Associated Press. I then directed the issuance by the Commanding Officer, Alamogordo Air Base of a news release as shown on the enclosure.[4] With the assistance of the Office of Censorship we were able to limit the news stories to the approved release supplemented in the

4. Identified in the source copy as a clipping from the *Albuquerque Tribune* for July 16, 1945:

Alamogordo, N. M., July 16—William O. Eareckson, commanding officer of the Alamogordo Army Air Base, made the following statement today:

"Several inquiries have been received concerning a heavy explosion which occurred on the Alamogordo Air Base reservation this morning.

"A remotely located ammunition magazine containing a considerable amount of high explosive and pyrotechnics exploded.

"There was no loss of life or injury to anyone, and the property damage outside of the explosives magazine itself was negligible.

"Weather conditions affecting the content of gas shells exploded by the blast may make it desirable for the Army to evacuate temporarily a few civilians from their homes."

local papers by brief stories from the many eyewitnesses not connected with our project. One of these was a blind woman who saw the light.

11. Brigadier General Thomas F. Farrell was at the control shelter located 10,000 yards south of the point of explosion. His impressions are given below:

"The scene inside the shelter was dramatic beyond words. In and around the shelter were some twenty-odd people concerned with last minute arrangements prior to firing the shot. Included were: Dr. Oppenheimer, the Director who had borne the great scientific burden of developing the weapon from the raw materials made in Tennessee and Washington and a dozen of his key assistants—Dr. Kistiakowsky, who developed the highly special explosives; Dr. Bainbridge, who supervised all the detailed arrangements for the test; Dr. Hubbard, the weather expert, and several others. Besides these, there were a handful of soldiers, two or three Army officers and one Naval officer. The shelter was cluttered with a great variety of instruments and radios.

"For some hectic two hours preceding the blast, General Groves stayed with the Director, walking with him and steadying his tense excitement. Every time the Director would be about to explode because of some untoward happening General Groves would take him off and walk with him in the rain, counselling with him and reassuring him that everything would be all right. At twenty minutes before zero hours, General Groves left for his station at the base camp, first because it provided a better observation point and second, because of our rule that he and I must not be together in situations where there is an element of danger, which existed at both points.

"Just after General Groves left, announcements began to be broadcast of the interval remaining before the blast. They were sent by radio to the other groups participating in and observing the test. As the time interval grew smaller and changed from minutes to seconds, the tension increased by leaps and bounds. Everyone in that room knew the awful potentialities of the thing that they thought was about to happen. The scientists felt that their figuring must be right and that the bomb had to go off but there was in everyone's mind a strong measure of doubt. The feeling of many could be expressed by 'Lord, I believe; help Thou mine unbelief.' We were reaching into the unknown and we did not know what might come of it. It can be safely said that most of those present—Christian, Jew and Atheist—were praying and praying harder than they had ever prayed before. If the shot were successful, it was a justification of the several years of intensive effort of tens of thousands of people—statesmen, scientists, engineers, manufacturers, soldiers, and many others in every walk of life.

"In that brief instant in the remote New Mexico desert the tremendous effort of the brains and brawn of all these people came suddenly and startlingly to the fullest fruition. Dr. Oppenheimer, on whom had rested a very heavy burden, grew tenser as the last seconds ticked off. He scarcely breathed. He held on to a post to steady himself. For the last few seconds, he stared directly ahead and then when the announcer shouted 'Now!' and there came this tremendous burst of light followed shortly thereafter by the deep growling roar of the explosion, his face relaxed into an expression of tremendous relief. Several of the observers standing back of the shelter to watch the lighting effects were knocked flat by the blast.

"The tension in the room let up and all started congratulating each other. Everyone sensed 'This is it!' No matter what might happen now all knew that the impossible scientific job had been done. Atomic fission would no longer be hidden in the cloisters of the theoretical physicists' dreams. It was almost full grown at birth. It was a great new force to be used for good or for evil. There was a feeling in that shelter that those concerned with its nativity should dedicate their lives to the mission that it would always be used for good and never for evil.

"Dr. Kistiakowsky,[5] the impulsive Russian, threw his arms around Dr. Oppenheimer and embraced him with shouts of glee. Others were equally enthusiastic. All the pent-up emotions were released in those few minutes and all seemed to sense immediately that the explosion had far exceeded the most optimistic expectations and wildest hopes of the scientists. All seemed to feel that they had been present at the birth of a new age—The Age of Atomic Energy—and felt their profound responsibility to help in guiding into right channels the tremendous forces which had been unlocked for the first time in history.

"As to the present war, there was a feeling that no matter what else might happen, we now had the means to insure its speedy conclusion and save thousands of American lives. As to the future, there had been brought into being something big and something new that would prove to be immeasurably more important than the discovery of electricity or any of the other great discoveries which have so affected our existence.

"The effects could well be called unprecedented, magnificent, beautiful, stupendous and terrifying. No man-made phenomenon of such tremendous power had ever occurred before. The lighting effects beggared description. The whole country was lighted by a searing light with the intensity many times that of the midday sun. It was golden, purple, violet, gray and blue. It lighted every peak, crevasse and ridge of the nearby mountain range with

5. At this point is the following manuscript interpolation by Groves: "An American and Harvard Professor for many years."

a clarity and beauty that cannot be described but must be seen to be imagined. It was that beauty the great poets dream about but describe most poorly and inadequately. Thirty seconds after the explosion came first, the air blast pressing hard against the people and things, to be followed almost immediately by the strong, sustained, awesome roar which warned of doomsday and made us feel that we puny things were blasphemous to dare tamper with the forces heretofore reserved to The Almighty. Words are inadequate tools for the job of acquainting those not present with the physical, mental and psychological effects. It had to be witnessed to be realized."

12. My impressions of the night's high points follow:

After about an hour's sleep I got up at 0100 and from that time on until about five I was with Dr. Oppenheimer constantly. Naturally he was nervous, although his mind was working at its usual extraordinary efficiency. I devoted my entire attention to shielding him from the excited and generally faulty advice of his assistants who were more than disturbed by their excitement and the uncertain weather conditions. By 0330 we decided that we could probably fire at 0530. By 0400 the rain had stopped but the sky was heavily overcast. Our decision became firmer as time went on. During most of these hours the two of us journeyed from the control house out into the darkness to look at the stars and to assure each other that the one or two visible stars were becoming brighter. At 0510 I left Dr. Oppenheimer and returned to the main observation point which was 17,000 yards from the point of explosion. In accordance with our orders I found all personnel not otherwise occupied massed on a bit of high ground.

At about two minutes of the scheduled firing time all persons lay face down with their feet pointing towards the explosion. As the remaining time was called from the loud speaker from the 10,000 yard control station there was complete silence. Dr. Conant said he had never imagined seconds could be so long. Most of the individuals in accordance with orders shielded their eyes in one way or another. There was then this burst of light of a brilliance beyond any comparison. We all rolled over and looked through dark glasses at the ball of fire. About forty seconds later came the shock wave followed by the sound, neither of which seemed startling after our complete astonishment to the extraordinary lighting intensity. Dr. Conant reached over and we shook hands in mutual congratulations. Dr. Bush, who was on the other side of me, did likewise. The feeling of the entire assembly was similar to that described by General Farrell, with even the uninitiated feeling profound awe. Drs. Conant and Bush and myself were struck by an even stronger feeling that the faith of those who had been responsible for the initiation and the carrying on of this Herculean project had been justified. I personally thought of Blondin crossing Niagara Falls on his tight rope, only to me this tight rope had lasted for almost three years and of my repeated confident-appearing assurances that such a thing was possible and that we would do it.

13. A large group of observers were stationed at a point about 27 miles north of the point of explosion. Attached is a memorandum written shortly after the explosion by Dr. E. O. Lawrence which may be of interest.

14. While General Farrell was waiting about midnight for a commercial airplane to Washington at Albuquerque—120 miles away from the site—he overheard several airport employees discussing their reaction to the blast. One said that he was out on the parking apron; it was quite dark; the light lasted several seconds. Another remarked that if a few exploding bombs could have such an effect, it must be terrible to have them drop on a city.

15. My liaison officer at the Alamogordo Air Base, 60 miles away, made the following report:

"There was a blinding flash of light that lighted the entire northwestern sky. In the center of the flash, there appeared to be a huge billow of smoke. The original flash lasted approximately 10 to 15 seconds. As the first flash died down, there arose in the approximate center of where the original flash had occurred an enormous ball of what appeared to be fire and closely resembled a rising sun that was three-fourths above a mountain. The ball of fire lasted approximately 15 seconds, then died down and the sky resumed an almost normal appearance.

"Almost immediately, a third, but much smaller, flash and billow of smoke of a whitish-orange color appeared in the sky, again lighting the sky for approximately 4 seconds. At the time of the original flash, the field was lighted well enough so that a newspaper could easily have been read. The second and third flashes were of much lesser intensity.

"We were in a glass-enclosed control tower some 70 feet above the ground and felt no concussion or air compression. There was no noticeable earth tremor although reports overheard at the Field during the following 24 hours indicated that some believed that they had both heard the explosion and felt some earth tremor."

16. I have not written a separate report for General Marshall as I feel you will want to show this to him. I have informed the necessary people here of our results. Lord Halifax after discussion with Mr. Harrison and myself stated that he was not sending a full report to his government at this time. I informed him that I was sending this to you and that you might wish to show it to the proper British representatives.

17. We are all fully conscious that our real goal is still before us. The battle test is what counts in the war with Japan.

18. May I express my deep personal appreciation for your congratulatory

cable to us and for the support and confidence which I have received from you ever since I have had this work under my charge.

19. I know that Colonel Kyle will guard these papers with his customary extraordinary care.

L. R. Groves

18. Leslie R. Groves to George C. Marshall, July 30, 1945

30 July 1945

MEMORANDUM TO THE CHIEF OF STAFF

1. The following additional conclusions have been drawn from the test in New Mexico with respect to the probable effects of the combat bomb which will be exploded about 1800 feet in the air:

a. Measured from the point on the ground directly below the explosion the blast should be lethal to at least 1000 feet. Between 2500 and 3500 feet, blast effects should be extremely serious to personnel. Heat and flame should be fatal to about 1500 to 2000 feet.

b. At 10 miles for a few thousandths of a second the light will be as bright as a thousand suns; at the end of a second, as bright as one or possibly two suns. The effect on anyone about a half mile away who looks directly at the explosion would probably be permanent sight impairment; at one mile, temporary blindness; and up to and even beyond ten miles, temporary sight impairment. To persons who are completely unshielded, gamma rays may be lethal to 3500 feet and neutrons to about 2000 feet.

c. No damaging effects are anticipated on the ground from radioactive materials. These effects at New Mexico resulted from the low altitude from which the bomb was set off.

d. Practically all structures in an area of one or two square miles should be completely demolished and a total area of six to seven square miles should be so devastated that the bulk of the buildings would have to have major repairs to make them habitable.

This document is found in Manhattan Engineer District-Top Secret, Manhattan Project File, Folder 4, Trinity Test, National Archives, Washington, D.C.

e. At New Mexico tanks could have gone through the immediate explosion area at normal speeds within thirty minutes after the blast. With the explosion at the expected 1800 feet, we think we could move troops through the area immediately preferably by motor but on foot if desired. The units should be preceded by scouts with simple instruments. The nearest exposed personnel should not be nearer to the blast than six miles plus the necessary allowance for bombing inaccuracy and they would require a high order of discipline and special but simple instructions. As an extra precaution, extra special dark glasses might be issued to all commanders of units as large as a platoon. If dropped on the enemy lines, the expected effect on the enemy would be to wipe out his resistance over an area 2000 feet in diameter; to paralyse it over an area a mile in diameter; and to impede it seriously over an area five miles in diameter. Troops which were in deep cave shelters at distances of over a mile should not be seriously affected. Men in slit trenches within 800 feet should be killed by the blast.

2. The energy of the test explosion has been broken down as follows:
Total theoretical energy contained in the bomb at 100% efficiency was [sensitive information deleted]. Of this amount, 21,000 to 24,000 tons were converted into actual energy made up of:
Blast—10,500 tons minimum, 13,500 maximum
Light—2500 tons
Waste Heat—8000 tons, about 4000 of which went into the air and 4000 into the ground. If the explosion had been at the combat altitude of 1800 feet, most of the 4000 that went into the ground would have been converted into blast, making the total blast from 14,000 to 17,000 tons.

3. There is a definite possibility, [sensitive information deleted] as we increase our rate of production at the Hanford Engineer Works, with the type of weapon tested that the blast will be smaller due to detonation in 19advance of the optimum time. But in any event, the explosion should be on the order of thousands of tons. The difficulty arises from an undesirable isotope which is created in greater quantity as the production rate increases.

4. The final components of the first gun type bomb have arrived at Tinian, those of the first implosion type should leave San Francisco by airplane early on 30 July. I see no reason to change our previous readiness predictions on the first three bombs. In September, we should have three or four bombs. One of these will be made from 235 material and will have a smaller effectiveness, about two-thirds that of the test type, but by November, we should be able to bring this up to full power. There should be either four or three bombs in October, one of the lesser size. In November there should be at least five bombs and the rate will rise to seven in December

and increase decidedly in early 1946. By some time in November, we should have the effectiveness of the 235 implosion type bomb equal to that of the tested plutonium implosion type.

5. By mid-October we could increase the number of bombs slightly by changing our design now to one using both materials in the same bomb. I have not made this change because of the ever present possibilities of difficulties in new designs. We could, if it were wise, change our plans and develop the combination bomb. But if this is to be done, it would entail an initial ten-day production setback which would be caught up in about a month's time; unless the decision to change were made before 1 August, in which case it would probably not entail any delay. From what I know of the world situation, it would seem wiser not to make this change until the effects of the present bomb are determined.

L. R. GROVES
Major General, U.S.A.

19. Truman-Stalin Conversation, Potsdam, July 24, 1945

Present*

United States†	Soviet Union
President Truman	Generalissimo Stalin
	Mr. Pavlov

[Harry] Truman gives the following account of the conversation in *Year of Decisions*, p. 416: "On July 24 I casually mentioned to Stalin that we had

This text appears in Foreign Relations of the United States, *Potsdam*, vol. 2 (Washington, D.C.: U.S. Government Printing Office, 1960), pp. 1376–79.

*This conversation took place at the Cecelienhof Palace immediately after the adjournment of the Eighth Plenary Meeting. [No official record of it has ever been found.]

†[James F.] Byrnes (*All in One Lifetime* [New York: Harper, 1958], p. 300) mentions [Charles] Bohlen's presence as Truman's interpreter at this conversation, and, in a conference Department of State historians had with Truman and members of his staff on January 24, 1956, Truman supplied the information that both Bohlen and a Soviet interpreter were present at this conversation. In a conversation with a Department of State historian on January 26, 1960, however, Bohlen stated categorically that Truman walked over to Stalin alone in order to give the conversation a more casual flavor, and that Pavlov did the interpreting.

a new weapon of unusual destructive force. The Russian Premier showed no special interest. All he said was that he was glad to hear it and hoped we would make 'good use of it against the Japanese.'"

[James F.] Byrnes[1] gives the following information on the conversation in *Speaking Frankly*, p. 263: "At the close of the meeting of the Big Three on the afternoon of July 24, the President walked around the large circular table to talk to Stalin. After a brief conversation the President rejoined me and we rode back to the 'Little White House' together. He said he had told Stalin that, after long experimentation, we had developed a new bomb far more destructive than any other known bomb, and that we planned to use it very soon unless Japan surrendered. Stalin's only reply was to say that he was glad to hear of the bomb and he hoped we would use it."

[William D.] Leahy's account is as follows (*I Was There*, p. 429): "At the plenary session on July 24, Truman walked around to Stalin and told him quietly that we had developed a powerful weapon, more potent than anything yet seen in war. The President said later that Stalin's reply indicated no especial interest and that the Generalissimo did not seem to have any conception of what Truman was talking about. It was simply another weapon and he hoped we would use it effectively."

[Winston] Churchill, who was also an eye-witness to the conversation, gives the following information (*Triumph and Tragedy*, pp. 669–70): "Next day, July 24, after our plenary meeting had ended and we all got up from the round table and stood about in twos and threes before dispersing, I saw the President go up to Stalin, and the two conversed alone with only their interpreters. I was perhaps five yards away, and I watched with the closest attention the momentous talk. I knew what the President was going to do. What was vital to measure was its effect on Stalin. I can see it all as if it were yesterday. He seemed to be delighted. . . . As we were waiting for our cars I found myself near Truman. 'How did it go?' I asked. 'He never asked a question,' he replied."

---

1. Books cited here and following paragraphs: Byrnes, *Speaking Frankly* (1947; rpt. Westport, Conn.: Greenwood Press, 1974); Leahy, *I Was There*, ed. Richard H. Kohn (1950; rpt. Salem, N.H.: Ayer Co.); Churchill, *Triumph and Tragedy*, vol. 6 of *The Second World War* (Boston: Houghton Mifflin, 1953).

20. Chicago Scientists' Petition to the President, July 17, 1945

(Drafted by Leo Szilard and signed by 68 members of the Metallurgical Laboratory in Chicago)

Discoveries of which the people of the United States are not aware may affect the welfare of this nation in the near future. The liberation of atomic power which has been achieved places atomic bombs in the hands of the Army. It places in your hands, as Commander-in-Chief, the fateful decision whether or not to sanction the use of such bombs in the present phase of the war against Japan.

We, the undersigned scientists, have been working in the field of atomic power. Until recently we have had to fear that the United States might be attacked by atomic bombs during this war and that her only defense might lie in a counterattack by the same means. Today, with the defeat of Germany, this danger is averted and we feel impelled to say what follows:

The war has to be brought speedily to a successful conclusion and attacks by atomic bombs may very well be an effective method of warfare. We feel, however, that such attacks on Japan could not be justified, at least not until the terms which will be imposed after the war on Japan were made public in detail and Japan were given an opportunity to surrender.

If such, public announcement gave assurance to the Japanese that they could look forward to a life devoted to peaceful pursuits in their homeland and if Japan still refused to surrender, our action might then, in certain circumstances, find itself forced to resort to the use of atomic bombs. Such a step, however, ought not to be made at any time without seriously considering the moral responsibilities which are involved.

The development of atomic power will provide the nations with new weapons of destruction. The atomic bombs at our disposal represent only the first step in this direction, and there is almost no limit to the destructive power which will become available in the course of their future development. Thus a nation which sets the precedent of using these newly liberated forces of nature for purposes of destruction may have to bear the responsibility of opening the door to an era of devastation on an unimaginable scale.

If after the war a situation is allowed to develop in the world which

This document appears in S. R. Weart and Gertrud Weiss, eds., *Leo Szilard: His Version of the Facts*, vol. 2 (Cambridge, Mass.: MIT Press, 1978), pp. 211–12. Copyright © 1978 by MIT Press. Reprinted by permission of MIT Press.

permits rival powers to be in uncontrolled possession of these new means of destruction, the cities of the United States as well as the cities of other nations will be in continuous danger of sudden annihilation. All the resources of the United States, moral and material, may have to be mobilized to prevent the advent of such a world situation. Its prevention is at present the solemn responsibility of the United States—singled out by virtue of her lead in the field of atomic power.

The added material strength which this lead gives to the United States brings with it the obligation of restraint and if we were to violate this obligation our moral position would be weakened in the eyes of the world and in our own eyes. It would then be more difficult for us to live up to our responsibility of bringing the unloosened forces of destruction under control.

In view of the foregoing, we, the undersigned, respectfully petition: first, that you exercise your power as Commander-in-Chief to rule that the United States shall not resort to the use of atomic bombs in this war unless the terms which will be imposed upon Japan have been made public in detail and Japan knowing these terms has refused to surrender; second, that in such an event the question whether or not to use atomic bombs be decided by you in the light of the consideration presented in this petition as well as all the other moral responsibilities which are involved.

---

21. White House Press Release on Hiroshima, August 6, 1945

Statement by the President of the United States *

Sixteen hours ago an American airplane dropped one bomb on Hiroshima, an important Japanese Army base. That bomb had more power than 20,000 tons of T.N.T. It had more than two thousand times the blast power of the British "Grand Slam" which is the largest bomb ever yet used in the history of warfare.

The Japanese began the war from the air at Pearl Harbor. They have

This text appears in Foreign Relations of the United States, *Potsdam*, vol. 2 (Washington, D.C.: U.S. Government Printing Office, 1960), pp. 1380–81.

* Released August 6, 1945. A draft of this statement had been discussed with the British.

been repaid many fold. And the end is not yet. With this bomb we have now added a new and revolutionary increase in destruction to supplement the growing power of our armed forces. In their present form these bombs are now in production and even more powerful forms are in development.

It is an atomic bomb. It is a harnessing of the basic power of the universe. The force from which the sun draws its power has been loosed against those who brought war to the Far East.

Before 1939, it was the accepted belief of scientists that it was theoretically possible to release atomic energy. But no one knew any practical method of doing it. By 1942, however, we knew that the Germans were working feverishly to find a way to add atomic energy to the other engines of war with which they hoped to enslave the world. But they failed. We may be grateful to Providence that the Germans got the V-1's and the V-2's late and in limited quantities and even more grateful that they did not get the atomic bomb at all.

The battle of the laboratories held fateful risks for us as well as the battles of the air, land and sea, and we have now won the battle of the laboratories as we have won the other battles.

Beginning in 1940, before Pearl Harbor, scientific knowledge useful in war was pooled between the United States and Great Britain, and many priceless helps to our victories have come from that arrangement. Under that general policy the research on the atomic bomb was begun. With American and British scientists working together we entered the race of discovery against the Germans.

The United States had available the large number of scientists of distinction in the many needed areas of knowledge. It had the tremendous industrial and financial resources necessary for the project and they could be devoted to it without undue impairment of other vital war work. In the United States the laboratory work and the production plants, on which a substantial start had already been made, would be out of reach of enemy bombing, while at that time Britain was exposed to constant air attack and was still threatened with the possibility of invasion. For these reasons Prime Minister Churchill and President Roosevelt agreed that it was wise to carry on the project here. We now have two great plants and many lesser works devoted to the production of atomic power. Employment during peak construction numbered 125,000 and over 65,000 individuals are even now engaged in operating the plants. Many have worked there for two and a half years. Few know what they have been producing. They see great quantities of material going in and they see nothing coming out of these plants, for the physical size of the explosive charge is exceedingly small. We have spent two billion dollars on the greatest scientific gamble in history—we won.

But the greatest marvel is not the size of the enterprise, its secrecy, nor its cost, but the achievement of scientific brains in putting together infinitely complex pieces of knowledge held by many men in different fields of science into a workable plan. And hardly less marvelous has been the capacity of industry to design, and of labor to operate, the machines and methods to do things never done before so that the brain child of many minds came forth in physical shape and performed as it was supposed to do. Both science and industry worked under the direction of the United States Army, which achieved a unique success in managing so diverse a problem in the advancement of knowledge in an amazingly short time. It is doubtful if such another combination could be got together in the world. What has been done is the greatest achievement of organized science in history. It was done under high pressure and without failure.

We are now prepared to obliterate more rapidly and completely every productive enterprise the Japanese have above ground in any city. We shall destroy their docks, their factories, and their communications. Let there be no mistake; we shall completely destroy Japan's power to make war.

It was to spare the Japanese people from utter destruction that the ultimatum of July 26 was issued at Potsdam. Their leaders promptly rejected that ultimatum. If they do not now accept our terms they may expect a rain of ruin from the air, the like of which has never been seen on this earth. Behind this air attack will follow sea and land forces in such numbers and power as they have not yet seen and with the fighting skill of which they are already well aware.

The Secretary of War, who has kept in personal touch with all phases of this project, will immediately make public a statement giving further details.[1]

His statement will give facts concerning the sites of Oak Ridge near Knoxville, Tennessee, and at Richland near Pasco, Washington, and an installation near Santa Fe, New Mexico. Although the workers at the sites have been making materials to be used in producing the greatest destructive force in history they have not themselves been in danger beyond that of many other occupations, for the utmost care has been taken of their safety.

The fact that we can release atomic energy ushers in a new era in man's understanding of nature's forces. Atomic energy may in the future supplement the power that now comes from coal, oil, and falling water, but at pres-

---

1. For the text of Stimson's statement of August 6, 1945, see Raymond Dennett and Robert K. Turner, eds., *Documents on American Foreign Relations, July 1, 1945–December 31, 1945* (Princeton, N.J.: Princeton University Press, 1948), p. 413

ent it cannot be produced on a basis to compete with them commercially. Before that comes there must be a long period of intensive research.

It has never been the habit of the scientists of this country or the policy of this Government to withhold from the world scientific knowledge. Normally, therefore, everything about the work with atomic energy would be made public.

But under present circumstances it is not intended to divulge the technical processes of production or all the military applications, pending further examination of possible methods of protecting us and the rest of the world from the danger of sudden destruction.

I shall recommend that the Congress of the United States consider promptly the establishment of an appropriate commission to control the production and use of atomic power within the United States. I shall give further consideration and make further recommendations to the Congress as to how atomic power can become a powerful and forceful influence towards the maintenance of world peace.

III

Atomic Energy in a Postwar World

At the end of World War II, the United States had spent over two billion dollars to develop an enormous atomic industry operated by the U.S. Army. The Manhattan Engineer District had nurtured a new science and produced a weapon that won the war and stunned the world. Yet, in 1945, scientists associated with the Manhattan Project also saw a potential for the peaceful atom.

The military and scientific establishments fought for control of atomic energy after the war. The military services proposed that they be represented on an atomic energy commission, an independent agency in which the military would have a strong voice. The May-Johnson Bill introduced in Congress in 1946 represented the military view that weapons were of the highest importance and that the secret of the bomb could be preserved only through military control.

Most scientists disagreed. The Federation of Atomic Scientists, made up of many of those who had worked on the bomb, lobbied in Washington for civilian control. Working privately in small groups, prominent scientists visited individual congressmen. During the war, military control had restricted exchange of information and actually hindered progress on the bomb, they said. More importantly, the scientists persuasively argued that civilian decision makers, not the military, should control nuclear weapons. As a safety measure, they wanted the agency in charge of atomic weapons to be independent of the armed forces.

The scientists won the legislative battle. The Atomic Energy Act of 1946, *sponsored by Senator Brien McMahon of Connecticut, established an independent, civilian-controlled Atomic Energy Commission. A Military Liaison Committee did oversee the military activities of the Atomic Energy Commission, but the five civilian commissioners retained complete responsibility for weapons, which could not be delivered to the armed forces without a direct presidential order.*

Civilian control at home gave no assurances for what happened in the international arena. Those who had witnessed the horrible destructive power of the bomb at Alamogordo or had seen photographs of Hiroshima and Nagasaki sought to ban the bomb altogether. In the spring of 1946 a special panel within the Department of State drafted a proposal for the international control of atomic energy. The Acheson-Lilienthal report took its name from the two best known members of the panel, Undersecretary Dean G. Acheson and David E. Lilienthal, former head of the Tennessee Valley Authority but soon to be the first chairman of the Atomic Energy Commission. The principal author of the report, however, was Oppenheimer. The proposal suggested a series of steps by which the nations of the world would surrender their possession of atomic weapons. During the transition, the United States would maintain its monopoly and all nations would be subject to international inspection.

In June 1946, shortly after Secretary of State George C. Marshall outlined his plan for the economic recovery of Europe, President Truman set aside the Acheson-Lilienthal Report and asked Bernard M. Baruch, a prominent New York banker and public servant, to devise a new plan as the U.S. delegate to the United Nations Atomic Energy Commission. Baruch proposed the creation of an International Atomic Development Authority. The United States would agree to place its atomic bomb secrets with the authority provided that the great powers could establish a suitable international control and inspection system. In addition, Baruch offered to cease manufacturing weapons and destroy all existing bombs, indeed, to eliminate war altogether. The plan, Baruch announced, was "the last, best hope of earth."

The sticking point for the Soviet Union was Baruch's insistence that the Russians could not block the plan through a veto in the United Nations Security Council, especially on the issue of vote punishment for offensive nations. The Soviet Union did not want a majority vote of the Security Council to determine its atomic future. But the United States recognized that it could not maintain its nuclear monopoly for long, and only a gradual reduction in its nuclear arsenal would maintain the balance of power. Many felt that the ways to eliminate nuclear weapons had to be explored.

The positions taken by the United States and the Soviet Union did nothing to ease suspicions and mutual distrust. The opportunity for international control of atomic weapons at their birth was gone, frozen for over a de-cade by the icy storms of the Cold War. The collapse of negotiations at the United Nations a year after the introduction of the Baruch Plan and Presi-dent Harry S. Truman's failure to achieve any disarmament agreement with the Soviet Union had driven the issue out of the international forum until 1952. Early that year, the General Assembly established a new disarmament commission, but, as before, disagreements over disclosure of weapons and an effective inspection system doomed its efforts.

Shortly after Eisenhower took office in January 1953, he received a re-port from an advisory group headed by Oppenheimer. Truman's Secretary of State, Dean Acheson, had asked Oppenheimer to reexamine the question of disarmament. The report argued that there were too many nuclear weapons. The United States and the Soviet Union were like "two scorpions in a bottle, each capable of killing the other, but only at the risk of his own life," Oppen-heimer wrote. The American people had to understand the dangers of atomic weapons. What was required, the scientist explained, was "candor on the part of the officials, the representatives, of the people of their coun-try." The idea for Operation Candor was launched.

Operation Candor became a heated subject for debate within the Eisen-hower administration. Some advisers supported Oppenheimer in advocating a frank discussion to increase the public's knowledge of nuclear danger. Ad-miral Lewis Strauss, the chairman of the Atomic Energy Commission, was horrified that too much candor would aid Soviet espionage but do little for the American public. Eisenhower settled the debate by borrowing from both sides. Alarmed that a nuclear war would leave everyone dead on both sides, the president said he didn't "want to scare the country to death. Can't we find some hope?" The death of Joseph Stalin in March and the explosion of a Soviet hydrogen device in August gave the appeal even more promise and urgency.

Over a series of discussions, Eisenhower developed a new approach. Each nation, he proposed, should give the United Nations a specified amount of fissionable material to form an atomic pool for peaceful uses, pri-marily nuclear power. Because the plan emerged over the breakfast table, insiders soon called the approach that replaced Operation Candor, "Opera-tion Wheaties."

Eisenhower arranged to announce his Atoms for Peace plan before the United Nations after consulting with Prime Minister Winston Churchill and his advisers in Bermuda. On December 8, 1953, Eisenhower stood be-fore the General Assembly. The Voice of America would carry his words around the world.

The first part of the speech suggested that a nuclear or thermonuclear holocaust demanded an easing of East-West relationships, but the heart of Eisenhower's message was an appeal for peace. The powers had to find a way "to serve the needs rather than the fears of mankind." By establishing a stockpile of uranium under an International Atomic Energy Agency, the atom might "serve the peaceful pursuits of mankind." Concluding, Eisenhower promised that the United States would "devote its entire heart and mind to find the way by which the miraculous inventiveness of man shall not be dedicated to his death, but consecrated to his life."

The audience of thirty-five hundred people sat silently until the final words, then burst into enthusiastic applause unprecedented in United Nations history. The delegations from the Soviet bloc countries joined the cheers. The Atoms for Peace proposal had offered hope to a fearful world.

Eisenhower's speech, full of idealism, did not address the thorny issues of nuclear proliferation, of nuclear disarmament, or of verifiable inspection systems. Rather, the president sought to avoid the old barrier and went his way to a simple international agreement before proceeding with controversial proposals on general disarmament. But within a year the proposal was foundering on mutual American and Soviet suspicions. The United States wanted to bring the Soviets to some terms before their atomic stockpile grew too large. The Russians sought further delays. Not until October 1956 was the International Atomic Energy Agency established with the participation of seventy-two nations, including the Soviet Union. Although it has allocated fissionable materials to other nations for industrial purposes, its main contribution has been the forum for the establishment of international standards for radiation protection.

Valuable memoirs on these years are Roger M. Anders, ed., Forging the Atomic Shield: Excerpts from the Office Diary of Gordon E. Dean *(Chapel Hill: University of North Carolina Press, 1987); Lewis L. Strauss,* Men and Decisions *(Garden City, N.Y.: Doubleday, 1962); and Dwight D. Eisenhower,* Mandate for Change, 1953–1956: The White House Years *(Garden City, N.Y.: Doubleday, 1963). See also U.S. Congress, Senate, Committee on Foreign Relations,* Atoms for Peace Manual, *84 Congress, 1 sess., 1955, Senate Report 55; and more recently Herbert S. Parmet,* Eisenhower and the American Crusades *(New York: Macmillan, 1972).*

On the postwar years, see both Richard G. Hewlett and Oscar E. Anderson, The New World, 1939–1946 *(University Park: Pennsylvania State University Press, 1962) and Richard G. Hewlett and Francis Duncan,* Atomic Shield, 1947–1952, *in the same series (1969). On the international control of atomic energy after 1945, see Walter LaFeber,* America, Russia, and the Cold War, 1945–1966 *(New York: Wiley, 1967) and the United Nations Department of Political and Security Council Affairs,* The United Nations and

Disarmament, 1945–1970 *(New York: United Nations, 1970)*. *On the postwar role of the atomic bomb in American foreign policy, see Gregg Herken,* The Winning Weapon: The Atomic Bomb in the Cold War, 1945–1950 *(New York: Alfred A. Knopf, 1980)*.

The importance of the Eisenhower administration in the development of atomic energy and the peaceful uses of the atom are examined in the prizewinning third volume of the history of the Atomic Energy Commission by Richard G. Hewlett and Jack M. Holl, Atoms for Peace and War, 1953–1961: Eisenhower and the Atomic Energy Commission *(Berkeley: University of California Press, 1989); George T. Mazuzan and J. Samuel Walker,* Controlling the Atom: The Beginnings of Nuclear Regulation, 1946–1962 *(Berkeley: University of California Press, 1984); and Richard Pfau,* No Sacrifice Too Great: The Life of Lewis L. Strauss *(Charlottesville: University Press of Virginia, 1984)*. *A provocative study of the best known Soviet spy is Robert Chadwell Williams,* Klaus Fuchs *(Cambridge, Mass.: Harvard University Press, 1987)*.

22. Henry L. Stimson to Harry S. Truman, September 11, 1945

Dear Mr. President:

In handing you today my memorandum about our relations with Russia in respect to the atomic bomb, I am not unmindful of the fact that when in Potsdam I talked with you about the question whether we could be safe in sharing the atomic bomb with Russia while she was still a police state and before she put into effect provisions assuring personal rights of liberty to the individual citizen.

I still recognize the difficulty and am still convinced of the ultimate importance of a change in Russian attitude toward individual liberty but I have come to the conclusion that it would not be possible to use our possession of the atomic bomb as a direct lever to produce the change. I have become convinced that any demand by us for an internal change in Russia as a condition of sharing in the atomic weapon would be so resented that it would make the objective we have in view less probable.

This text appears in Foreign Relations of the United States, *1945*, vol. 2 (Washington, D.C.: U.S. Government Printing Office, 1971), pp. 40–44.

I believe that the change in attitude toward the individual in Russia will come slowly and gradually and I am satisfied that we should not delay our approach to Russia in the matter of the atomic bomb until that process has been completed. My reasons are set forth in the memorandum I am handing you today. Furthermore, I believe that this long process of change in Russia is more likely to be expedited by the closer relationship in the matter of the atomic bomb which I suggest and the trust and confidence that I believe would be inspired by the method of approach which I have outlined.

Faithfully yours,
Henry L. Stimson
Secretary of War.

The President,
The White House

MEMORANDUM FOR THE PRESIDENT

Subject: The advent of the atomic bomb has stimulated great military and probably even greater political interest throughout the civilized world. In a world atmosphere already extremely sensitive to power, the introduction of this weapon has profoundly affected political considerations in all sections of the globe.

In many quarters it has been interpreted as a substantial offset to the growth of Russian influence on the continent. We can be certain that the Soviet Government has sensed this tendency and the temptation will be strong for the Soviet political and military leaders to acquire this weapon in the shortest possible time. Britain in effect already has the status of a partner with us in the development of this weapon. Accordingly, unless the Soviets are voluntarily invited into the partnership upon a basis of cooperation and trust, we are going to maintain the Anglo-Saxon bloc over against the Soviet in the possession of this weapon. Such a condition will almost certainly stimulate feverish activity on the part of the Soviet toward the development of this bomb in what will in effect be a secret armament race of a rather desperate character. There is evidence to indicate that such activity may have already commenced.

If we feel, as I assume we must, that civilization demands that some day we shall arrive at a satisfactory international arrangement respecting the control of this new force, the question then is how long we can afford to enjoy our momentary superiority in the hope of achieving our immediate peace council objectives.

Whether Russia gets control of the necessary secrets of production in a minimum of say four years or a maximum of twenty years is not nearly as important to the world and civilization as to make sure that when they do get it they are willing and cooperative partners among the peace-loving nations of the world. It is true if we approach them now, as I would propose, we may be gambling on their good faith and risk their getting into production of bombs a little sooner than they would otherwise.

To put the matter concisely, I consider the problem of our satisfactory relations with Russia as not merely connected with but as virtually dominated by the problem of the atomic bomb. Except for the problem of the control of the bomb, those relations, while vitally important, might not be immediately pressing. The establishment of relations of mutual confidence between her and us could afford to wait the slow progress of time. But with the discovery of the bomb they became immediately emergent. Those relations may be perhaps irretrievably embittered by the way in which we approach the solution of the bomb with Russia. For if we fail to approach them now and merely continue to negotiate with them, having this weapon rather ostentatiously on our hip, their suspicions and their distrust of our purposes and motives will increase. It will inspire them to greater efforts in an all-out effort to solve the problem. If the solution is achieved in that spirit, it is much less likely that we will ever get the kind of covenant we may desperately need in the future. This risk, is, I believe, greater than the other, inasmuch as our objective must be to get the best kind of international bargain we can—one that has some chance of being kept and saving civilization not for five or for twenty years, but forever.

The chief lesson I have learned in a long life is that the only way you can make a man trustworthy is to trust him; and the surest way to make him untrustworthy is to distrust him and show your distrust.

If the atomic bomb were merely another though more devastating military weapon to be assimilated into our pattern of international relations, it would be one thing. We could then follow the old custom of secrecy and nationalistic military superiority relying on international caution to prescribe the future use of the weapon as we did with gas. But I think the bomb instead constitutes merely a first step in a new control by man over the forces of nature too revolutionary and dangerous to fit into the old concepts. I think it really caps the climax of the age between man's growing technical power for destructiveness and his psychological power of self-control and group control—his moral power. If so, our method of approach to the Russians is a question of the most vital importance in the evolution of human progress.

Since the crux of the problem is Russia, any contemplated action leading to the control of this weapon should be primarily directed *to* Russia. It is my judgment that the Soviets would be more apt to respond sincerely to a direct and forthright approach made by the United States on the subject than would be the case if the approach were made as a part of a general international scheme, or if the approach were made after a succession of express or implied threats or near threats in our peace negotiations.

My idea of an approach to the Soviets would be a direct proposal after discussion with the British that we would be prepared in effect to enter an arrangement with the Russians, the general purpose of which would be to control and limit the use of the atomic bomb as an instrument of war and so far as possible to direct and encourage the development of atomic power for peaceful and humanitarian purposes. Such an approach might more specifically lead to the proposal that we would stop work on the further improvement in, or manufacture of, the bomb as a military weapon, provided the Russians and the British would agree to do likewise. It might also provide that we would be willing to impound what bombs we now have in the United States provided the Russians and the British would agree with us that in no event will they or we use a bomb as an instrument of war unless all three Governments agree to that use. We might also consider including in the arrangement a covenant with the U.K. and the Soviets providing for the exchange of benefits of future development whereby atomic energy may be applied on a mutually satisfactory basis for commercial or humanitarian purposes.

I would make such an approach just as soon as our immediate political considerations make it appropriate.

I emphasize perhaps beyond all other considerations the importance of taking this action with Russia as a proposal of the United States—backed by Great Britain but peculiarly the proposal of the United States. Action of any international group of nations, including many small nations who have not demonstrated their potential power or responsibility in this war would not, in my opinion, be taken seriously by the Soviets. The loose debates which would surround such proposal, if put before a conference of nations, would provoke but scant favor from the Soviets. As I say, I think this is the most important point in the program.

After the nations which have won this war have agreed to it, there will be ample time to introduce France and China into the covenants and finally to incorporate the agreement into the scheme of the United Nations. The use of this bomb has been accepted by the world as the result of the initiative and productive capacity of the United States, and I think this factor is a most potent lever toward having our proposals accepted by the Soviets, whereas I am most skeptical of obtaining any tangible results by way of any international debate. I urge this method as the most realistic means of accomplishing this vitally important step in the history of the world.

Henry L. Stimson
"Secretary of War."

23. The McMahon Bill, December 20, 1945 (Atomic Energy Act of 1946)

A BILL

For the development and control of atomic energy
Be it enacted by the Senate and House of Representatives of the United States of America in congress assembled,

DECLARATION OF POLICY

Section 1. (a) Findings and Declaration.—Research and experimentation in the field of nuclear fission have attained the stage at which the release of atomic energy on a large scale is practical. The significance of the atomic bomb for military purposes is evident. The effect of the use of atomic energy for civilian purposes upon the social, economic, and political structures of today cannot now be determined. It is reasonable to anticipate, however, that tapping this new source of energy will cause profound changes in our present way of life. Accordingly, it is hereby declared to be the policy of the people of the United States that the development and utilization of atomic energy shall be directed toward improving the public welfare, increasing the standard of living, strengthening free competition among private enterprises so far as practicable, and cementing world peace.

PL 79-585. This text appears in Senate Special Committee on Atomic Energy, *Atomic Energy Act of 1946. Hearings on S. 1717 January 22–April 4, 1946* (Washington, D.C., 1946), pp. 1–9.

(b) Purpose of Act.—It is the purpose of this Act to effectuate these policies by providing, among others, for the following major programs;

(1) A program of assisting and fostering private research and development on a truly independent basis to encourage maximum scientific progress;

(2) A program for the free dissemination of basic scientific information and for maximum liberality in dissemination of related technical information;

(3) A program of federally conducted research to assure the Government of adequate scientific and technical accomplishments;

(4) A program for Government control of the production, ownership, and use of fissionable materials to protect the national security and to insure the broadest possible exploitation of the field;

(5) A program for simultaneous study of the social, political, and economic effects of the utilization of atomic energy; and

(6) A program of administration which will be consistent with international agreements made by the United States, and which will enable the Congress to be currently informed so as to take further legislative action as may hereafter be appropriate.

ATOMIC ENERGY COMMISSION

Sec. 2. (a) There is hereby established an Atomic Energy Commission (herein called the Commission), which shall be composed of five members. Three members shall constitute a quorum of the Commission. The President shall designate one member as a Chairman of the Commission.

(b) Members of the Commission shall be appointed by the President, by and with the advice and consent of the Senate, and shall serve at the pleasure of the President. In submitting nominations to the Senate, the President shall set forth the experience and qualifications of each person so nominated. Each member, except the Chairman, shall receive compensation at the rate of $15,000 per annum; the Chairman shall receive compensation at the rate of $20,000 per annum. No member of the Commission shall engage in any other business, vocation, or employment than that of serving as a member of the Commission.

(c) The principal office of the Commission shall be in the District of Columbia, but the Commission may exercise any or all of its powers in any place. The Commission shall hold such meetings, conduct such hearings, and receive such reports as will enable it to meet its responsibilities for carrying out the purpose of this Act.

RESEARCH

Sec. 3. (a) Research Assistance.—The Commission is directed to exercise its powers in such a manner as to insure the continued conduct of research and developmental activities in the fields specified below by private or public institutions or persons and to assist in the acquisition of an ever-expanding fund of theoretical and practical knowledge in such fields. To this end the Commission is authorized and directed to make contracts, agreements, arrangements, grants-in-aid, and loans—

(1) for the conduct of research and developmental activities relating to (a) nuclear processes; (b) the theory and production of atomic energy, including processes and devices related to such production; (c) utilization of fissionable and radioactive materials for medical or health purposes; (d) utilization of fissionable and radioactive materials for all other purposes, including industrial uses; and (e) the protection of health during research and production activities; and

(2) for studies of the social, political, and economic effects of the availability and utilization of atomic energy.

The Commission may make partial advance payments on such contracts and arrangements. Such contracts or other arrangements may contain provisions to protect health, to minimize danger from explosion, and for reporting and inspection of work performed thereunder as the Commission may determine, but shall not contain any provisions or conditions which prevent the dissemination of scientific or technical information, except to the extent already required by the Espionage Act.

(b) Federal Atomic Research.—The Commission is authorized and directed to conduct research and developmental activities through its own facilities in the fields specified in (a) above.

PRODUCTION OF FISSIONABLE MATERIALS

Sec. 4. (a) Definition.—The term "production of fissionable materials" shall include all methods of manufacturing, producing, refining, or processing fissionable materials, including the process of separating fissionable material from other substances in which such material may be contained, whether by thermal diffusion, electromagnetic separation, or other processes.

(b) Authority to Produce.—The Commission shall be the exclusive producer of fissionable materials, except production incident to research or developmental activities subject to restrictions provided in subparagraph (d) below. The quantities of fissionable material to be produced in any quarter shall be determined by the President.

(c) Prohibition.—It shall be unlawful for any person to produce any fissionable material except as may be incident to the conduct of research or developmental activities.

(d) Research and Development on Production Processes. (1) The Commission shall establish by regulation such requirements for the reporting of research and developmental activities on the production of fissionable materials as will assure the Commission of full knowledge of all such activities, rates of production, and quantities produced.

(2) The Commission shall provide for the frequent inspection of all such activities by employees of the Commission.

(3) No person may in the course of such research or developmental activities possess or operate facilities for the production of fissionable material in quantities or at a rate sufficient to construct a bomb or other military weapon unless all such facilities are the property of and subject to the control of the Commission. The Commission is authorized, to the extent that it deems such action consistent with the purposes of this Act, to enter into contracts for the conduct of such research or developmental activities involving the use of the Commission's facilities.

(e) Existing Contracts.—The Commission is authorized to continue in effect and modify such contracts for the production of fissionable materials as may have been made prior to the effective date of this Act, except that, as rapidly as practicable, and in any event not more than one year after the effective date of this Act, the Commission shall arrange for the exclusive operation of facilities employed in the manufacture of fissionable materials by employees of the Commission.

CONTROL OF MATERIALS

Sec. 5. (a)(1) Definition.—The term "fissionable materials" shall include plutonium, uranium 235, and such other materials as the Commission may from time to time determine to be capable of releasing substantial quantities of energy through nuclear fission of the materials.

(2) Privately Owned Fissionable Materials.—Any person owning any right, title, or interest in or to any fissionable material shall forthwith transfer all such right, title, or interest to the Commission.

(3) Prohibition.—It shall be unlawful for any person to (a) own any fissionable material; or (b) after sixty days after the effective date of this Act and except as authorized by the Commission possess any fissionable material; or (c) export from or import into the United States any fissionable material, or directly or indirectly be a party to or in any way a beneficiary of, any contract, arrangement or other activity pertaining to the production, refining, or processing of any fissionable material outside of the United States.

(4) Distribution of Fissionable Materials.—The Commission is authorized and directed to distribute fissionable materials to all applicants requesting such materials for the conduct of research or developmental activities either independently or under contract or other arrangement with the Commission. If sufficient materials are not available to meet all such requests, and applications for licenses under section 7, the Commission shall allocate fissionable materials among all such applicants in the manner best calculated to encourage independent research and development by making adequate fissionable materials available for such purposes. The Commission shall refuse to distribute or allocate any materials to any applicant, or shall recall any materials after distribution or allocation from any applicant, who is not equipped or who fails to observe such safety standards to protect health and to minimize danger from explosion as may be established by the Commission.

(b) Source Materials.—

(1) Definition.—The term "source materials" shall include any ore containing uranium, thorium, or beryllium, and such other materials peculiarly essential to the production of fissionable materials as may be determined by the Commission with the approval of the President.

(2) License for Transfers Required.—No person may transfer possession or title to any source material after mining, extraction, or removal from its place of origin, and no person may receive any source material without a license from the Commission.

(3) Issuance of Licenses.—Any person desiring to transfer or receive possession of any source material shall apply for a license therefor in accordance with such procedures as the Commission may by regulation establish. The Commission shall establish such standards for the issuance or refusal of licenses as it may deem necessary to assure adequate source materials for

production, research or developmental activities pursuant to this Act or to prevent the use of such materials in a manner inconsistent with the national welfare.

(4) Reporting.—The Commission is authorized to issue such regulations or orders requiring reports of ownership, possession, extraction, refining, shipment or other handling of source materials as it may deem necessary.

(c) Byproduct Materials.—

(1) Definition.—The term "byproduct material" shall be deemed to refer to all materials (except fissionable material) yielded in the processes of producing fissionable material.

(2) Distribution.—The Commission is authorized and directed to distribute, with or without charge, byproduct materials to all applicants seeking such materials for research or developmental work, medical therapy, industrial uses, or such other useful applications as may be developed. If sufficient materials to meet all such requests are not available, the Commission shall allocate such materials among applicants therefor, giving preference to the use of such materials in the conduct of research and developmental activity and medical therapy. The Commission shall refuse to distribute or allocate any byproduct materials to any applicant, or recall any materials after distribution or allocation from any applicant, who is not equipped or who fails to observe such safety standards to protect health as may be established by the Commission.

(d) General Provisions.—(1) The Commission is authorized to—

(i) acquire or purchase fissionable or source materials within the United States or elsewhere;

(ii) take, requisition, or condemn within the United States any fissionable or source material and make just compensation therefor. The Commission shall determine such compensation. In the exercise of such rights of eminent domain and condemnation, proceedings may be instituted under the Act of August 1, 1888 (U. S. C. 1940, title 40, sec. 257), or any other applicable Federal statute. Upon or after the filing of the condemnation petition, immediate possession may be taken and the property may be treated by the Commission in the same manner as other similar property owned by it;

(iii) conduct exploratory operation, investigations, inspections to determine the location, extent, mode of occurrence, use, or condition of source materials with or without the consent of the owner of any interest therein, making just compensation for any damage or injury occasioned thereby.

(2) The Commission shall establish by regulation a procedure by which any person who is dissatisfied with its action in allocating, refusing to allocate

or in rescinding any allocation of fissionable, source, or byproduct materials to him may obtain a review of such determination by a board of appeal consisting of two or more members appointed by the Commission and at least one member of the Commission.

MILITARY APPLICATIONS OF ATOMIC POWER

Sec. 6. (a) The Commission is authorized and directed to—

(1) conduct experiments and do research and developmental work in the military application of atomic power; and

(2) have custody of all assembled or unassembled atomic bombs, bomb parts, or other atomic military weapons, presently or hereafter produced, except that upon the express finding of the President that such action is required in the interests of national defense, the Commission shall deliver such quantities of weapons to the armed forces as the President may specify.

(b) The Commission shall not conduct any research or developmental work in the military application of atomic power if such research or developmental work is contrary to any international agreement of the United States.

(c) The Commission is authorized to engage in the production of atomic bombs, bomb parts, or other applications of atomic power as military weapons, only to the extent that the express consent and direction of the President of the United States has been obtained, which consent and direction shall be obtained for each quarter.

(d) It shall be unlawful for any person to manufacture, produce, or process any device or equipment designed to utilize fissionable materials as a military weapon, except as authorized by the Commission.

ATOMIC ENERGY DEVICES

Sec. 7. (a) License Required.—It shall be unlawful for any person to operate any equipment or device utilizing fissionable materials without a license issued by the Commission authorizing such operation.

Issuance of Licenses.—Any person desiring to utilize fissionable materials in any such device or equipment shall apply for a license therefor in accordance with such procedures as the Commission may by regulation establish. The Commission is authorized and directed to issue such a license on a nonexclusive basis and to supply appropriate quantities of fissionable materials to the extent available to any applicant (1) who is equipped to observe such safety standards to protect health and to minimize danger from explosion as the Commission may establish; and (2) who agrees to make available to the Commission such technical information and data concerning the operation of such device as the Commission may determine necessary to encourage the use of such devices by as many licensees as possible. Where any license might serve to maintain or foster the growth of monopoly, restraint of trade, unlawful competition, or other trade position inimical to the entry of new, freely competitive enterprises, the Commission is authorized and directed to refuse to issue such license or to establish such conditions to prevent these results as the Commission, in consultation with the Attorney General, may determine. The Commission shall report promptly to the Attorney General any information it may have of the use of such devices which appears to have these results. No license may be given to a foreign government or to any person who is not under and within the jurisdiction of the United States.

(c) Byproduct Power.—If in the production of fissionable materials the production processes yield energy capable of utilization, such energy may be used by the Commission, transferred to other Government agencies, sold to public or private utilities under contract providing for reasonable resale prices, or sold to private consumers at reasonable rates and on as broad a basis of eligibility as the Commission may determine to be possible.

(d) Reports to Congress.—Whenever in its opinion industrial, commercial, or other uses of fissionable materials have been sufficiently developed to be of practical value, the Commission shall prepare a report to the Congress stating all the facts, the Commission's estimate of the social, political, and economic effects of such utilization, and the Commission's recommendations for necessary or desirable supplemental legislation. Until such a report has been filed with the Commission and the period of ninety days has elapsed after such filing within which period the Commission may adopt supplemental legislation, no license for the use of atomic energy devices shall be issued by the Commission.

PROPERTY OF THE COMMISSION

Sec. 8. (a) The President shall direct the transfer to the Commission of the following property owned by the United States or any of its agencies, or any interest in such property held in trust for or on behalf of the United States:

(1) All fissionable materials; all bombs and bomb parts; all plants, facilities, equipment, and materials for the processing or production of fissionable

materials, bombs, and bomb parts; all processes and technical information of any kind, and the source thereof (including data, drawings, specifications, patents, patent applications, and other sources, relating to the refining or production of fissionable materials; and all contracts, agreements, leases, patents, applications for patents, inventions and discoveries (whether patented or unpatented), and other rights of any kind concerning any such items;

(2) All facilities and equipment, and materials therein, devoted primarily to atomic energy research and development; and

(3) All property in the custody and control of the Manhattan engineer district.

(b) In order to render financial assistance to those States and local governments in which the activities of the Commission are carried on and in which the Commission, or its agents, have acquired properties previously subject to State and local taxation, the Commission is authorized to make payments to State and local governments in lieu of such taxes. Such payments may be in the amounts, at the times, and upon the terms the Commission deems appropriate, but the Commission shall be guided by the policy of not exceeding the taxes which would have been payable for such property in the condition in which it was acquired, except where special burdens have been cast upon the State or local government by activities of the Commission, the Manhattan engineer district, or their agents, and in such cases any benefits accruing to the States and local governments by reason of these activities shall be considered in the determination of such payments. The Commission and any corporation created by it, and the property and income of the Commission or of such corporation, are hereby expressly exempted from taxation in any manner or form by any State, county, municipality, or any subdivision thereof.

DISSEMINATION OF INFORMATION

Sec. 9. (a) Basic Scientific Information.—Basic scientific information in the fields specified in section 3 may be freely disseminated. The term "basic scientific information" shall include, in addition to theoretical knowledge of nuclear and other physics, chemistry, biology, and therapy, all results capable of accomplishment, as distinguished from the processes or techniques of accomplishing them.

(b) Related Technical Information.—The Commission shall establish a Board of Commission. The Board shall, under the direction and supervision of the Commission, provide for the dissemination of related technical information with the utmost liberality as freely as may be consistent with the foreign and domestic policies established by the President and shall have authority to—

(1) establish such information services, publications, libraries, and other registers of available information as may be helpful in effectuating this policy;

(2) designate by regulation the types of related technical information the dissemination of which will effectuate the foregoing policy. Such designations shall constitute an administrative determination that such information is not of value to the national defense and that any person is entitled to receive such information, within the meaning of the Espionage Act. Failure to make any such designation shall not, however, be deemed a determination that such undesignated information is subject to the provisions of said Act;

(3) by regulation or order, require reports of the conduct of independent research or development activities in the fields specified in section 3 and of the operation of atomic energy devices under licenses issued pursuant to section 7;

(4) provide for such inspections of independent research and development activities of the types specified in section 3 and of the operation of atomic energy devices as the Commission or the Board may determine; and

(5) whenever it will facilitate the carrying out of the purposes of the Act, adopt by regulation administrative interpretations of the Espionage Act except that any such interpretation shall, before adoption, receive the express approval of the President.

PATENTS

Sec. 10. (a) Whenever any person invents a device or method for the production, refining, or processing of fissionable material: (i) he may file a patent application to cover such invention, sending a copy thereof to the Commission; (ii) if the Commissioner of Patents determines that the invention is patentable, he shall issue a patent in the name of the Commission; and (iii) the Commission shall make just compensation to such person. The Commission shall appoint a Patent Royalty Board consisting of one or more employees and at least one member of the Commission, and the Commissioner of Patents. The Patent Royalty Board shall determine what constitutes just compensation in each such case and whether such compensation is to be paid in periodic payments rather than in a lump sum. Any person to whom any such patent has heretofore been issued shall forthwith transfer all right, title, and interest in and to such patent to the Commission and shall receive therefor just compensation as provided above.

(b) (1) Any patent now or hereafter issued covering any process or device utilizing or peculiarly necessary to the utilization of fissionable materials, or peculiarly necessary to the conduct of research or developmental activities in the fields specified in section 3, is hereby declared to be affected with the public interest and its general availability for such uses is declared to be necessary to affectuate the purposes of this Act.

(2) Any person to whom any such patent has been issued, or any person desiring to use any device or process covered by such patent for such uses, may apply to the Patent Royalty Board, for determination by such Board of a reasonable royalty fee for such use of the patented process or device intended to be used under the Commission's license.

(3) In determining such reasonable royalty fee, the Patent Royalty Board shall take into consideration any defense, general or special, that might be pleaded by a defendant in an action for infringement, the extent to which, if any, such patent was developed through federally financed research, the degree of utility, novelty, and importance of the patent, the cost to the patentee of developing such process or device, and a reasonable rate of return on such research investment by the patentee.

(4) No court, Federal, State, or Territorial, shall have jurisdiction or power to stay, restrain, or otherwise enjoin any such use of any such patented device or process by any person on the ground of infringement of such patent. In any action for infringement of any such patent filed in any such court, the court shall have authority only to order the payment of reasonable royalty fees and attorney's fees and court costs as damages for any such infringement. If the Patent Royalty Board has not previously determined the reasonable royalty fee for the use of the patented device or process involved in any case, the court in such case shall, before entering judgment, obtain from the Patent Royalty Board a report containing its recommendation as to the reasonable royalty fee it would have established had application been made to it as provided in subparagraphs 2 and 3 above.

ORGANIZATIONAL AND GENERAL AUTHORITY

Sec. 11. (a) Organization.—There are hereby established within the Commission a Division of Research, a Division of Production, a Division of Materials, and a Division of Military Application. Each division shall be under the direction of a Directory who shall be appointed by the President, by and with the advice and consent of the Senate, and shall receive compensation at the rate of $15,000 per annum. The Commission shall delegate to each such division such of its powers under this Act as in its opinion from time to time will promote the effectuation of the purposes of this Act in an efficient manner. Nothing in this paragraph shall prevent the Commission from establishing such additional divisions or other subordinate organizations as it may deem desirable.

(b) General Authority.—In the performance of its functions the Commission is authorized to—

(1) establish advisory boards to advise with and make recommendations to the Commission on legislation, policies, administration, and research;

(2) establish by regulation or order such standards and instructions to govern the possession and use of fissionable and byproduct materials as the Commission may deem necessary or desirable to protect health or to minimize danger from explosion;

(3) make such studies and investigations, obtain such information, and hold such hearings as the Commission may deem necessary or proper to assist it in exercising any authority provided in this Act, or in the administration or enforcement of this Act, or any regulations or orders issued thereunder. For such purposes the Commission is authorized to require any person to permit the inspection and copying of any records or other documents, to administer oaths and affirmations, and by subpena to require any person to appear and testify, or to appear and produce documents, or both, at any designated place. Witnesses subpenaed under this subsection shall be paid the same fees and mileage as are paid witnesses in the district courts of the United States;

(4) create or organize corporations, the stock of which shall be wholly owned by the United States and controlled by the Commission, to carry out the provisions of this Act;

(5) appoint and fix the compensation of such officers and employees as may be necessary to carry out the functions of the Commission. All such officers and employees shall be appointed in accordance with the civil-service laws and their compensation fixed in accordance with the Classification Act of 1923, as amended, except that expert administrative, technical, and professional personnel may be employed and their compensation fixed without regard to such laws. The Commission shall make adequate provision for administrative review by a board consisting of one

or more employees and at least one member of the Commission of any determination to dismiss any scientific or professional employee; and

(6) acquire such materials, property, equipment, and facilities, establish or construct such buildings and facilities, modify such building and facilities from time to time, and construct, acquire, provide, or arrange for such facilities and services for the housing, health, safety, welfare, and recreation of personnel employed by the Commission as it may deem necessary.

ENFORCEMENT

Sec. 12. (a) Any person who willfully violates, attempts to violate, or conspires to violate, any of the provisions of this Act or any regulations or orders issued thereunder shall, upon conviction thereof, be punishable by a fine of not more than $10,000, or by imprisonment for a term of not exceeding five years, or both.

(b) Whenever in the judgment of the commission any person has engaged or is about to engage in any acts or practices which constitute or will constitute a violation of any provision of this Act, it may make application to the appropriate court for an order enjoining such acts or practices, or for an order enforcing compliance with such provision, and upon a showing by the Commission that such person has engaged or is about to engage in any such acts or practices a permanent or temporary injunction, restraining order, or other order shall be granted without bond.

(c) In case of contumacy by, or refusal to obey a subpena served upon, any person pursuant to section 11 (b) (3), the district court for any district in which such person is found or resides or transacts business, upon application by the Commission, shall have jurisdiction to issue an order requiring such person to appear and give testimony or to appear and produce documents, or both; and any failure to obey such order of the court may be punished by such court as a contempt thereof.

REPORTS

Sec. 13. The Commission shall, on the first days of January, April, July, and October, submit reports to the President, to the Senate and to the House of Representatives. Such reports shall summarize and appraise the activities of the Commission and of each division and board thereof, and specifically shall contain financial statements; lists of licenses issued, of property acquired, of research contracts and arrangements entered into, and of the amounts of fissionable material and the persons to whom allocated; the Commission's program for the following quarter including lists of research contracts and arrangements proposed to be entered into; conclusions drawn from studies of the social, political, and economic effects of the release of atomic energy; and such recommendations for additional legislation as the Commission may deem necessary or desirable.

DEFINITIONS

Sec. 14. As used in this Act—

(a) The term "atomic energy" shall include all forms of energy liberated in the artificial transmutation of atomic species.

(b) The term "Government agency" means any executive department, board, bureau, commission, or other agency in the executive branch of the Federal Government, or any corporation wholly owned (either directly or through one or more corporations) by the United States.

(c) The term "person" means any individual, corporation, partnership, firm, association, trust, estate, public or private institution, group, any government other than the United States, any political subdivision of any such government, and any legal successor, representative, agent, or agency of the foregoing, or other entity.

(d) The term "United States" includes all Territories and possessions of the United States.

APPROPRIATIONS

Sec. 15. There are hereby authorized to be appropriated such sums as may be necessary and appropriate to carry out the provisions and purposes of this Act. Funds appropriated to the Commission shall, if obligated during the fiscal year for which appropriated, remain available for expenditure for four years following the expiration of the fiscal year for which appropriated. After such four-year period, the unexpended balances of appropriations shall be carried to the surplus fund and covered into the Treasury.

SEPARABILITY OF PROVISIONS

Sec. 16. If any provision of this Act, or the application of such provision to any person or circumstances, is held invalid, the remainder of this Act or the application of such provision to persons of circumstances other than those to which it is held invalid, shall not be affected thereby.

SHORT TITLE

Sec. 17. This Act may be cited as the "Atomic Energy Act of 1946."

24. The Baruch Plan, June 1946

Memorandum by President Truman to the United States Representative on the Atomic Energy Commission (Baruch)

Confidential   Washington, June 7, 1946.

Memorandum for Mr. Baruch: Because you requested it I herewith attach a statement of the United States policy with reference to atomic energy. This statement is solely for your guidance in your deliberations as the Representative of the United States on the Atomic Energy Commission of the United Nations.

The statement is general in character because I want you to have authority to exercise your judgment as to the method by which the stated objectives can be accomplished.

If as negotiations progress you conclude that there should be changes in this statement of policy, I will expect you to advise me and to frankly give me your views.

I know that you will keep me advised as to the negotiations. However, I want you to know that I am relying upon you to exercise your own discretion in those negotiations, subject only to the general statement of policy attached, unless you should receive from me through the Secretary of State a further statement of policy.

Harry S. Truman

This text appears in Foreign Relations of the United States, 1946, vol. 1 (Washington, D.C.: U.S. Government Printing Office, 1972), pp. 846–51.

(Annex)

June 7, 1946.

Statement of United States Policy

The proposals in this paper, put forth as a basis of discussion, grow out of three basic conclusions:

1. It is believed that an international agreement leaving the development of atomic energy in national hands, subject to an obligation not to develop atomic energy for war purposes and relying solely on an international inspection system to detect evasions, will not provide adequate security and indeed may be a source of insecurity.

2. It is believed

(a) That a treaty merely outlawing possession or use of the atomic bomb would not be an effective fulfillment of the directions under which the Commission is to proceed; therefore, that an international atomic development authority be set up, with adequate powers;

(b) That in connection with the greatest safeguards which can be established through a competent international authority in this field, there should be a clear statement of the consequences of violations of the system of control, including definitions of the acts which would constitute such violations and the penalties and concerted action which would follow such violations;

(c) that one of the objectives of the plan should be that when the system of control is fully in operation there would be no stockpiles of bombs in existence;

(d) That the plan might also include a parallel statement as to a system of control for biological warfare.

3. It is further believed that the aim of preventing atomic warfare can only be achieved by entrusting to an international organization

(a) Managerial control of all atomic energy activities intrinsically dangerous to world security;

(b) Power to control, inspect, and license all other activities and stages.

If an international agency is given sole responsibility for the dangerous activities, leaving the non-dangerous open to nations and their citizens and if the international agency is given and carries forward affirmative development responsibility, furthering among other things the beneficial uses of atomic energy and enabling itself to comprehend and therefore detect the misuse of atomic energy, these afford the best prospect of security.

For purposes of discussion, the following measures are proposed as representing the fundamental features of a plan which would give effect to the conclusions just stated. In this paper the proposed international agency is referred to as the Atomic Development Authority.

1. General—The Atomic Development Authority should seek to set up a thorough plan of control through various "forms of ownership, dominion, licenses, operation, inspection, research and management by competent personnel."

It is believed that the plan of control in all its aspects must be adequate not only in concept—a combination of responsibility for developments as well as control—but in type of organization and in choice of personnel to guarantee the most effective control required to provide for the security of the nations.

2. Raw Materials—The Atomic Development Authority when set up should have as one of its earliest purposes to bring under its complete dominion world supplies of uranium and thorium. The precise pattern of control for various types of deposits of such materials will have to depend upon the geological, mining, refining, and economic facts involved in different situations.

The Authority should conduct continuous surveys so that it will have the most complete knowledge of the world geology of uranium and thorium. The agency should also constantly investigate new methods for recovering these materials where they occur in small quantities so that as their recovery from such sources becomes practical, means of control can be devised.

3. Primary Production Plants—The Atomic Development Authority should exercise complete managerial control of the production of fissionable materials. This means that it should control and operate all plants producing fissionable materials in dangerous quantities and own and control the product of these plants.

4. Atomic Explosives—The Authority should be given exclusive authority to conduct research in the field of atomic explosives. Research activities in the field of atomic explosives are essential in order that the Authority may keep in the forefront of knowledge in the field of atomic energy and fulfill the objective of preventing illicit manufacture of bombs. Only by preserving its position as the best informed agency will the Authority be able to tell where the line between the intrinsically dangerous and the non-dangerous should be drawn. If it turns out at some time in the future, as a result of new discoveries, that other materials or other processes lend themselves to dangerous atomic developments, it is important that the Authority should be the first to know. At that time measures would have to be taken to extend the boundaries of safeguards.

5. Strategic Distribution of Activities and Materials—The activities entrusted exclusively to the Authority because they are intrinsically dangerous to security should be distributed throughout the world. Similarly, stockpiles of raw materials and fissionable materials should not be centralized.

6. Non-Dangerous Activities—Atomic research (except in explosives), the use of research reactors, the production of radioactive tracers by means of non-dangerous reactors, the use of such tracers, and to some extent the production of power should be open to nations and their citizens under reasonable licensing arrangements from the Authority. Denatured materials necessary for these activities should be furnished, under lease or other suitable arrangement by the Atomic Development Authority.

It should be an essential function of the Atomic Development Authority to promote to the fullest possible extent the peace-time benefits that can be obtained from the use of atomic energy.

It is necessary at all times to take advantage of the opportunity for promotion of decentralized and diversified national and private developments and of avoiding unnecessary concentration of functions in the Authority. It should, therefore, be a primary function to the Authority to encourage development by nations and private enterprise in the broad field of non-dangerous activities.

7. Definition of Dangerous and Non-Dangerous Activities—Although a reasonable dividing line can be drawn between the dangerous and the non-dangerous, it is not hard and fast. Machinery should, therefore, be provided to assure constant examination and re-examination of the question, and to permit revision of the dividing line as changing conditions and new discoveries may require.

7 (a) Any plant dealing with uranium or thorium after it once reaches the potential of dangerous use must be not only subject to the most rigorous and competent inspection by the international Authority, but its actual operation shall be under the management, supervision and control of the international Authority.

8. Inspection Activities—By assigning intrinsically dangerous activities exclusively to management by the Atomic Development Authority, the difficulties of inspection are thereby reduced to manageable proportions. For if the Atomic Development Authority is the only agency which may lawfully conduct the dangerous activities in the field of raw materials, primary production plants, and research in explosives, then visible operation by others than the Authority will constitute a danger signal.

The plan does not contemplate any systematic or large-scale inspection procedures covering the whole of industry. The delegation of authority for making inspections will have to be carefully drawn so that the inspection may be adequate for the needs and responsibilities of the Authority and yet not go beyond this point. Many of the inspection activities of the Authority should grow out of and be incidental to its other functions. An important measure of inspection will be those associated with the tight control of raw materials, for this is one of the keystones of the plan. The continuing activities of prospecting, survey and research in relation to raw materials will be designed not only to serve the affirmative development functions of the agency but also to assure that no surreptitious operations are conducted in the raw materials field by nations or their citizens. Inspection will also occur in connection with the licensing functions of the Authority. Finally, a means should be provided to enable the international organization to make special "spot" investigations of any suspicious national or private activities.

9. Personnel—The personnel of the Atomic Development Authority should be recruited on a basis of proven competence but also so far as possible on an international basis, giving much weight to geographical and national distribution. Although the problem of recruitment of the high-quality personnel required for the top executive and technical positions will be difficult, it will certainly be far less difficult than the recruitment of the similarly high-quality personnel that would be necessary for any purely policing organization.

10. Negotiation Stage—The first step in the creating of the system of control is the spelling out in comprehensive terms of the functions, responsibilities, authority, and limitations of the Atomic Development Authority. Once a Charter for the agency has been written, and adopted, the Authority and the system of control for which it will be responsible will require time to become fully organized and effective. The plan of control will therefore have to come into effect in successive stages. These should be specifically fixed in the Charter or means should be otherwise set forth in the Charter for transitions from one stage to another, as contemplated in the resolution of the U.N. Assembly which created this Commission.

11. Disclosures—In the deliberation of the United Nations Commission on Atomic Energy, the United States must be prepared to make available the information essential to a reasonable understanding of the proposals which it advocates. Further disclosures must be dependent, in the interests of all, upon the effective ratification of this treaty. If and when the Authority is actually created, the United States must then also be prepared to make available other information essential to that organization or the performance of its functions. And as the successive stages of international control are reached, the United States must further be prepared to yield, to the extent required by each stage, national control of activities in this field to the international agency.

12. International Control—There will be questions about the extent of control allowed to national bodies, should an international body be established and in this respect, it is believed that any control by an atomic energy authority set up by any state should to the extent necessary for the effective operation of the international control system be subordinate to direction and absolute dominion on the part of the international authority. It will readily be seen at this time that this is not an endorsement or disapproval of the creation of national authorities, or a definition of their jurisdiction. This problem will be before the Commission and it should deal with a clear separation of duties and responsibilities of such state authorities, with the purpose of preventing possible conflicts of jurisdiction.

Harry S. Truman

25. Dwight D. Eisenhower's "Atoms for Peace" Address to the United Nations General Assembly, December 8, 1953

Madame President,[1] Members of the General Assembly:

When Secretary General Hammarskjold's invitation to address this General Assembly reached me in Bermuda, I was just beginning a series of conferences with the Prime Ministers and Foreign Ministers of Great Britain and of France. Our subject was some of the problems that beset the world.

During the remainder of the Bermuda Conference, I had constantly in mind that ahead of me lay a great honor. That honor is mine today as I stand here, privileged to address the General Assembly of the United Nations.

At the same time that I appreciate the distinction of addressing you, I have a sense of exhilaration as I look upon this assembly.

Never before in history has so much hope for so many people been gathered together in a single organization. Your deliberations and decisions during these somber years have already realized part of those hopes.

But the great tests and the great accomplishments still lie ahead. And in the confident expectation of those accomplishments, I would use the office

This text appears in the *Congressional Record*, vol. 100, January 7, 1954, pp. 61–63.
1. Mme. Vijaya Pandit, president of the United Nations General Assembly.

which, for the time being, I hold, to assure you that the Government of the United States will remain steadfast in its support of this body. This we shall do in the conviction that you will provide a great share of the wisdom, of the courage and the faith which can bring to this world lasting peace for all nations and happiness and well-being for all men.

Clearly, it would not be fitting for me to take this occasion to present to you a unilateral American report on Bermuda. Nevertheless, I assure you that in our deliberations on that lovely island we sought to invoke those same great concepts of universal peace and human dignity which are so cleanly etched in your Charter.

Neither would it be a measure of this great opportunity merely to recite, however hopefully, pious platitudes.

I therefore decided that this occasion warranted my saying to you some of the things that have been on the minds and hearts of my legislative and executive associates and on mine for a great many months—thoughts I had originally planned to say primarily to the American people.

I know that the American people share my deep belief that if a danger exists in the world, it is a danger shared by all—and equally, that if hope exists in the mind of one nation, that hope should be shared by all.

Finally, if there is to be advanced any proposal designed to ease, even by the smallest measure, the tensions of today's world, what more appropriate audience could there be than the members of the General Assembly of the United Nations?

Language of Atomic Warfare

I feel impelled to speak today in a language that, in a sense, is new—one, which I, who spent so much of my life in the military profession, would have preferred never to use.

That new language is the language of atomic warfare.

The atomic age has moved forward at such a pace that every citizen of the world should have some comprehension, at least in comparative terms, of the extent of this development, of the utmost significance to every one of us. Clearly, if the people of the world are to conduct an intelligent search for peace, they must be armed with the significant facts of today's existence.

My recital of atomic danger and power is necessarily stated in United States terms, for these are the only incontrovertible facts that I know. I need hardly point out to this assembly, however, that this subject is global, not merely national in character.

On July 16, 1945, the United States set off the world's first atomic test explosion. Since that date in 1945, the United States of America has conducted forty-two test explosions.

Atomic bombs today are more than twenty-five times as powerful as the weapons with which the atomic age dawned, while hydrogen weapons are in the ranges of millions of tons of TNT equivalent.

Today, the United States' stockpile of atomic weapons, which, of course, increases daily, exceeds by many times the explosive equivalent of the total of all bombs and all shells that came from every plane and every gun in every theatre of war through all the years of World War II.

A single air group, whether afloat or land based, can now deliver to any reachable target a destructive cargo exceeding in power all the bombs that fell on Britain in all of World War II.

In size and variety the development of atomic weapons has been no less remarkable. This development has been such that atomic weapons have virtually achieved conventional status within our armed services. In the United States services, the Army, the Navy, the Air Force and the Marine Corps are all capable of putting this weapon to military use.

But the dread secret and the fearful engines of atomic might are not ours alone.

In the first place, the secret is possessed by our friends and Allies, Great Britain and Canada, whose scientific genius made a tremendous contribution to our original discoveries and the designs of atomic bombs.

The secret is also known by the Soviet Union.

The Soviet Union has informed us that, over recent years, it has devoted extensive resources to atomic weapons. During this period, the Soviet Union has exploded a series of atomic devices, including at least one involving thermo-nuclear reactions.

Monopoly Is Now Ended

If at one time the United States possessed what might have been called a monopoly of atomic power, that monopoly ceased to exist several years ago. Therefore, although our earlier start has permitted us to accumulate what is today a great quantitative advantage, the atomic realities of today comprehend two facts of even greater significance.

First, the knowledge now possessed by several nations will eventually be shared by others, possibly all others.

Second, even a vast superiority in numbers of weapons, and a consequent capability of devastating retaliation, is no preventive, of itself, against the fearful material damage and toll of human lives that would be inflicted by surprise aggression.

The free world, at least dimly aware of these facts, has naturally embarked on a large program of warning and defense systems. That program will be accelerated and expanded.

But let no one think that the expenditure of vast sums for weapons and systems of defense can guarantee absolute safety for the cities and the citizens of any nation. The awful arithmetic of the atomic bomb does not permit of such an easy solution. Even against the most powerful defense, an aggressor in possession of the effective minimum number of atomic bombs or a surprise attack could probably place a sufficient number of his bombs on the chosen targets to cause hideous damage.

Should such an atomic attack be launched against the United States, our reaction would be swift and resolute. But for me to say that the defense capabilities of the United States are such that they could inflict terrible losses upon an aggressor—for me to say that the retaliation capabilities of the United States are so great that such an aggressor's land would be laid waste—all this, while fact, is not the true expression of the purpose and the hope of the United States.

To pause there would be to confirm the hopeless finality of a belief that two atomic colossi are doomed malevolently to eye each other indefinitely across a trembling world. To stop there would be to accept helplessly the probability of civilization destroyed—the annihilation of the irreplaceable heritage of mankind handed down to us generation from generation—and the condemnation of mankind to begin all over again the age-old struggle upward from savagery toward decency and right and justice.

No Victory in Desolation

Surely no sane member of the human race could discover victory in such desolation. Could anyone wish his name to be coupled by history with such human degradation and destruction?

Occasional pages of history do record the faces of the "Great Destroyers" but the whole book of history reveals mankind's never-ending quest for peace and mankind's God-given capacity to build.

It is with the book of history, and not with isolated pages, that the United States will ever wish to be identified. My country wants to be constructive, not destructive. It wants agreements, not wars, among nations. It wants, itself, to live in freedom and in the confidence that the people of every other nation enjoy equally the right of choosing their own way of life.

So my country's purpose is to help us move out of the dark chamber of horrors into the light, to find a way by which the minds of men, the hopes of men, the souls of men everywhere, can move forward toward peace and happiness and well-being.

In this quest, I know that we must not lack patience.

I know that in a world divided, such as ours today, salvation cannot be attained by one dramatic act.

I know that many steps will have to be taken over many months before the world can look at itself one day and truly realize that a new climate of mutually peaceful confidence is abroad in the world.

But I know, above all else, that we must start to take these steps—now.

The United States and its Allies, Great Britain and France, have, over the past months, tried to take some of these steps. Let no one say that we shun the conference table.

On the record has long stood the request of the United States, Great Britain and France, to negotiate with the Soviet Union the problems of a divided Germany.

On that record has long stood the request of the same three nations to negotiate an Austrian peace treaty.

On the same record still stands the request of the United Nations to negotiate the problems of Korea.

Conference with the Russians

Most recently, we have received from the Soviet Union what is in effect an expression of willingness to hold a four-power meeting. Along with our Allies, Great Britain and France, we were pleased to see that this note did not contain the unacceptable preconditions previously put forward.

As you already know from our joint Bermuda communique, the United States, Great Britain and France have agreed promptly to meet with the Soviet Union.

The Government of the United States approaches this conference with hopeful sincerity. We will bend every effort of our minds to the single purpose of emerging from that conference with tangible results toward peace—the only true way of lessening international tension.

We never have, we never will, propose to suggest that the Soviet Union surrender what is rightfully theirs.

We will never say that the peoples of Russia are an enemy with whom we have no desire ever to deal or mingle in friendly and fruitful relationship.

On the contrary, we hope that this coming conference may initiate a relationship with the Soviet Union which will eventually bring about a free intermingling of the peoples of the East and of the West—the one sure, human way of developing the understanding required for confident and peaceful relations.

Instead of the discontent which is now settling upon Eastern Germany, occupied Austria and the countries of Eastern Europe, we seek a harmonious family of free European nations, with none a threat to the other, and least of all a threat to the peoples of Russia.

Beyond the turmoil and strife and misery of Asia, we seek peaceful opportunity for these peoples to develop their natural resources and to elevate their lot.

These are not idle words of shallow vision. Behind them lies a store of nations lately come to independence, not as a result of war but through free grant or peaceful negotiation. There is a record already written of assistance gladly given by nations of the West to needy peoples and to those suffering the temporary effects of famine, drought and natural disaster.

These are deeds of peace. They speak more loudly than promises or protestations of peaceful intent.

Would Explore Every Channel

But I do not wish to rest either upon the reiteration of past proposals or the restatement of past deeds. The gravity of the time is such that every new avenue of peace, no matter how dimly discernible, should be explored.

There is at least one new avenue of peace which has not yet been well explored—an avenue now laid out by the General Assembly of the United Nations.

In its resolution of November 18, 1953, this General Assembly suggested—and I quote—"that the Disarmament Commission study the desirability of establishing a subcommittee consisting of representatives of the powers principally involved, which should seek, in private, an acceptable solution—and report such a solution to the General Assembly and to the Security Council not later than 1 September, 1954."

The United States, heeding the suggestion of the General Assembly of the United Nations, is instantly prepared to meet privately with such other countries as may be "principally involved," to seek "an acceptable solution" to the atomic armaments race which overshadows not only the peace but the very life of the world.

We shall carry into these private or diplomatic talks a new conception.

The United States would seek more than the mere reduction or elimination of atomic materials for military purposes.

It is not enough to take this weapon out of the hands of the soldiers. It must be put into the hands of those who will know how to strip its military casing and adapt it to the arts of peace.

The United States knows that if the fearful trend of atomic military build-up can be reversed, this greatest of destructive forces can be developed into a great boon for the benefit of all mankind.

The United States knows that peaceful power from atomic energy is no dream of the future. That capability, already proved, is here now—today. Who can doubt, if the entire body of the world's scientists and engineers had adequate amounts of fissionable material with which to test and develop their ideas, that this capability would rapidly be transformed into universal, efficient and economic usage.

To hasten the day when fear of the atom will begin to disappear from the minds of people and the governments of the East and West there are certain steps that can be taken now.

I therefore make the following proposals:

The governments principally involved to the extent permitted by elementary prudence, to begin now and continue to make joint contributions from their stockpiles of normal uranium and fissionable materials to an international atomic energy agency. We would expect that such an agency would be set up under the aegis of the United Nations.

The ratios of contributions, the procedures and other details would properly be within the scope of the "private conversations" I have referred to earlier.

The United States is prepared to undertake these explorations in good faith. Any partner of the United States acting in the same good faith will find the United States a not unreasonable or ungenerous associate.

Undoubtedly initial and early contributions to this plan would be small in quantity. However, the proposal has the great virtue that it can be undertaken without irritations and mutual suspicions incident to any attempt to set up a completely acceptable system of world-wide inspection and control.

The atomic energy agency could be made responsible for the impounding, storage and protection of the contributed fissionable and other material. The ingenuity of our scientists will provide special, safe conditions under which such a bank of fissionable material can be made essentially immune to surprise seizure.

The more important responsibility of this atomic energy agency would be to devise methods whereby this fissionable material would be allocated to serve the peaceful pursuits of mankind. Experts would be mobilized to apply atomic energy to the needs of agriculture, medicine and other peaceful activities. A special purpose would be to provide abundant electrical energy in the power-starved areas of the world. Thus the contributing powers would be dedicating some of their strength to serve the needs rather than the fears of mankind.

Outlines Plan for Congress

The United States would be more than willing—it would be proud—to take up with others "principally involved" the development of plans whereby such peaceful use of atomic energy would be expedited.

Of those "principally involved" the Soviet Union must, of course, be one.

I would be prepared to submit to the Congress of the United States, and with every expectation of approval, any such plan that would:

First, encourage world-wide investigation into the most effective peacetime uses of fissionable material;

Second, begin to diminish the potential destructive power of the world's atomic stockpiles;

Third, allow all peoples of all nations to see that, in this enlightened age, the great powers of the earth, both of the East and of the West, are interested in human aspirations first rather than building up the armaments of war;

Fourth, open up a new channel for peaceful discussion and initiate at least a new approach to the many difficult problems that must be solved in both private and public conversations if the world is to shake off the inertia imposed by fear and is to make positive progress toward peace.

Against the dark background of the atomic bomb, the United States does not wish merely to present strength, but also the desire and the hope for peace.

The coming months will be fraught with fateful decisions. In this Assembly, in the capitals and military headquarters of the world; in the hearts of men everywhere, be they governed or governors, may they be the decisions which will lead this world out of fear and into peace.

To the making of these fateful decisions, the United States pledges before you—and therefore before the world—its determination to help solve the fearful atomic dilemma—to devote its entire heart and mind to find the way by which the miraculous inventiveness of man shall not be dedicated to his death, but consecrated to his life.

I again thank the delegates for the great honor they have done me in inviting me to appear before them and in listening to me so courteously. Thank you.

---

26. Lewis L. Strauss, "My Faith in the Atomic Future," August 1955

As told to James Monahan

Many people regard the atomic discoveries of recent years as part of a nightmare that disrupts the peaceful dreams of civilized man. I do not believe history will see them in that light. We have gained control over natural forces that can advance civilization, even within a single generation, to a point which man has never attained before. I believe firmly that our knowledge of the atom is intended by the Creator for the service and not the destruction of mankind.

The Atomic Energy Act of 1946 was a farsighted law. But I had certain specific reservations about it. Nuclear energy, which I believed could change the world, was straitjacketed in Government regulations. Research, development, patents, manufacturing and possession of fissionable materials were denied to private enterprise. Atomic energy was an absolute Government monopoly.

Atomic weapons development is necessarily a Government responsibility. But I was convinced that developments in agriculture, industry and power production would not be realized fully until the field was opened to the genius and enterprise of American industry.

Actually, the restrictions might have been relaxed sooner but for the attitude of Soviet Russia. Beginning in 1946, when the United States held a virtual world monopoly on nuclear weapons, we proposed international control, subject to rigid inspections and enforcement, which would have limited

This article appears in *The Reader's Digest*, August 1955, pp. 17–21. Copyright © 1955 by the Reader's Digest Association, Inc. Reprinted by permission.

the use of atomic energy to peaceful purposes. At that time we even offered to share our knowledge and resources with all nations.

The Soviets did everything possible to delay, confuse and destroy that plan. Actually, they were launching their own secret atomic program. We detected their clandestine weapon test in 1949, and were at once engaged in the costly and perilous contest for supremacy in nuclear weapons. Every thinking person knows now that our present great and versatile stockpile is the major safeguard of the free world.

Meanwhile, atomic energy became associated in popular thinking with death and destruction. Yet the custodians of atomic energy under President Truman and President Eisenhower never lost sight of its benign potentials. Progress was phenomenal in the production of radioisotopes.* They were produced by AEC reactors as early as 1947. They were distributed freely to institutions here and abroad, and within a few years revolutionized some areas of medical research and the diagnosis and treatment of certain diseases. Scarce, high-priced radium for the treatment of cancer was rendered virtually obsolete by radioactive cobalt and other elements which are equally effective sources of gamma rays and yet are now available to institutions at a small fraction of radium's cost.

Several different types of nuclear reactors for the generation of electrical power were designed by the AEC. But most authorities put the date of their construction in the remote future.

When I returned to the AEC as chairman in 1953 I was deeply impressed by the growing conviction in the White House and the Congress that the time had come for full-scale development of atomic energy outside the military area.

President Eisenhower, in his address to the General Assembly of the United Nations on December 8, 1953, stated: "The United States pledges before the world its determination to help solve the fearful atomic dilemma, to devote its entire heart and mind to find the way by which the miraculous inventiveness of man shall not be dedicated to his death, but consecrated to his life."

Two months later the President sent to Congress the message which resulted in the Atomic Energy Act of 1954. The new law had two great aims—to make international coöperation possible, and to enable private enterprise to develop the atom for peaceful purposes.

---

* An isotope is the "twin" of an element. It is chemically identical, but differs slightly in atomic weight. Radioisotopes also differ from their stable twins by giving off radiation. Some radioisotopes (radium and uranium, for example) occur in nature. Today it is possible to produce radioactive "twins" of any stable element (carbon, sodium, phosphorus, etc.) in the atomic pile or reactor.

The progress of the past 18 months—only a moment in history—has been extraordinary. For example, the AEC announced its program to develop power-producing reactors, and invited private companies to participate. The quick response was totally unexpected. The Duquesne Light Co. is building our first full-scale nuclear-power plant at Shippingport, Pa. At least four or five others will be constructed in the near future in Massachusetts, Michigan, Nebraska, Illinois and New York.

These pioneer nuclear-power plants cannot be economically competitive with conventional plants at present. Yet the participating companies are paying about *90 percent* of the total costs! This, I maintain, could only happen under free enterprise in an expanding economy.

Indeed, two or three proposed plants will be constructed *entirely* without financial help from the Government. Mr. Hudson R. Searing, president of one of these companies, Consolidated Edison of New York, recently told stockholders that nuclear power "is the only way we can see of bringing about lower electricity costs over the long pull."

It is pointless to speculate on how soon nuclear power will be cheaper than power produced by falling water or the burning of coal or oil. We do know that our resources of fossil fuels are limited, and that coal and oil will be needed for functions which atomic energy cannot perform. We know that there is a great disparity in electricity costs between those areas where water power, coal and oil are plentiful and regions like New England where such resources are scarce or nonexistent. We also must remember that there are many countries which are not blessed with such abundant resources as our own. So to me the present question of "economic" nuclear power is academic. I believe that it will be available before long, and that it logically will be used first where it is needed most.

Nuclear power for the propulsion of ships and aircraft will also come sooner than is generally realized. Few people have grasped the significance of the *Nautilus* and her sister submarine, the *Seawolf.* With the feasibility and safety of the marine propulsion reactor established beyond doubt, the job now is up to the designers and builders of surface vessels. The time to begin is *now.* That was the thought behind the President's recent recommendation of a nuclear-powered merchant ship.

I am convinced that the radioactive isotopes will continue to be the wonders of the atomic age. Today, they are being used by many industries to control processes, detect flaws and test the durability or wearing quality of all sorts of materials. New uses for them are found every day.

Used as "tracers" or as radiation sources, these atomic particles can search out the innermost secrets of nature and give man greater control over his environment. For example, plant geneticists have already used radiation from isotopes to produce a new strain of rust-resistant oats, wilt-resistant tomato seedlings, and a peanut plant with 30 percent greater yield. These and similar developments will mean millions—perhaps billions—of dollars to farmers.

By incorporating small amounts of radiophosphorus in fertilizers, and then using instruments to trace the uptake from the soil through roots, stem, leaves and blossom, agricultural experts can now determine the right amount of fertilizer to use in the most economical manner, and at the proper time in the growing cycle.

For nearly a century science has tried in vain to solve the fundamental secret of photosynthesis, the process whereby nature traps solar energy in the green leaf and converts water and carbon dioxide into the sugars and starches on which all higher life subsists. Using radioactive carbon as the tracer, researchers today seem to be on the point of solving (and perhaps duplicating) this mysterious process. If successful, that achievement might lead to the synthetic production of basic foodstuffs from simple and abundant chemicals—the solution to the world's pressing food problems.

Since 1946 the American people have spent more than 12 billion dollars on atomic energy. We will probably continue spending about two billion a year. Most of this money is invested in our stockpile of nuclear weapons, which represents the security of the free world. We have no choice but to maintain that security—until the whole world joins us in arriving at a safe solution to the "atomic dilemma." I firmly believe that can be accomplished.

But our nuclear stockpile also represents a national resource of incalculable value. With nuclear weapons you can "beat swords into plowshares and spears into pruning hooks" even more realistically than the Scriptures envisioned. The material is immediately convertible to peaceful uses. That is what President Eisenhower had in mind when he told the United Nations that the weapon "must be put in the hands of those who know how to strip its military casing and adapt it to the arts of peace."

Young people have asked me if I sincerely think that we shall enjoy the benefits of the atom before the world is overtaken by the destructive power that is within man's grasp. With all my heart I can answer: Yes!

We are living in an era that seems designed to test the courage and faith of free men. Yet I do not believe that any great discovery of the atom's magnitude came from man's intelligence alone. A Higher Intelligence decided that man was ready to receive it. My faith tells me that the Creator did not intend man to evolve through the ages to this stage of civilization only now to devise something that would destroy life on this earth.

My old chief, former President Herbert Hoover, to whose Quaker convictions all warfare is revolting, listened to President Eisenhower's U.N. speech and said: "I pray it may be accepted by all the world." We pray that Divine Providence will guide men of all nations to grasp this opportunity to "shake off the inertia imposed by fear and make positive progress toward peace."

IV

The Hydrogen Bomb

*I*n August 1949 the first successful Soviet test of an atomic bomb, nick-named "Joe I" after Stalin, ended the American monopoly on nuclear weapons and inaugurated the era of proliferation. The American response was to launch a crash program to develop an even more fearful thermonuclear, or hydrogen, weapon based upon the fusing of light elements at extremely high temperatures, as opposed to the fission of heavy ones. Although the idea of a fusion weapon had been broached by Edward Teller in 1942, relatively little work had been done.

Despite opposition by Oppenheimer and the General Advisory Committee of the Atomic Energy Commission, Truman decided to approve the development of the new weapon upon the advice of the secretaries of state and defense. The achievement of a successful hydrogen bomb did not occur until 1954, a consequence less of the opposition of Oppenheimer than of the extraordinary difficulty in coming up with a feasible design of the weapon. Whereas the early atomic bombs had a yield of approximately twenty kilotons of TNT equivalent, the hydrogen bomb yield immediately reached the fifteen megaton range; the largest weapon was exploded in 1961 by the Soviet Union, with a yield of nearly sixty megatons.

On the development of the hydrogen bomb, see Herbert York, The Advisors: Oppenheimer, Teller and the Superbomb (San Francisco: W. H. Freeman, 1976), which includes the previously classified General Advisory Committee report, and Richard G. Hewlett and Francis Duncan, Atomic Shield, 1947–1952 (University Park: Pennsylvania State University Press, 1969). Hans Bethe wrote his version of the hydrogen bomb project in Los Alamos Science (Fall 1982). In 1979 a journalist named Howard Morland attempted to publish a description of the hydrogen bomb "secret" in the Progressive magazine, and the U.S. government took him to court and delayed publication. For the details of this case, see Alexander De Volpi et al.,

Born Secret: The H-Bomb, The *Progressive* Case, and National Security *(Oxford: Pergamon Press, 1981). On the position of the Atomic Energy Commission, see* The Journals of David E. Lilienthal, *vol. 2,* The Atomic Energy Years, 1945–1950 *(New York: Harper & Row, 1964). An important overview of the history of atomic weapons is McGeorge Bundy,* Danger and Survival: Choices About the Bomb in the First Fifty Years *(New York: Random House, 1988). A more specialized work in the history of science is the first volume of a projected multi-volume work, J. L. Heilbron and Robert W. Seidel,* Lawrence and His Laboratory: A History of the Lawrence Berkeley Laboratory, *(Berkeley: University of California Press, 1989).*

---

27. "Political Implications of Detonation of Atomic Bomb by the U.S.S.R.," August 16, 1949.

Top Secret

The Problem

1. To determine the political implications if this Government could know with certainty when the U.S.S.R. detonates an atomic bomb.

Analysis and Conclusions

The Department of State obviously cannot pass on the question whether scientific techniques or equipment can be developed to detect the explosion by the U.S.S.R. of an atomic bomb, and it cannot express judgment as between competing demands for research and development funds. It is clear, however, that "only if a high degree of certainty can be placed on systems of

U.S. Department of State, Policy Planning Staff PPS/58, August 16, 1949. The text appears in Foreign Relations of the United States, *1949,* vol. 1 (Washington, D.C.: U.S. Government Printing Office, 1976), pp. 514–15.

detection, would this Government be warranted in basing policy decisions on intelligence derived from them."

Definite knowledge by this Government of the explosion by the U.S.S.R. of its first bomb is considered by the Department to be important for the following reasons:

(1) It would have a steadying effect on the American people and give them a sense of security if this Government could give assurance that the U.S.S.R. probably could not, without our knowledge, have a bomb or bombs for any length of time. With this knowledge, the Government would be able to combat intelligently defeatist or irrational attitudes arising from uncertainty as to whether the U.S.S.R. was capable of using atomic bombs, and would be in a position to refute with conviction false claims or rumors.

(2) It would be of the utmost importance for us to know when the U.S.S.R. has successfully tested a bomb in order to anticipate and counter possible changes in Soviet foreign policy which might result therefrom, and to know whether a shift in its foreign policy was the result of the possession of atomic bombs. We cannot know whether the U.S.S.R. would make the knowledge public if it did possess the atomic bomb; however, we would be in a position to know the truth of what the U.S.S.R. said publicly.

(3) The Soviet possession of a bomb or bombs may require a reevaluation of U.S. policy in the United Nations in our efforts to obtain effective international control.

(4) Most of the free nations of the world are inclined at present to cooperate with the United States in view of the threat of Soviet aggression. A belief that we are now the sole possessor of atomic bombs and that the U.S.S.R. has none probably tends to increase their desire to collaborate with us and also their sense of safety in doing so. This tendency would probably be reinforced even further by certain knowledge that the U.S.S.R. does not possess the bomb and that we would have means of knowing if it and when it did come into possession of the bomb. However, it is realized that knowledge that the U.S.S.R. did in fact possess the bomb also might tend to incline third countries toward a position of neutrality between the United States and the U.S.S.R.

(5) If at some later time we should learn with certainty that the U.S.S.R. did possess the atomic bomb, this knowledge would be of importance in reevaluating the necessity for precautionary measures to reduce U.S. vulnerability to atomic attack. However, this is a matter of primary concern to the NME. (National Military Establishment)

28. Statement by the President on Announcing the First Atomic Explosion in the USSR, September 23, 1949

I believe the American people, to the fullest extent consistent with national security, are entitled to be informed of all developments in the field of atomic energy. That is my reason for making public the following information.

We have evidence that within recent weeks an atomic explosion occurred in the U.S.S.R.

Ever since atomic energy was first released by man, the eventual development of this new force by other nations was to be expected. This probability has always been taken into account by us.

Nearly 4 years ago I pointed out that "scientific opinion appears to be practically unanimous that the essential theoretical knowledge upon which the discovery is based is already widely known. There is also substantial agreement that foreign research can come abreast of our present theoretical knowledge in time." And, in the Three-Nation Declaration of the president of the United States and the prime ministers of the United Kingdom and of Canada, dated November 15, 1945, it was emphasized that no single nation could in fact have a monopoly of atomic weapons.

This recent development emphasizes once again, if indeed such emphasis were needed, the necessity for that truly effective enforceable international control of atomic energy which this Government and the large majority of the members of the United Nations support.

This text appears in *Public Papers of the Presidents of the United States: Harry S. Truman, 1949* (Washington, D.C.: U.S. Government Printing Office, 1964), p. 485.

29. USAEC General Advisory Committee Minutes, October 28–30, 1949

MINUTES:

Seventeenth Meeting of the General Advisory Committee
to the U.S. Atomic Energy Commission,
October 28–30, 1949,
Washington, D.C.

FIRST SESSION
October 28, 1949

This Session was convened by the Chairman at 2:00 P.M. Those present were Dr. Oppenheimer, Dr. Buckley, Dr. DuBridge, Dr. Fermi, Dr. Rabi, Dr. Smith, the Secretary, and Mr. George F. Kennan, Counsellor of the Department of State. Dr. Seaborg was absent for all sessions of this meeting.

Mr. Kennan presented his views of the Russian situation and replied to a number of questions from Committee members. He left the meeting at 2:45 P.M.

The remainder of the Session, which Dr. Rowe joined, was devoted to continuing the discussion of our present policy and the role of atomic energy in present and future implementation of developing policy.

This Session was adjourned by 4:00 P.M. to permit the members to have informal discussions with Dr. H. A. Bethe of Cornell University and Dr. R. Serber of the Radiation Laboratory, Berkeley.

SECOND SESSION
October 29, 1949

This Session was convened at 9:30 A.M. Dr. Oppenheimer, Dr. Buckley, Dr. Conant, Dr. DuBridge, Dr. Fermi, Dr. Rabi, Dr. Rowe, Dr. Smith, the Secretary, and Mr. Tomei were present.

As the result of agreement that the proposal to initiate a program for the development of a super-bomb was the chief issue under the broad objective of examining whether the Commission is doing everything reasonably possible for it to do for the common defense and security, the discussion was

This text appears in United States Atomic Energy Commission General Advisory Committee Records, United States Department of Energy, Germantown, Maryland.

directed toward a consideration of the details and possible consequences of such a program.

At 10:00 A.M., Commissioners Lilienthal, Dean, Smyth, and Strauss, the General Manager, the Deputy General Manager, the Director of Production, the Director of Reactor Development, the General Counsel, and the Executive Officer of the Program Council of the Commission joined the meeting, followed at 10:10 A.M. by the Director of Biology and Medicine and the Director of Military Application, and at 10:20 A.M. by the Director of Research. Mr. Tomei was absent for the remainder of this Session. Discussion of the super-bomb program continued with this group.

At 11:00 A.M., the Chairman of the Joint Chiefs of Staff, the Chairman and the Executive Secretary of the Military Liaison Committee, the Chairman of the Weapons Systems Evaluation Group, Lt. Gen. L. Norstad, and Rear Adm. W. S. Parsons joined the Committee to discuss the military factors relevant to the Commission's program. Commissioners Lilienthal, Dean, Smyth and Strauss, the General Manager, and the Director of Military Application of the Commission were also present. The purpose of this meeting was to explore the military implications of the Commission's present or future programs.

This Session was adjourned at 12:30 P.M.

THIRD SESSION

The Committee met with Dr. Malcolm C. Henderson, Deputy Director of Intelligence, AEC, at 2:10 P.M. to obtain background information on Russian activities as known to his office. Commissioners Lilienthal, Dean, Smyth, and Strauss were also present. Dr. Henderson left the meeting at 3:30 P.M. Discussion continued until adjournment of the Session at 5:00 P.M.

FOURTH SESSION

This Session was convened at 8:15 P.M. All Committee members (except Dr. Seaborg), the Secretary, and Mr. Tomei were present.

Further exploration of the super-bomb program was conducted. Before adjournment at 10:10 P.M. it was understood that drafts of specific positions would be the first order of business at the next Session.

FIFTH SESSION
October 30, 1949

This Session was convened at 9:05 A.M. Dr. Oppenheimer, Dr Buckley, Dr. Conant, Dr. DuBridge, Dr. Fermi, Dr. Rabi, Dr. Rowe, Dr. Smith, the Secretary, and Mr. Tomei were present.

Preliminary positions on the super-bomb program were outlined, and other possibilities for extension of the Commission's program—reactors for production of freely available neutrons, increased fissionable material production, and the use of atomic weapons for tactical purposes—were discussed.

At 10:10 A.M., Commissioners Lilienthal, Dean, Pike, Smyth, and Strauss, the General Manager, and the General Counsel of the Commission joined the group for a summary by Dr. Oppenheimer of the views of the Committee at this time and for additional discussion. Mr. Tomei was absent until 12:15 P.M.

During this portion of this Session, the Committee agreed that the Commission should feel free to use as it wished the reports and written views it would receive from the Committee. It was the implication of this action that the members of the GAC would not discuss the Committee's views until freed to do so by action of the Commission. In addition, it was agreed that each member would avoid discussion of his personal opinion at least for a period of one week, the initial day of the proposed Los Alamos conference. This was done in order to assure the maximum freedom of action to the Commission in the matters under discussion.

The guests of the Committee left at 12:15 P.M.

After taking action to confirm the date of the next meeting as December 1, 2, and 3, 1949, the place as Washington, and the dates of the subsequent meeting as January 31 and February 1, 1950, the Committee constituted drafting groups for the preparation of papers. The papers thus prepared were edited and modified to consist of the following:

1. A report by the Chairman addressed to the Chairman of the AEC, consisting of a covering letter and two Parts, approved by all members of the Committee (except Dr. Seaborg, who was not present). The "majority" to which reference is made (Part II, page 4, last paragraph) included neither Dr. Fermi nor Dr. Rabi.
2. One attachment to the report approved and signed by Dr. Conant, Dr. Rowe, Dr. Smith, Dr. DuBridge, Dr. Buckley, and Dr. Oppenheimer.
3. A second attachment approved and signed by Dr. Fermi and Dr. Rabi.

It was agreed that in view of the nature of the documents they should bear a Top Secret classification.

The Minutes of the Sixteenth Meeting of the Committee were unanimously approved and this Meeting was adjourned at 3:00 P.M.

J. H. Manley
Secretary

30. USAEC General Advisory Committee Report on the "Super," October 30, 1949

GENERAL ADVISORY COMMITTEE
to the
U.S. ATOMIC ENERGY COMMISSION
Washington 25, D.C.

October 30, 1949

Dear Mr. Lilienthal:

At the request of the Commission, the seventeenth meeting of the General Advisory Committee was held in Washington on October 29 and 30, 1949 to consider some aspects of the question of whether the Commission was making all appropriate progress in assuring the common defense and security. Dr. Seaborg's absence in Europe prevented his attending this meeting. For purposes of background, the Committee met with the Counsellor of the State Department, with Dr. Henderson of AEC Intelligence, with the Chairman of the Joint Chiefs of Staff, the Chairman of the Military Liaison Committee, the Chairman of the Weapons Systems Evaluation Group, General Norstadt and Admiral Parsons. In addition, as you know, we have had intimate consultations with the Commission itself.

The report which follows falls into three parts. The first describes certain recommendations for action by the Commission directed toward the common defense and security. The second is an account of the nature of the super project and of the super as a weapon, together with certain comments on which the Committee is unanimously agreed. Attached to the report, but not a part of it, are recommendations with regard to action on the super project which reflect the opinions of Committee members.

The Committee plans to hold its eighteenth meeting in the city of Washington on December 1, 2, and 3, 1949. At that time we hope to return to many of the questions which we could not deal with at this meeting.

J. R. Oppenheimer
Chairman

This text appears in United S. Atomic Energy Commission General Advisory Committee Records, United States Department of Energy, Germantown, Maryland.

UNITED STATES ATOMIC ENERGY COMMISSION
WASHINGTON, D.C. 20545
HISTORICAL DOCUMENT NUMBER 349

David E. Lilienthal
Chairman
U.S. Atomic Energy Commission
Washington 25, D.C.

PART I

(1) PRODUCTION. With regard to the present scale of production of fissionable material, the General Advisory Committee has a recommendation to make to the Commission. We are not satisfied that the present scale represents either the maximum or the optimum scale. We recognize the statutory and appropriate role of the National Military Establishment in helping to determine that. We believe, however, that before this issue can be settled, it will be desirable to have from the Commission a careful analysis of what the capacities are which are not now being employed. Thus we have in mind that an acceleration of the program on beneficiation of low grade ores could well turn out to be possible. We have in mind that further plants, both separation and reactor, might be built, more rapidly to convert raw material into fissionable material. It would seem that some notion of the costs, yields and time scales for such undertakings would have to precede any realistic evaluation of what we should do. We recommend that the Commission undertake such studies at high priority. We further recommend that projects should not be dismissed because they are expensive but that their expense be estimated.

(2) TACTICAL DELIVERY. The General Advisory Committee recommends to the Commission an intensification of efforts to make atomic weapons available for tactical purposes, and to give attention to the problem of integration of bomb and carrier design in this field.

(3) NEUTRON PRODUCTION. The General Advisory Committee recommends to the Commission the prompt initiation of a project for the production of freely absorbable neutrons. With regard to the scale of this project the figure [deleted] per day may give a reasonable notion. Unless obstacles appear, we suggest that the expediting of design be assigned to the Argonne National Laboratory.

With regard to the purposes for which these neutrons may be required, we need to make more explicit statements. The principal purposes are the following:

(a)   The production of U-233.
(b)   The production of radiological warfare agents.
(c)   Supplemental facilities for the test of reactor components.
(d)   The conversion of U-235 to plutonium.
(e)   A secondary facility for plutonium production.
(f)   The production of tritium (1) for boosters, (2) for super bombs.

We view these varied objectives in a quite different light. We have a great interest in the U-233 program both for military and for civil purposes. We strongly favor, subject to favorable outcome of the 1951 Eniwetok tests, the booster program. With regard to radiological warfare, we would not wish to alter the position previously taken by our Committee. With regard to the conversion to plutonium, we would hardly believe that this alone could justify the construction of these reactors, though it may be important should unanticipated difficulties appear in the U-233 and booster programs. With regard to the use of tritium in the super bomb, it is our unanimous hope that this will not prove necessary. It is the opinion of the majority that the super program itself should not be undertaken and that the Commission and its contractors understand that construction of neutron producing reactors is not intended as a step in the super program.

PART II

Super Bombs

The General Advisory Committee has considered at great length the question of whether to pursue with high priority the development of the super bomb. No member of the Committee was willing to endorse this proposal. The reasons for our views leading to this conclusion stem in large part from the technical nature of the super and of the work necessary to establish it as a weapon. We therefore here transmit "an elementary" account of these matters.

The basic principle of design of the super bomb is the ignition of the thermo-nuclear DD reaction by the use of a fission bomb, and of high temperatures, pressure, and neutron densities which accompany it. In overwhelming probability, tritium is required as an intermediary, more easily ignited than the deuterium itself and, in turn, capable of igniting the deuterium. The steps which need to be taken if the super bomb is to become a reality include:

(1) The provision of tritium in amounts perhaps of several [deleted] per unit.
(2) Further theoretical studies and criticisms aimed at reducing the very great uncertainties still inherent in the behavior of this weapon under extreme conditions of temperature, pressure and flow.
(3) The engineering of designs which may on theoretical grounds appear hopeful, particularly with regard to the [deleted] problems presented.
(4) Carefully instrumented test programs to determine whether the deuterium-tritium mixture will be ignited by the fission bomb, . . .

It is notable that there appears to be no experimental approach short of actual test which will substantially add to our conviction that a given model will or will not work, and it is also notable that because of the unsymmetric and extremely unfamiliar conditions obtaining, some considerable doubt will surely remain as to the soundness of theoretical anticipation. Thus we are faced with a development which cannot be carried to the point of conviction without the actual construction and demonstration of the essential elements of the weapon in question. This does not mean that further theoretical studies would be without avail. It does mean that they could not be decisive. A final point that needs to be stressed is that many tests may be required before a workable model has been evolved or before it has been established beyond reasonable doubt that no such model can be evolved. Although we are not able to give a specific probability rating for any given model, we believe that an imaginative and concerted attack on the problem has a better than even chance of producing the weapon within five years.

A second characteristic of the super bomb is that once the problem of initiation has been solved, there is no limit to the explosive power of the bomb itself except that imposed by requirements of delivery. This is because one can continue to add deuterium—an essentially cheap material—to make larger and larger explosions, the energy release and radioactive products of which are both proportional to the amount of deuterium itself. Taking into account the probable limitations of carriers likely to be available for the delivery of such a weapon, it has generally been estimated that the weapon would have an explosive effect some hundreds of times that of present fission bombs. This would correspond to a damage area of the order of hundreds of square miles, to thermal radiation effects extending over a comparable area, and to very grave contamination problems which can easily be made more acute, and may possibly be rendered less acute, by surrounding the deuterium with uranium or other material. It needs to be borne in mind that for delivery by ship, submarine or other such carrier, the limitations here outlined no longer apply and that the weapon is from a technical point of view without limitation with regard to the damage that it can inflict.

It is clear that the use of this weapon would bring about the destruction of innumerable human lives; it is not a weapon which can be used exclusively for the destruction of material installations of military or semi-military purposes. Its use therefore carries much further than the atomic bomb itself the policy of exterminating civilian populations. It is of course true that super bombs which are not as big as those here contemplated could be made, provided the initiating mechanism works. In this case, however, there appears to be no chance of their being an economical alternative to the fission weapons themselves. It is clearly impossible with the vagueness of design and the uncertainty as to performance as we have them at present to give anything like a cost estimate of the super. If one uses the strict criteria of damage area per dollar and if one accepts the limitations on air carrier capacity likely to obtain in the years immediately ahead, it appears uncertain to us whether the super will be cheaper or more expensive than the fission bomb.

PART III

Although the members of the Advisory Committee are not unanimous in their proposals as to what should be done with regard to the super bomb, there are certain elements of unanimity among us. We all hope that by one means or another, the development of these weapons can be avoided. We are all reluctant to see the United States take the initiative in precipitating this development. We are all agreed that it would be wrong at the present moment to commit ourselves to an all-out effort toward its development.

We are somewhat divided as to the nature of the commitment not to develop the weapon. The majority feel that this should be an unqualified commitment. Others feel that it should be made conditional on the response of the Soviet government to a proposal to renounce such development. The Committee recommends that enough be declassified about the super bomb so that a public statement of policy can be made at this time. Such a statement might in our opinion point to the use of deuterium as the principal source of energy. It need not discuss initiating mechanisms nor the role which we believe tritium will play. It should explain that the weapon cannot be explored without developing it and proof-firing it. In one form or another the statement should express our desire not to make this development. It should explain the scale and general nature of the destruction which its use would entail. It should make clear that there are no known or foreseen non-military applications of this development. The separate views of the members of the Committee are attached to this report for your use.

J. R. Oppenheimer

October 30, 1949

We have been asked by the Commission whether or not they should immediately initiate an "all-out" effort to develop a weapon whose energy release is 100 to 1000 times greater and whose destructive power in terms of area of damage is 20 to 100 times greater than those of the present atomic bomb. We recommend strongly against such action.

We base our recommendation on our belief that the extreme dangers to mankind inherent in the proposal wholly outweigh any military advantage that could come from this development. Let it be clearly realized that this is a super weapon; it is in a totally different category from an atomic bomb. The reason for developing such super bombs would be to have the capacity to devastate a vast area with a single bomb. Its use would involve a decision to slaughter a vast number of civilians. We are alarmed as to the possible global effects of the radioactivity generated by the explosion of a few super bombs of conceivable magnitude. If super bombs will work at all, there is no inherent limit in the destructive power that may be attained with them. Therefore, a super bomb might become a weapon of genocide.

The existence of such a weapon in our armory would have far-reaching effects on world opinion: reasonable people the world over would realize that the existence of a weapon of this type whose power of destruction is essentially unlimited represents a threat to the future of the human race which is intolerable. Thus we believe that the psychological effect of the weapon in our hands would be adverse to our interest.

We believe a super bomb should never be produced. Mankind would be far better off not to have a demonstration of the feasibility of such a weapon, until the present climate of world opinion changes.

It is by no means certain that the weapon can be developed at all and by no means certain that the Russians will produce one within a decade. To the argument that the Russians may succeed in developing this weapon, we would reply that our undertaking it will not prove a deterrent to them. Should they use the weapon against us, reprisals by our large stock of atomic bombs would be comparably effective to the use of a super.

In determining not to proceed to develop the super bomb, we see a unique opportunity of providing by example some limitations on the totality of war and thus of limiting the fear and arousing the hopes of mankind.

James B. Conant
Hartley Rowe
Cyril Stanley Smith
L. A. DuBridge
Oliver E. Buckley
J. R. Oppenheimer

October 30, 1949

An Opinion on the Development of the "Super"

A decision on the proposal that an all-out effort be undertaken for the development of the "Super" cannot in our opinion be separated from considerations of broad national policy. A weapon like the "Super" is only an advantage when its energy release is from 100–1000 times greater than that of ordinary atomic bombs. The area of destruction therefore would run from 150 to approximately 1000 square miles or more.

Necessarily such a weapon goes far beyond any military objective and enters the range of very great natural catastrophes. By its very nature it cannot be confined to a military objective but becomes a weapon which in practical effect is almost one of genocide.

It is clear that the use of such a weapon cannot be justified on any ethical ground which gives a human being a certain individuality and dignity even if he happens to be a resident of an enemy country. It is evident to us that this would be the view of peoples in their countries. Its use would put the United States in a bad moral position relative to the peoples of the world.

Any postwar situation resulting from such a weapon would leave unresolvable enmities for generations. A desirable peace cannot come from such an inhuman application of force. The postwar problems would dwarf the problems which confront us at present.

The application of this weapon with the consequent great release of radioactivity would have results unforeseeable at present, but would certainly render large areas unfit for habitation for long periods of time.

The fact that no limits exist to the destructiveness of this weapon makes its very existence and the knowledge of its construction a danger to humanity as a whole. It is necessarily an evil thing considered in any light.

For these reasons we believe it important for the President of the United States to tell the American public, and the world, that we think it wrong on fundamental ethical principles to initiate a program of development of such a weapon. At the same time it would be appropriate to invite the nations of the world to join us in a solemn pledge not to proceed in the development or construction of weapons of this category. If such a pledge were accepted even without control machinery, it appears highly probable that an advanced stage of development leading to a test by another power could be detected by available physical means. Furthermore, we have in our possession, on our stockpile of atomic bombs, the means for adequate "military" retaliation for the production or use of "super."

E. Fermi
I. I. Rabi

31. Lewis Strauss to Harry S. Truman, November 25, 1949

25 November 1949

Dear Mr. President:

As you know, the thermonuclear (super) bomb was suggested by scientists working at Los Alamos during the war. The current consideration of the super bomb was precipitated, I believe, by a memorandum which I addressed to my fellow Commissioners following your announcement on September 23rd of an atomic explosion in Russia. I participated in the discussions which were antecedent to the letter to you from the Commission on November 9th, but did not join in the preparation of the letter as I was then on the Pacific Coast. It was my belief that a comprehensive recommendation should be provided for you, embodying the judgment of the Commission (in the areas where it is competent), together with the views of the Departments of State and Defense. My colleagues, however, felt that you would prefer to obtain these views separately.

Excerpt from Lewis Strauss, *Men and Decisions* (New York: Doubleday and Company, 1962), pp. 219–22. Copyright © by Lewis L. Strauss and the Lewis L. Strauss Literary Trust. Reprinted by permission of Doubleday and Company, Inc.

Differences on the broad question of policy between my associates as individuals were included in the Commission's letter to you, and it was correctly stated that the views of Commissioner Dean and mine were in substantial accord on the main issue. It is proper, I believe, that I should state them on my own responsibility and in my own words.

I believe that the United States must be as completely armed as any possible enemy. From this, it follows that I believe it unwise to renounce, unilaterally, any weapon which an enemy can reasonably be expected to possess. I recommend that the President direct the Atomic Energy Commission to proceed with the development of the thermonuclear bomb, at highest priority subject only to the judgment of the Department of Defense as to its value as a weapon, and of the advice of the Department of State as to the diplomatic consequences of its unilateral renunciation or its possession. In the event that you may be interested, my reasoning is appended in a memorandum.

/s/ Lewis L. Strauss

25 November 1949

This is a memorandum to accompany a letter of even date to the President to supply the reasoning for my recommendation that he should direct the Atomic Energy Commission to proceed at highest priority with the development of the thermonuclear weapon.

Premises
(1) The production of such a weapon appears to be feasible (i.e., better than a 50-50 chance).
(2) Recent accomplishments by the Russians indicate that the production of a thermonuclear weapon is within their technical competence.
(3) A government of atheists is not likely to be dissuaded from producing the weapon on "moral" grounds. ("Reason and experience both forbid us to expect that national morality can prevail in exclusion of religious principle." G. Washington, September 17, 1796.)
(4) The possibility of producing the thermonuclear weapon was suggested more than six years ago, and considerable theoretical work has been done which may be known to the Soviets—the principle has certainly been known to them.
(5) The time in which the development of this weapon can be perfected is perhaps of the order of two years, so that a Russian enterprise started some years ago may be well along to completion.

(6) It is the historic policy of the United States not to have its forces less well armed than those of any other country (viz., the 5:5:3 naval ratio, etc. etc.).

(7) Unlike the atomic bomb which has certain limitations, the proposed weapon may be tactically employed against a mobilized army over an area of the size ordinarily occupied by such a force.

(8) The Commission's letter of November 9th to the President mentioned the "possibility that the radioactivity released by a small number (perhaps ten) of these bombs would pollute the earth's atmosphere to a dangerous extent." Studies requested by the Commission have since indicated that the number of such weapons necessary to pollute the earth's atmosphere would run into many hundreds. Atmospheric pollution is a consequence of present atomic bombs if used in quantity.

Conclusions

(1) The danger in the weapon does not reside in its physical nature but in human behavior. Its unilateral renunciation by the United States could very easily result in its unilateral possession by the Soviet Government. I am unable to see any satisfaction in that prospect.

(2) The Atomic Energy Commission is competent to advise the President with respect to the feasibility of making the weapon; its economy in fissionable material as compared with atomic bombs; the possible time factor involved; and a description of its characteristics compared to atomic bombs. Judgment, however, as to its strategic or tactical importance for the armed forces should be furnished by the Department of Defense, and views as to the effect on friendly nations or of unilateral renunciation of the weapon is a subject for the Department of State. My opinion as an individual, however, based upon discussion with military experts is to the effect that the weapon may be critically useful against a large enemy force both as a weapon of offense and as a defensive measure to prevent landings on our own shores.

(3) I am impressed with the arguments which have been made to the effect that this is a weapon of mass destruction on an unprecedented scale. So, however, was the atomic bomb when it was first envisaged and when the National Academy of Sciences in its report of November 6, 1941, referred to it as "of superlatively destructive power." Also on June 16, 1945, the Scientific Panel of the Interim Committee on Nuclear Power, comprising some of the present members of the General Advisory Committee, reported to the Secretary of War, "We believe the subject of thermonuclear reactions among light nuclei is one of the most important that needs study. There is a reasonable presumption

that with skillful research and development, fission bombs can be used to initiate the reactions of deuterium, tritium, and possibly other light nuclei. If this can be accomplished, the energy release of explosive units can be increased by a factor of 1000 or more over that of presently contemplated fission bombs." This statement was preceded by the recommendation, "Certainly we would wish to see work carried out on the problems mentioned below."

The General Advisory Committee to the Atomic Energy Commission, in its recent communication to the Commission recommending against the development of the super bomb, noted that it "strongly favors" the booster program, which is a program to increase the explosive power and hence the damage area and deadliness of atomic bombs. These positions and those above appear not to be fully consistent and indicate that the scientific point of view is not unanimous.

(4) Obviously the current atomic bomb as well as the proposed thermonuclear weapon are horrible to contemplate. All war is horrible. Until, however, some means is found of eliminating war, I cannot agree with those of my colleagues who feel that an announcement should be made by the President to the effect that the development of the thermonuclear weapon will not be undertaken by the United States at this time. This is because: (a) I do not think the statement will be credited in the Kremlin; (b) that when and if it should be decided subsequent to such a statement to proceed with the production of the thermonuclear bomb, it might in a delicate situation, be regarded as an affirmative statement of hostile intent; and (c) because primarily until disarmament is universal, our arsenal must be not less well equipped than with the most potent weapons that our technology can devise.

Recommendation

In sum, I believe that the president should direct the Atomic Energy Commission to proceed with all possible expedition to develop the thermonuclear weapon.

32. Statement by the President on the Hydrogen Bomb, January 31, 1950

It is part of my responsibility as Commander in Chief of the Armed Forces to see to it that our country is able to defend itself against any possible aggressor. Accordingly, I have directed the Atomic Energy Commission to continue its work on all forms of atomic weapons, including the so-called hydrogen or superbomb. Like all other work in the field of atomic weapons, it is being and will be carried forward on a basis consistent with the overall objectives of our program for peace and security.

This we shall continue to do until a satisfactory plan for international control of atomic energy is achieved. We shall also continue to examine all those factors that affect our program for peace and this country's security.

33. Excerpts from President Eisenhower's Press Conference, March 31, 1954

The following questions by news correspondents and answers by Lewis L. Strauss, Chairman, AEC, came at the conclusion of his prepared statement, which is attached.[1]

Q. Mr. Chairman, you said that this particular explosion was not out of control. But is it possible that in any series of tests that a hydrogen explosion or series of them could get out of control?

A. I am informed by the scientists that that is impossible.

Q. Admiral Strauss, yesterday, at his news conference, Secretary of Defense Wilson said the results of the March 1st test—is the one he was

(Selection 32). This text appears in *Public Papers of the Presidents of the United States: Harry S. Truman, 1950* (Washington, D.C.: Government Printing Office, 1965), p. 138.

(Selection 33). This text appeared in Eisenhower Press Conference, March 31, 1954; *New York Times*, April 1, 1954. Reprinted by permission of the *New York Times*.

1. Statement not reproduced.

referring to—was unbelievable. Would you care to comment on that?

A. No, I don't think I should comment on that. The use of that adjective, I think, was played up beyond the point where the Secretary intended it. I don't know what is meant by "unbelievable" and I would rather not comment.

Q. Mr. Chairman, do you intend to imply by the last paragraph in this statement that the work on the weapon phase of the atom is reaching a completion; that we are approaching a point where pursuit of this will no longer yield very large profits, and that we will, therefore, turn our research power to the peaceful applications?

A. . . . I think the answer to that is this: The Military have certain requirements. The Commission is engaged in attempting to fill those requirements. The ability of the Commission to devote attention and fissionable material to peaceful requirements, peaceful needs, is always junior to the defense needs, by definition of the Act itself. The result of these tests has brought us very much nearer to the day of the satisfaction of military requirements, put us within sight of them, so that we can see the ability to proceed aggressively with the peacetime development of power to an extent that we were not able to before the tests.

Q. Admiral Strauss, can you go beyond this statement and describe the area of the blast, the effectiveness of the blast, and give a general description of what actually happened when the H-bomb went off?

A. The area of the blast, would be about—
THE PRESIDENT: Why not depend on these pictures they are all going to see?
(Mr Strauss) I understand you are going to see a film, a picture, of the 1952 shot. The area, if I were to describe it specifically, would be translatable into the number of megatons involved, which is a matter of military secrecy.

The effects, you said the effectiveness—I don't know exactly what you meant by that, sir, so I don't know how to answer it.

Q. Well, I don't mean in the percentage of the effectiveness of or the efficiency of the blast itself. But many people in Congress, I think many elsewhere, have been reaching out and grasping for some information as to what happens when the H-bomb goes off, how big is the area of destruction in its various stages; and what I am asking you for now is some enlightenment on that subject.

A. Well, the nature of an H-bomb, Mr. Wilson, is that, in effect, it can be made to be as large as you wish, as large as the military requirement demands, that is to say, an H-bomb can be made as—large enough to take out a city.

(A chorus of "What?")

To take out a city, to destroy a city.

Q. How big a city?

A. Any city.

Q. Any city, New York?

A. The metropolitan area, yes.

(With reference to the foregoing, Mr. Strauss added later that he meant "put out of commission," not "to destroy.")

Q. Mr. Chairman, may I ask this specific question: If you were to make a comparison, duplicating the explosion that occurred at Eniwetok, with this building in which we are right now as the center, what would be left of this city of Washington?

A. Well I couldn't say, because the precise measurements of these two shots have not been completely calibrated. It may be as many as a month or two before I know the answer to it. It would be a very extensive—

Q. Will you provide that answer at some time, sir?

A. I won't make a definite commitment, but I would certainly like to.

. . . .

34. Hans Bethe, Comments on the History of the H-bomb, 1954[1]

The H-bomb was suggested by Teller in 1942. Active work on it was pursued in the summer of 1942 by Oppenheimer, Teller, myself, and others. The idea did not develop from Teller's "quiet work" at Los Alamos during the war.

When Los Alamos was started in Spring 1943, several groups of scientists were included who did work on this problem specifically. However, it was realized that this was a long-range project and that the main efforts of Los

This text appears in *Los Alamos Science* (Fall 1982): 43–53.

1. Hans A. Bethe, a distinguished theoretical physicist from Cornell University, was long associated with weapons development at Los Alamos.

Alamos must be concentrated on making A-bombs. Teller, working on the H-bomb at Los Alamos, discovered a major difficulty (testimony by Oppenheimer). This discovery made it clear that it would be a very hard problem to make a "classical super" work, as this type of H-bomb was called. I shall refer to the classical super as Method A.

The work on thermonuclear weapons at Los Alamos never stopped. At this stage of the development, the main requirements were for theoretical work and for a few experimental physics measurements. Both of these types of work went ahead. On the basis of the monthly reports of the theoretical division of Los Alamos, it has been estimated that between 1946 and 1949 the work of that division was about equally divided between fission weapon design and problems related to thermonuclear weapons. (In this respect I was mistaken when testifying in the Oppenheimer case. I said then, from memory, that a relatively small fraction of the scientists of the division, though consisting of especially able men, were working on thermonuclear problems. Actually, the fraction was large.)

Two new methods of designing a thermonuclear weapon were invented (Methods B and C). Both inventions were due to Teller. Method B was invented in 1946, Method C in 1947. Method B was actively worked on by Richtmyer, Nordheim, and others. However, at that time, there seemed to be no way of putting Method B into practice, as Dr. Bradbury has mentioned in his statement to "The New Mexican." Teller himself wrote a most pessimistic report on the feasibility of this method in September 1947.

Method C is different from all the others in that thermonuclear reactions are used only in a minor way, for weapons of relatively small yield. This method seemed quite promising from the start, as early as the Summer of 1948 it was added to the devices to be tested in the Greenhouse tests.

Theoretical work on the "classical super," Method A, proceeded continually, since this method was considered the most important of all thermonuclear devices. New plans for calculations were made frequently, mostly by consultation between Teller and the senior staff of the theoretical division. However, as Teller stated in 1946, "The required scientific effort is clearly much larger than that needed for the first fission weapon." In particular, the theoretical computations required were of such complication that they could not be handled in any reasonable time by any of the computing machines then available. Some greatly simplified calculations were done but it was realized that they left out many important factors and were therefore quite unreliable. Work was therefore concentrated on preparing full-scale calculations "for the time when adequate fast computing machines become available"—a sentence which recurs in many of the theoretical reports of this period. The plans for such a calculation on Method A were laid in September

1948, and the mathematical work was virtually completed by December 1949—all before the directive of President Truman—but it was not until mid-1952 that adequate computing machines finally became available, and by that time the most capable of them were fully engaged on the new and more promising proposal (Method D) discussed below.

When Dr. Teller and Admiral Strauss proposed in the Fall of 1949 to start a full-scale development of H-bombs, the method in their minds, as well as in the minds of the opponents of the program, was Method A. To accomplish Method A, two major problems had to be solved which I shall call Part 1 and Part 2. Part 1 seemed to be reasonably well in hand according to calculations made by Teller's group from 1944 to 1946 although nobody had been able to perform a really convincing calculation, as discussed in the paragraph above. Teller now believed that he had a solution for Part 2. In principle, the accomplishment of Part 2 had never been seriously in doubt, although the question of whether or not any particular device would behave in the way required could not be settled without experiment.

The Greenhouse thermonuclear experiment was designed to test Part 2. After President Truman made the decision to go ahead with a full-scale thermonuclear program, Los Alamos made plans to add to the Greenhouse test series an experiment intended to test a particular proposal relating to Part 2. Teller played a large part in the specification of this device, and as it turned out it behaved very well. However, as on previous occasions, Teller did not do so well in directing the detailed theoretical work of his group. Only as late as January 1951, a month or so before the test device had to be shipped to the Pacific, was the full theoretical prediction of the (probably successful) behavior of the device available. But even while complete theoretical proof was lacking, most of us connected with the work at Los Alamos were confident that the Greenhouse experiment would work. As far as I could make out, at a meeting at Los Alamos in October 1950 which I attended as a guest, this was also the opinion of the GAC including Dr. Oppenheimer. Shepley and Blair instead report on page 116 that Dr. Oppenheimer expected the device to fail. (The correct story on Oppenheimer's attitude will be discussed below.)

A very large fraction of the members of the Los Alamos Laboratory, not just a "small handful of his" (Teller's) "associates" were extremely busy from Spring 1950 to Spring 1951 with the preparation of Teller's thermonuclear experiment. They did this in addition to preparing the Nevada tests of early 1951.

The major feature of the year 1950 was, however, the discovery that Part 1 of Method A was by no means under control. While Teller and most of the Los Alamos Laboratory were busy preparing the Greenhouse test, a number of persons in the theoretical division had continued to consider the various problems posed by Part 1. In particular, Dr. Ulam on his own initiative had decided to check the feasibility of aspects of Part 1 without the aid of high-speed computing equipment. He, and Dr. Everett who assisted him, soon found that the calculations of Teller's group of 1946 were wrong. Ulam's calculations showed that an extraordinarily large amount of tritium would be necessary. In the Summer of 1950 further calculations by Ulam and Fermi showed further difficulties with Part 1.

That Ulam's calculations had to be done at all was proof that the H-bomb project was not ready for a "crash" program when Teller first advocated such a program in the Fall of 1949. Nobody will blame Teller because the calculations of 1946 were wrong, especially because adequate computing machines were not then available. But he was blamed at Los Alamos for leading the Laboratory, and indeed the whole country, into an adventurous program on the basis of calculations which he himself must have known to have been very incomplete. The technical skepticism of the GAC on the other hand had turned out to be far more justified than the GAC itself had dreamed in October 1949.

We can now appreciate better the attitude of the GAC, and indeed of most of the members of Los Alamos, to the Greenhouse thermonuclear test. They did not expect it to fail, but they considered it as irrelevant because there appeared to be no solution to Part 1 of the problem. The correct description of this attitude is given by Oppenheimer in his own testimony.

The lack of a solid theoretical foundation was the only reason why the Los Alamos work might have seemed to some to have gotten off to a slow start in 1950. Purely theoretical work may seem slow in a project intended to develop "hardware," but there was simply no basis for building hardware until the theory had been clarified. As far as the mental attitude of Los Alamos in early 1950, it was almost the exact opposite of that described by Shepley and Blair. I visited Los Alamos around April 1, 1950 and tried to defend the point of view of the GAC in their decision of October 1949. I encountered almost universal hostility. The entire Laboratory seemed enthusiastic about the project and was working at high speed. That they continued to work with full energy on Teller's Greenhouse test, after Ulam's calculations had made the success of the whole program very doubtful, shows how far they were willing to go in following Teller's lead.

Teller himself was desperate between October 1950 and January 1951. He proposed a number of complicated schemes to save Method A, none of which seemed to show much promise. It was evident that he did not know of any solution. In spite of this, he urged that the Laboratory be put essentially at his disposal for another year or more after the Greenhouse test, at which time there should then be another test on some device or other. After the failure of the major part of his program in 1950, it would have been folly of the Los Alamos Laboratory to trust Teller's judgment, at least until he could present a definite idea which showed practical promise. This attitude was strongly held by most of those on the permanent staff of the Laboratory who were responsible for its operation. As might be expected, the many discussions of aspects of this situation bred considerable emotion.

Between January and May 1951, the "new concept" was developed. (This I shall call Method D.) In addition, it should be remembered that between January and May both tests in Nevada and the Greenhouse series of tests took place, and this required many senior members of the Laboratory to be at the test sites for prolonged periods of time and the attention of many others was engaged on study of results of these tests.

But what are the actual facts about this alleged delay in work on the new concept? In January Teller obviously did not know how to save the thermonuclear program. On March 9, 1951, according to Bradbury's press statement, Teller and Ulam published a paper which contained one-half of the new concept. As Bradbury has pointed out, Ulam as well as Teller should be given credit for this. Ulam, by the way, made his discovery while studying some aspects of fission weapons. This shows once more how the important ideas may not come from a straightforward attack on the main problem.

Within a month, the very important second half of the new concept occurred to Teller, and was given preliminary checks by de Hoffman. This immediately became the main focus of attention of the thermonuclear design program.

It is worth noting that the entire new concept was developed before the thermonuclear Greenhouse test which took place on May 8, 1951. The literature is full of statements that the success of Greenhouse was the direct cause of the new concept. This is historically false. Teller may have been influenced by thinking about the Greenhouse design when developing the new concept, but the success of Greenhouse (which was anticipated) had no influence on either the creation of the new concept, or on its quick adoption by the Laboratory or later by the GAC. The new concept stood on its own.

As early as the end of May 1951, I received from the Associate Director of Los Alamos a detailed proposal for the future program of the Laboratory in which Teller's new concept figured most prominently. By early June, when I visited Los Alamos for two weeks, everybody in the theoretical division was talking about the new concept.

Not only was the acceptance of the new concept not slow; but the realization of the development was a sensationally rapid accomplishment, in the same class as the achievement of Los Alamos during the war.

It is difficult to describe to a non-scientist the novelty of the new concept. It was entirely unexpected from the previous development. It was also not anticipated by Teller, as witness his despair immediately preceding the new concept. I believe that this very despair stimulated him to an invention that even he might not have made under calmer conditions. The new concept was to me, who had been rather closely associated with the program, about as surprising as the discovery of fission had been to physicists in 1939. Before 1939 scientists had a vague idea that it might be possible to release nuclear energy but nobody could think even remotely of a way to do it. If physicists had tried to discover a way to release nuclear energy before 1939, they would have worked on anything else rather than the field which finally led to the discovery of fission, namely radio-chemistry. At that time, concentrated work on any "likely" way of releasing nuclear energy would have led nowhere. Similarly, concentrated work on Method A would never have led to Method D. The Greenhouse test had a vague connection with Method D but one that nobody, including Teller, could have foreseen or did foresee when that test was planned. By a misappraisal of the facts many persons not closely connected with the development have concluded that the scientists who had shown good judgment concerning the technical feasibility of Method A were now suddenly proved wrong, whereas Teller, who had been wrong in interpreting his own calculations, was suddenly right. The fact was that the new concept had created an entirely new technical situation. Such miracles incidentally do happen occasionally in scientific history but it would be folly to count on their occurrence. One of the dangerous consequences of the H-bomb history may well be that government administrators, and perhaps some scientists, too, will imagine that similar miracles should be expected in other developments.

Before the end of the Summer of 1951, the Los Alamos Laboratory was putting full force behind attempts to realize the new concept. However, the continued friction of 1950 and early 1951 had strained a number of personal relations between Teller and others at Los Alamos. In addition, Teller insisted on an earlier test date than the Laboratory deemed possible. There was further disagreement between Teller and Bradbury on personalities, in particular on the person who was to direct the actual development of hardware. Bradbury had great experience in administrative matters like these. Teller had no experience and had in the past shown no talent for administration. He had given countless examples of not completing the work he had started; he was inclined to inject constantly new modifications into an already going program which becomes intolerable in an engineering development beyond a certain stage; and he had shown poor technical judgment. Everybody recognizes that Teller more than anyone else contributed ideas at every stage of the H-bomb program, and this fact should never be obscured. However, as an article in "Life" of September 6, 1954, clearly portrays, nine out of ten of Teller's ideas are useless. He needs men with more judgment, even if they be less gifted, to select the tenth idea which often is a stroke of genius.

It has been loosely said that the people at Los Alamos couldn't "get along" with Teller and it might be worthwhile to clarify this point. Both during the difficulties of the wartime period and again in 1951, Teller was on excellent terms with the vast majority of the scientists at Los Alamos with whom he came in contact in the course of the technical work. On both occasions, however, friction arose between him and some of those responsible for the organization and operation of the Laboratory. In each case, Teller, who was essentially alone in his opinion, was convinced that things were hopelessly bad and that nothing would go right unless things were arranged quite differently. In each case, the Laboratory accomplished its mission with distinction. In September 1951, when the program for a specific test of the new concept was being planned, Teller was strongly urged to take the responsibility for directing the theoretical work on the design of Mike. But he felt sure the test date should be a few months earlier; he didn't like some of the people with whom he would have to work; he was convinced they weren't up to the job; the Laboratory was not organized properly and didn't have the right people. Teller decided to leave and left. The Mike shot went off exactly on schedule and was a full success.

It took much more than the idea of the new concept to design Mike. Major difficulties occurred in the theoretical design in early 1952, which happened to be a period when I was again at Los Alamos. They were all solved by the splendid group of scientists at Los Alamos.

At this time more than one-half of all the development work of the Los Alamos Laboratory went into thermonuclear weapons and into the preparation of the Mike test in particular. All but a small percentage of the theoretical division were thinking about this subject. In addition, there was a group of theorists working in Princeton under the direction of Professor John A. Wheeler in collaboration with the theoretical group at Los Alamos. Shepley and Blair, however, have to say of this period (on page 141) "Progress on the thermonuclear program still lagged."

Teller "helped" at this time by intensive agitation against Los Alamos and for a second laboratory. This agitation was very disturbing to the few leading scientists at Los Alamos who knew about it. Much precious time was spent in trying to counteract Teller's agitation by bringing the true picture to Washington. I myself wrote a history of the thermonuclear development to Chairman Dean of the AEC which was mentioned in the Oppenheimer testimony. This loss of time could be ill afforded at a time when the technical preparations for Mike were in a crisis.

Nevertheless, the theoretical design of Mike was completed by June 1952 in good time to make the device ready for test on November 1. Not only this, but, in the same period, much work was done leading to the conceptual design of the devices which were later tested in the Castle series in the Spring 1954. The approximate date for the Castle tests was also set at that time, and it was planned then that it should lead to a deliverable H-bomb if the experimental Mike shot was successful. It is necessary always to plan approximately two years ahead. Between Summer 1952 and Spring 1954, theoretical calculations on the proposed thermonuclear weapons proceeded; they were followed and in some areas paralleled by mechanical design of the actual device and finally followed by manufacture of the "hardware".

In July 1952, the new laboratory at Livermore was officially established by the AEC. Its existence did not, and in fact could not, accelerate the Los Alamos work because in all essentials the work for Castle had been planned before Livermore was established. In August 1952 an additional device was conceived at Los Alamos which might possibly have been slightly influenced by ideas then beginning to be considered at Livermore. In addition, Los Alamos decided to make a few experimental small-scale shots in Nevada in the Spring of 1953, and this program may have been slightly stimulated by the existence of Livermore. Livermore did assist in the observation of the performance of some of the devices tested at Castle.

Concerning the performance of Livermore's own designs, I will only quote the statement of Dr. Bradbury to the press which says, "Every successful thermonuclear weapon tested so far" (1954) "has been developed by the Los Alamos Laboratory."[1] This statement has not been contradicted.

Was the H-bomb Necessary?

Although the GAC were seeking a solution rather than offering one, the proposal of its minority still seems worthwhile, even as seen from today's (1954) viewpoint. The proposal was to enter negotiations with Russia with the aim that both countries undertake an obligation not to develop the

1. Bethe added this comment in 1982: "In the intervening 28 years, Livermore has contributed greatly to nuclear weapons development. Some weapons programs are assigned to Livermore, some to Los Alamos, and the talents of the two laboratories complement each other."

H-bomb. If such an agreement could have been reached and had been kept, it would have gone far to avoid the peril in which the world now stands. At that time neither we nor the Russians presumably knew whether an H-bomb could be made. In this blissful state of ignorance we might have remained for a long time to come. Since the technical program was a very difficult one it could never be accomplished without a major effort. It is possible, perhaps likely, that the Russians would have refused to enter an agreement on this matter. If they had done so, this refusal would have been a great propaganda asset for us in the international field and would in addition have gone far to persuade the scientists of this country to cooperate in the H-bomb program with enthusiasm.

Many people will argue that the Russians might have accepted such an agreement, but then broken it. I do not believe so. Thermonuclear weapons are so complicated that nobody will be confident that he has the correct solution before he has tested such a device. But it is well known that any test of a bomb of such high yield is immediately detected. Therefore, without any inspection, each side would know immediately if the other side had broken the agreement.

It is difficult to tell whether or not the Russians would have developed the H-bomb independently of us. I am not sure what would have happened if we had followed the recommendations of the GAC majority and had merely announced that for such and such reasons, we would refrain from developing the H-bomb. Once we announced that we would go ahead, the Russians clearly had no choice but to do the same. In the field of atomic weapons, we have called the tune since the end of the war, both in quality and in quantity. Russia has to follow the tune or be a second-class power.

In summary I still believe that the development of the H-bomb is a calamity. I still believe that it was necessary to make a pause before the decision and to consider this irrevocable step most carefully. I still believe that the possibility of an agreement with Russia not to develop the bomb should have been explored. But once the decision was made to go ahead with the program, and once there was a sound technical program, I cooperated with it to the best of my ability. I did and still do this because it seems to me that once one is engaged in a race, one clearly must endeavor to win it. But one can try to forestall the race itself.

V

The Oppenheimer Case

*D*uring the winter of 1953–54, in the wake of the Korean War and several Soviet espionage cases involving atomic energy (notably those of the British physicist Klaus Fuchs and the Americans Ethel and Julius Rosenberg), the Cold War meant keeping atomic secrets from the Soviet Union. Both Moscow and Washington were directing major projects on developing a thermonuclear weapon as part of an escalating arms race. Los Alamos, together with the new Livermore national laboratory in California, was testing both atomic and hydrogen devices in the Nevada desert and on Pacific islands. As the power of the new weapons increased from kilotons to megatons in yield, and the size of deliverable weapons became ever smaller, many scientists had second thoughts about the power of the atom.

Among those scientists was J. Robert Oppenheimer, scientific director of the Manhattan Project and father of the atomic bomb, a man whose gaunt appearance and porkpie hat were as familiar to most Americans as a Norman Rockwell painting or Ike's grin. To the atomic scientists Oppie was a kind of father figure, a godlike administrator with a brilliant and incisive mind whose interests ranged from Sanskrit and French literature to Marxism and driving fast cars. Educated at Göttingen and Harvard, Oppie in the 1930s had commuted between the University of California at Berkeley and the California Institute of Technology at Pasadena, where he mesmerized a generation of young graduate students with his lectures on nuclear physics and quantum mechanics. Through his wife and his brother Frank, Oppie became involved in a number of then modish left-wing groups and causes in the California labor movement. He contributed money, he made new friends, he hosted meetings. In this he was not unlike many intellectuals of the 1930s for whom the forces of evil were represented by fascism and anti-Semitism, not communism.

From 1941 on Oppenheimer was deeply involved in the top secret *project to build an American atomic bomb. In 1942 Arthur Holly Compton named him to work with General Leslie Groves as chief scientific director of the Manhattan Project. Oppenheimer's initial work at Berkeley and his later move to Los Alamos also attracted the attention of Army security agents concerned about his left-wing past and Communist friends. Even as Los Alamos was being constructed on a remote New Mexico mesa in early 1943, Oppie was being tailed by FBI and Army security agents and his telephone wiretapped. In July 1943 General Groves obtained a security clearance for Oppenheimer over the objections of army security advisers, claiming that Oppie was absolutely essential for the project whatever his political past.*

With the success of the Manhattan Project, Oppenheimer's early political indiscretions and contacts were forgotten. After 1945 he left Los Alamos for Princeton, New Jersey, where he directed the Institute for Advanced Study. He served on the General Advisory Committee of the Atomic Energy Commission, a key panel of scientists, and became a familiar public speaker, author, and culture hero. To the public Oppie was a genius of a particular sort, an academic and intellectual who had unleashed the greatest power on earth—atomic energy—and struck a Faustian bargain with political and military authority.

In October 1949 Oppenheimer and the General Advisory Committee recommended against a crash program to build the hydrogen bomb. President Truman, under pressure from Edward Teller and other scientific and military advisers, rejected that recommendation within months and decided to initiate the project. Despite his sense that the hydrogen bomb made no military sense and was essentially a weapon of genocide, Oppie participated in its early development for several years as a consultant to the Atomic Energy Commission. His opposition to it seemed forgotten. It was not.

In the winter of 1953–54, during the period of political paranoia associated with Senator Joseph McCarthy (R-Wis), Oppenheimer was surprised to receive a letter from Kenneth Nichols of the Atomic Energy Commission informing him that his security clearance had been suspended. The letter included a number of accusations, innuendos, and charges relating to Oppie's early left-wing associations in California in the 1930s and 1940s and to his more recent opposition to the hydrogen bomb. Nichols informed Oppenheimer that if he wished he could request a hearing to defend himself against these charges that he was a security risk because of his politics and his ethics.

After writing a letter of response to Nichols, in April 1954 Oppenheimer confronted an Atomic Energy Commission three-man board of inquiry headed by Gordon Gray. His appearance promptly became a Gallilean drama of scientific truth and conscience pitted against irrational power and authority. The testimony of the witnesses, who included Groves, Teller, and most of the leading scientists in the Manhattan Project, largely defended Oppenheimer's character, patriotism, loyalty, and sense of integrity. Nevertheless, the Gray Board found Oppenheimer to be a security risk and recommended that his clearance not be reinstated.

The result was a public outcry and a cathartic debate within the entire scientific community about Oppenheimer's guilt or innocence regarding vague charges. The publication by the government of the testimony and charges only confirmed the deep division between the larger purposes of scientific inquiry and military power in the name of national security. Although many scientists rallied to Oppie's defense, they were not free to disclose classified details of either fission or fusion bomb development, details that could have shown that any delay in the hydrogen bomb project was due not to Oppenheimer's politics but to Edward Teller's scientific mistakes. Teller himself had testified against Oppenheimer and had been ostracized by many in the scientific community; only in 1982 did it become known, in the classified 1954 article by Hans Bethe included above, that Teller himself was partly responsible for the delays in hydrogen bomb development.

Oppenheimer died in 1967, only partly vindicated by his government that had demanded his brilliant talents for so long and then discarded him as a scapegoat no longer necessary for nuclear weapons research.

The transcript of the Oppenheimer hearing is available from the U.S. Government as In the Matter of J. Robert Oppenheimer: Transcript of Hearing Before Personnel Security Board, Washington, D.C., April 12, 1954, Through May 6, 1954 (Washington, D.C.: Government Printing Office, 1954; paperback ed. Cambridge, Mass.: MIT Press, 1971). On the case itself, see Philip M. Stern, with Harold P. Green, The Oppenheimer Case: Security on Trial (New York: Harper & Row, 1969). The best single volume on the case is John Major, The Oppenheimer Hearing (London: B. T. Batsford, 1971). On the Atomic Energy Commission's role in the case, see Richard G. Hewlett and Jack M. Holl, Atoms for Peace and War, 1953–1961: Eisenhower and the Atomic Energy Commission (Berkeley: University of California Press, 1989). The best treatment of Oppenheimer's early years may be found in Alice Kimball Smith and Charles Weiner, eds., Robert Oppenheimer: Letters and Recollections (Cambridge, Mass.: Harvard University Press, 1980). A more recent study is James W. Kunetka, Oppenheimer: The Years of Risk (Englewood Cliffs, N.J.: Prentice-Hall, 1982).

35. K. D. Nichols, General Manager, AEC, to
J. R. Oppenheimer, December 23, 1953

Section 10 of the Atomic Energy Act of 1946 places upon the Atomic Energy Commission the responsibility for assuring that individuals are employed by the Commission only when such employment will not endanger the common defense and security. In addition, Executive Order 10450 of April 27, 1953, requires the suspension of employment of any individual where there exists information indicating that his employment may not be clearly consistent with the interests of the national security.

As a result of additional investigation as to your character, associations, and loyalty, and review of your personnel security file in the light of the requirements of the Atomic Energy Act and the requirements of Executive Order 10450, there has developed considerable question whether your continued employment on Atomic Energy Commission work will endanger the common defense and security and whether such continued employment is clearly consistent with the interests of the national security. This letter is to advise you of the steps which you may take to assist in the resolution of this question.

The substance of the information which raises the question concerning your eligibility for employment on Atomic Energy Commission work is as follows:

It was reported that in 1940 you were listed as a sponsor of the Friends of the Chinese People, an organization which was characterized in 1944 by the House Committee on Un-American Activities as a Communist-front organization. It was further reported that in 1940 your name was included on a letterhead of the American Committee for Democratic and Intellectual Freedom as a member of its national executive committee. The American Committee for Democracy and Intellectual Freedom was characterized in 1942 by the House Committee on Un-American Activities as a Communist front which defended Communist teachers, and in 1943 it was characterized as subversive and un-American by a special subcommittee of the House Committee on Appropriations. It was further reported that in 1938 you were a member of the Western Council of the Consumers Union. The Consumers Union was cited in 1944 by the House Committee on Un-American

This document appears in *In the Matter of J. Robert Oppenheimer* (Washington, D.C.: U.S. Government Printing Office, 1954; Cambridge, Mass.: MIT Press, 1971), pp. 3–7.

Activities as a Communist-front headed by the Communist Arthur Kallet. It was further reported that you stated in 1943 that you were not a Communist, but had probably belonged to every Communist front organization on the west coast and had signed many petitions in which Communists were interested.

It was reported that in 1943 and previously you were intimately associated with Dr. Jean Tatlock, a member of the Communist Party in San Francisco, and that Dr. Tatlock was partially responsible for your association with Communist-front groups.

It was reported that your wife, Katherine Puening Oppenheimer, was formerly the wife of Joseph Dallet, a member of the Communist Party, who was killed in Spain in 1937 fighting for the Spanish Republican Army. It was further reported that during the period of her association with Joseph Dallet, your wife became a member of the Communist Party. The Communist Party has been designated by the Attorney General as a subversive organization which seeks to alter the form of Government of the United States by unconstitutional means, within the purview of Executive Order 9835 and Executive Order 10450. . . .

It was reported that you have associated with members and officials of the Communist Party including Isaac Folkoff, Steve Nelson, Rudy Lambert, Kenneth May, Jack Manley, and Thomas Addis.

It was reported that you were a subscriber to the *Daily People's World*, a west coast Communist newspaper, in 1941 and 1942. . . .

It was reported that prior to March 1, 1943, possibly 3 months prior, Peter Ivanov, secretary of the Soviet consulate, San Francisco, approached George Charles Eltenton for the purpose of obtaining information regarding work being done at the Radiation Laboratory for the use of Soviet scientists; that George Charles Eltenton subsequently requested Haakon Chevalier to approach you concerning this matter; that Haakon Chevalier thereupon approached you, either directly or through your brother, Frank Friedman Oppenheimer, in connection with this matter; and that Haakon Chevalier finally advised George Charles Eltenton that there was no chance whatsoever of obtaining the information. It was further reported that you did not report this episode to the appropriate authorities until several months after its occurrence; that when you initially discussed this matter with the appropriate authorities on August 26, 1943, you did not identify yourself as the person who had been approached, and you refused to identify Haakon Chevalier as the individual who made the approach on behalf of George Charles Eltenton; and that it was not until several months later, when you were ordered by a superior to do so, that you so identified Haakon Chevalier. It was further reported that upon your return to Berkeley following your separation from the Los Alamos project, you were visited by the Chevaliers on several occasions; and that your wife was in contact with Haakon and Barbara Chevalier in 1946 and 1947.

It was reported that in 1945 you expressed the view that "there is a reasonable possibility that it (the hydrogen bomb) can be made," but that the feasibility of the hydrogen bomb did not appear, on theoretical grounds, as certain as the fission bomb appeared certain, on theoretical grounds, when the Los Alamos Laboratory was started; and that in the autumn of 1949 the General Advisory Committee expressed the view that "an imaginative and concerted attack on the problem has a better than even chance of producing the weapon within 5 years." It was further reported that in the autumn of 1949 and subsequently, you strongly opposed the development of the hydrogen bomb; (1) on moral grounds, (2) by claiming that it was not feasible, (3) by claiming that there were insufficient facilities and scientific personnel to carry on the development and (4) that it was not politically desirable. It was further reported that even after it was determined, as a matter of national policy, to proceed with development of a hydrogen bomb, you continued to oppose the project and declined to cooperate fully in the project. It was further reported that you departed from your proper role as an adviser to the Commission by causing the distribution separately and in private, to top personnel at Los Alamos of the majority and minority reports of the General Advisory Committee on development of the hydrogen bomb for the purpose of trying to turn such top personnel against the development of the hydrogen bomb. It was further reported that you were instrumental in persuading other outstanding scientists not to work on the hydrogen-bomb project, and that the opposition to the hydrogen bomb, of which you are the most experienced, most powerful, and most effective member, has definitely slowed down its development.

In view of your access to highly sensitive classified information, and in view of these allegations which, until disproved, raise questions as to your veracity, conduct and even your loyalty, the Commission has no other recourse, in discharge of its obligations to protect the common defense and security, but to suspend your clearance until the matter has been resolved. Accordingly, your employment on Atomic Energy Commission work and your eligibility for access to restricted data are hereby suspended, effective immediately, pending final determination of this matter.

To assist in the resolution of this matter, you have the privilege of appearing before an Atomic Energy Commission personnel security board. To avail yourself of the privileges afforded you under the Atomic Energy Commission hearing procedures, you must, within 30 days following receipt of this letter, submit to me, in writing, your reply to the information outlined above and request the opportunity of appearing before the personnel security board. Should you signify your desire to appear before the board, you will be notified of the composition of the board and may challenge any member of it for cause. Such challenge should be submitted within 72 hours of the receipt of notice of composition of the board. . . .

If a written response is not received from you within 30 days it will be assumed that you do not wish to submit any explanation for further consideration. In that event, or should you not advise me in writing of your desire to appear before the personnel security board, a determination in your case will be made by me on the basis of the existing record. . . .

Very truly yours,

K. D. Nichols, General Manager.

36. J. R. Oppenheimer to K. D. Nichols, March 4, 1954

In the spring of 1936, I had been introduced by friends to Jean Tatlock, the daughter of a noted professor of English at the university; and in the autumn, I began to court her, and we grew close to each other. We were at least twice close enough to marriage to think of ourselves as engaged. Between 1939 and her death in 1944 I saw her very rarely. She told me about her Communist Party memberships; they were on again, off again affairs, and never seemed to provide for her what she was seeking. I do not believe that her interests were really political. She loved this country and its people and its life. She was, as it turned out, a friend of many fellow travelers and Communists, with a number of whom I was later to become acquainted.

I should not give the impression that it was wholly because of Jean Tatlock that I made leftwing friends, or felt sympathy for causes which hitherto would have seemed so remote from me, like the Loyalist cause in Spain, and the organization of migratory workers. I have mentioned some of the other contributing causes. I liked the new sense of companionship, and at the time felt that I was coming to be part of the life of my time and country.

This document appears in *In the Matter of J. Robert Oppenheimer* (Washington, D.C.: U.S. Government Printing Office, 1954; Cambridge, Mass.: MIT Press, 1971), pp. 8–11.

In 1937, my father died; a little later, when I came into an inheritance, I made a will leaving this to the University of California for fellowships to graduate students.

This was the era of what the Communists then called the United Front, in which they joined with many non-Communist groups in support of humanitarian objectives. Many of these objectives engaged my interest. I contributed to the strike fund of one of the major strikes of Bridges' union; I subscribed to the *People's World*; I contributed to the various committees and organizations which were intended to help the Spanish Loyalist cause. I was invited to help establish the teacher's union, which included faculty and teaching assistants at the university, and school teachers of the East Bay. I was elected recording secretary. My connection with the teacher's union continued until some time in 1941, when we disbanded our chapter. . . .

My own views were also evolving. Although Sidney and Beatrice Webb's book on Russia, which I had read in 1936, and the talk that I heard at that time had predisposed me to make much of the economic progress and general level of welfare in Russia, and little of its political tyranny, my views on this were to change. I read about the purge trials, though not in full detail, and could never find a view of them which was not damning to the Soviet system. In 1938 I met three physicists who had actually lived in Russia in the thirties. All were eminent scientists, Placzek, Weisskopf, and Schein; and the first two have become close friends. What they reported seemed to me so solid, so unfanatical, so true, that it made great impression; and it presented Russia, even when seen from their limited experience, as a land of purge and terror, of ludicrously bad management and of a long-suffering people. I need to make clear that this changing opinion of Russia, which was to be reinforced by the Nazi-Soviet Pact, and the behavior of the Soviet Union in Poland and in Finland, did not mean a sharp break for me with those who held to different views. At that time I did not fully understand—as in time I came to understand—how completely the Communist Party in this country was under the control of Russia. During and after the battle of France, however, and during the battle of England the next autumn, I found myself increasingly out of sympathy with the policy of disengagement and neutrality that the Communist press advocated. . . .

Because of these associations that I have described, and the contributions mentioned earlier, I might well have appeared at the time as quite close to the Communist Party—perhaps even to some people as belonging to it. As I have said, some of its declared objectives seemed to me desirable. But I never was a member of the Communist Party. I never accepted Communist dogma or theory; in fact, it never made sense to me. I had no clearly formulated political views. I hated tyranny and repression and every form of dictatorial control of thought. In most cases I did not in those days know who was and who was not a member of the Communist Party. No one ever asked me to join the Communist Party. . . .

In 1943 when I was alleged to have stated that "I knew several individuals then at Los Alamos who had been members of the Communist Party," I knew of only one; she was my wife, of whose disassociation from the party, and of whose integrity and loyalty to the United States I had no question. Later, in 1944 or 1945, my brother Frank, who had been cleared for work in Berkeley and at Oak Ridge, came to Los Alamos from Oak Ridge with official approval.

I knew of no attempt to obtain secret information at Los Alamos. Prior to my going there my friend Haakon Chevalier with his wife visited us on Eagle Hill, probably in early 1943. During the visit, he came into the kitchen and told me that George Eltenton had spoken to him of the possibility of transmitting technical information to Soviet scientists. I made some strong remark to the effect that this sounded terribly wrong to me. The discussion ended there. Nothing in our long standing friendship would have led me to believe that Chevalier was actually seeking information; and I was certain that he had no idea of the work on which I was engaged.

It has long been clear to me that I should have reported the incident at once. The events that led me to report it—which I doubt ever would have become known without my report—were unconnected with it. During the summer of 1943, Colonel Landsdale, the intelligence officer of the Manhattan District, came to Los Alamos and told me that he was worried about the security situation in Berkeley because of the activities of the Federation of Architects, Engineers, Chemists, and Technicians. This recalled to my mind that Eltenton was a member and probably a promoter of the FAECT. Shortly thereafter, I was in Berkeley and I told the security officer that Eltenton would bear watching. When asked why, I said that Eltenton had attempted, through intermediaries, to approach people on the project, though I mentioned neither myself nor Chevalier. Later, when General Groves urged me to give the details, I told him of my conversation with Chevalier. I still think of Chevalier as a friend. . . .

From the close of the war, when I returned to the west coast until finally in the spring of 1947 when I went to Princeton as the director of the Institute for Advanced Study, I was able to spend very little time at home and in teaching in California. In October 1945, at the request of Secretary of War Patterson, I had testified before the House Committee on Military Affairs in support of the May-Johnson bill, which I endorsed as an interim means of bringing about without delay the much needed transition from the wartime administration of the Manhattan District to postwar management of the atomic-energy enterprise. In December 1945, and later, I appeared at Senator McMahon's request in sessions of his Special Committee on Atomic Energy, which was considering legislation on the same subject. Under the chairmanship of Dr. Richard Tolman, I served on a committee set up by General Groves to consider classification policy on matters of atomic energy. For 2 months, early in 1946, I worked steadily as a member of a panel, the Board of Consultants to the Secretary of State's Committee on Atomic Energy, which, with the Secretary of State's Committee, prepared the so-called Acheson-Lilienthal report. After the publication of this report, I spoke publicly in support of it. A little later, when Mr. Baruch was appointed to represent the United States in the United Nations Atomic Energy Committee, I became one of the scientific consultants to Mr. Baruch, and his staff in preparation for and in the conduct of our efforts to gain support for the United States' plan. I continued as a consultant to General Osborn when he took over the effort.

At the end of 1946 I was appointed by the President as a member of the General Advisory Committee to the Atomic Energy Commission. At its first meeting I was elected Chairman, and was reelected until the expiration of my term in 1952. This was my principal assignment during these years as far as the atomic-energy program was concerned, and my principal preoccupation apart from academic work. . . .

The initial members of the General Advisory Committee were Conant, then president of Harvard, DuBridge, president of the California Institute of Technology, Fermi of the University of Chicago, Rabi of Columbia University, Rowe, vice president of the United Fruit Co., Seaborg of the University of California, Cyril Smith of the University of Chicago, and Worthington of the duPont Co. In 1948 Buckley, president of the Bell Telephone Laboratories, replaced Worthington; in the summer of 1950, Fermi, Rowe, and Seaborg were replaced by Libby of the University of Chicago, Murphree president of Standard Oil Development Co., and Whitman of the Massachusetts Institute of Technology. Later Smith resigned and was succeeded by von Neumann of the Institute for Advanced Study.

In these years from early 1947 to mid-1952 the Committee met some 30 times and transmitted perhaps as many reports to the Commission. Formulation of policy and the management of the vast atomic-energy enterprises were responsibilities vested in the Commission itself. The General Advisory Committee had the role, which was fixed for it by statute, to advise the Commission. In that capacity we gave the Commission our views on questions which the Commission put before us, brought to the Commission's attention on our initiative technical matters of importance, and encouraged and supported the work of the several major installations of the Commission. . . .

The super itself had a long history of consideration beginning, as I have said, with our initial studies in 1942 before Los Alamos was established. It continued to be the subject of study and research at Los Alamos throughout the war. After the war, Los Alamos itself was inevitably handicapped pending the enactment of necessary legislation for the atomic energy enterprise. With the McMahon Act, the appointment of the Atomic Energy Commission and the General Advisory Committee, we in the committee had occasion at our early meetings in 1947 as well as in 1948 to discuss the subject. In that period the General Advisory Committee pointed out the still extremely unclear status of the problem from the technical standpoint, and urged encouragement of Los Alamos' efforts which were then directed toward modest exploration of the super and of thermonuclear systems. No serious controversy arose about the super until the Soviet explosion of an atomic bomb in the autumn of 1949.

Shortly after that event, in October 1949, the Atomic Energy Commission called a special session of the General Advisory Committee and asked us to consider and advise on two related questions: First, whether in view of the Soviet success the Commission's program was adequate, and if not, in what way it should be altered or increased; second, whether a crash program for the development of the super should be a part of any new program. The committee considered both questions, consulting various officials from the civil and military branches of the executive departments who would have been concerned, and reached conclusions which were communicated in a report to the Atomic Energy Commission in October 1949.

This report, in response to the first question that had been put to us, recommended a great number of measures that the Commission should take to increase in many ways our overall potential in weapons.

As to the super itself, the General Advisory Committee stated its unanimous opposition to the initiation by the United States of a crash program of the kind we had been asked to advise on. The report of that meeting, and the Secretary's notes, reflect the reasons which moved us to this conclusion. The annexes, in particular, which dealt more with political and policy considerations—the report proper was essentially technical in character—indicated differences in the views of members of the committee. There were two annexes, one signed by Rabi and Fermi, the other by Conant, DuBridge, Smith, Rowe, Buckley and myself. (The ninth member of the committee, Seaborg, was abroad at the time.)

It would have been surprising if eight men considering a problem of extreme difficulty had each had precisely the same reasons for the conclusion in which we joined. But I think I am correct in asserting that the unanimous opposition we expressed to the crash program was based on the conviction, to which technical considerations as well as others contributed, that because of our overall situation at that time such a program might weaken rather than strengthen the position of the United States.

After the report was submitted to the Commission, it fell to me as chairman of the committee to explain our position on several occasions, once at a meeting of the Joint Congressional Committee on Atomic Energy. All this, however, took place prior to the decision by the President to proceed with the thermonuclear program.

This is the full story of my "opposition to the hydrogen bomb." It can be read in the records of the general transcript of my testimony before the joint congressional committee. It is a story which ended once and for all when in January 1950 the President announced his decision to proceed with the program. I never urged anyone not to work on the hydrogen bomb project. I never made or caused any distribution of the GAC reports except to the Commission itself. As always, it was the Commission's responsibility to determine further distribution.

In summary, in October 1949, I and the other members of the General Advisory Committee were asked questions by the Commission to which we had a duty to respond, and to which we did respond with our best judgment in the light of evidence then available to us. . . .

In this letter, I have written only of those limited parts of my history which appear relevant to the issue now before the Atomic Energy Commission. In order to preserve as much as possible the perspective of the story, I have dealt very briefly with many matters. I have had to deal briefly or not at all with instances in which my actions or views were adverse to Soviet or Communist interest, and of actions that testify to my devotion to freedom, or that have contributed to the vitality, influence and power of the United States.

In preparing this letter, I have reviewed two decades of my life. I have recalled instances where I acted unwisely. What I have hoped was, not that I could wholly avoid error, but that I might learn from it. What I have learned has, I think, made me more fit to serve my country.

Very truly yours,

J. Robert Oppenheimer

Princeton, N.J., March 4, 1954.

37. Testimony in the Matter of J. Robert Oppenheimer

a. General Leslie B. Groves

Q. General, did your security officers on the project advise against the clearance of Dr. Oppenheimer?

A. Oh, I am sure that they did. I don't recall exactly. They certainly were not in favor of his clearance. I think a truer picture is to say that they reported that they could not and would not clear him.

Q. General, you were in the Army actively for how many years?

A. I don't know. 1916 to 1948, and of course raised in it, also.

Q. And you rose to the rank of lieutenant general?

A. That is right.

Q. During your entire Army career, I assume you were dealing with matters of security?

A. I would say I devoted about 5 percent of my time to security problems.

Q. You did become thoroughly familiar with security matters.

A. I think that I was very familiar with security matters.

Q. In fact, it could be said that you became something of an expert in it?

A. I am afraid that is correct.

Q. I believe you said that you became pretty familiar with the file of Dr. Oppenheimer?

A. I think I was thoroughly familiar with everything that was reported about Dr. Oppenheimer; and that included, as it did on every other matter of importance, personally reading the original evidence if there was any original evidence. In other words, I would read the reports of the interviews with people. In other words, I was not reading the conclusions of any security officer. The reason for that was that in this project there were so many things that the security officer would not know the significance of that I felt I had to do it myself. Of course, I have been criticized for doing all those things myself and not having a staff of any kind; but, after all, it did work, and I did live through it.

Q. General, in the light of your experience with security matters and in the light of your knowledge of the file pertaining to Dr. Oppenheimer, would you clear Dr. Oppenheimer today?

These materials appear in *In the Matter of J. Robert Oppenheimer* (Washington, D.C.: U.S. Government Printing Office, 1954; Cambridge, Mass.: MIT Press, 1971), pp. 170–71, 330–31, 356–57, 468–69, 710.

A. I think before answering that I would like to give my interpretation of what the Atomic Energy Act requires. I have it, but I never can find it as to just what it says. Maybe I can find it this time.

Q. Would you like me to show it?

A. I know it is very deeply concealed in the thing.

Q. Do you have the same copy?

A. I have the original act.

Q. It is on page 14, I think, where you will find it, General. You have the same pamphlet I have.

A. Thank you. That is it. The clause to which I am referring is this: It is the last of paragraph (b) (i) on page 14. It says:

"The Commission shall have determined that permitting such person to have access to restricted data will not endanger the common defense or security," and it mentions that the investigation should include the character, associations, and loyalty.

My interpretation of "endanger"—and I think it is important for me to make that if I am going to answer your question—is that it is a reasonable presumption that there might be a danger, not a remote possibility, a tortured interpretation of maybe there might be something, but that there is something that might do. Whether you say that is 5 percent or 10 percent or something of that order does not make any difference. It is not a case of proving that the man is a danger. It is a case of thinking, well, he might be a danger, and it is perfectly logical to presume that he would be, and that there is no consideration whatsoever to be given to any of his past performances or his general usefulness or, you might say, the imperative usefulness. I don't care how important the man is, if there is any possibility other than a tortured one that his associations or his loyalty or his character might endanger.

In this case I refer particularly to associations and not to the associations as they exist today but the past record of the associations. I would not clear Dr. Oppenheimer today if I were a member of the Commission on the basis of this interpretation.

If the interpretation is different, then I would have to stand on my interpretation of it.

b. Hans Bethe.

Q. Do you have any opinion, Dr. Bethe, on the question of whether there has been in fact any delay in the development and the perfection of thermonuclear weapons by the United States?

A. I do not think that there has been any delay. I will try to keep this unclassified. I can't promise that I can make myself fully clear on this.

Q. Try to, will you?

A. I will try. When President Truman decided to go ahead with the hydrogen bomb in January 1950, there was really no clear technical program that could be followed. This became even more evident later on when new calculations were made at Los Alamos, and when these new calculations showed that the basis for technical optimism which had existed in the fall of 1949 was very shaky, indeed. The plan which then existed for the making of a hydrogen bomb turned out to be less and less promising as time went on.

Q. What interval are you now speaking of?

A. I am speaking of the interval of from January 1950 to early 1951. It was a time when it would not have been possible by adding more people to make any more progress. The more people would have to do would have to be work on the things which turned out to be fruitful.

Finally there was a very brilliant discovery made by Dr. Teller. It was one of the discoveries for which you cannot plan, one of the discoveries like the discovery of the relativity theory, although I don't want to compare the two in importance. But something which is a stroke of genius, which does not occur in the normal development of ideas. But somebody has to suddenly have an inspiration. It was such an inspiration which Dr. Teller had which put the program on a sound basis.

Only after there was such a sound basis could one really talk of a technical program. Before that, it was essentially only speculation, essentially only just trying to do something without having really a direction in which to go. Now things changed very much. After this brilliant discovery there was a program.

Q. Dr. Bethe, if the board and Mr. Robb would permit me, I would like to ask you somewhat a hypothetical question. Would your attitude about work on the thermonuclear program in 1949 have differed if at that time there had been available this brilliant discovery or brilliant inspiration, whatever you call it, that didn't come to Teller until the spring of 1951?

A. It is very difficult to answer this.

Q. Don't answer it if you can't.

A. I believe it might have been different.

Q. Why?

A. I was hoping that it might be possible to prove that thermonuclear reactions were not feasible at all. I would have thought that the greatest security for the United States would have lain in the conclusive proof of the impossibility of a thermonuclear bomb. I must confess that this was the main motive which made me start work on thermonuclear reactions in the summer of 1950.

With the new [Teller-Ulam idea?] I think the situation changed because it was then clear, or almost clear—at least very likely—that thermonuclear weapons were indeed possible. If thermonuclear weapons were possible, I felt that we should have that first and as soon as possible. So I think my attitude might have been different.

Q. One final question, Dr. Bethe. I should have asked you this. I have referred you to the press statements and the article that you published in the late winter and spring of 1950, expressing critical views of the H-bomb program. Did you ever discuss those moves, that is to make such statements and write such articles, with Dr. Oppenheimer?

A. I never did. In fact, after the President's decision, he would never discuss any matters of policy with me. There had been in fact a directive from President Truman to the GAC not to discuss the reasons of the GAC or any of the procedures, and Dr. Oppenheimer held to this directive very strictly.

Q. Did you consult him about the article?

A. I don't think I consulted him at all about the article. I consulted him about the statement that we made. As far as I remember, he gave no opinion.

Q. On the basis of your association with him, your knowledge of him over these many years, would you care to express an opinion about Dr. Oppenheimer's loyalty to the United States, about his character, about his discretion in regard to matters of security?

A. I am certainly happy to do this. I have absolute faith in Dr. Oppenheimer's loyalty. I have always found that he had the best interests of the United States at heart. I have always found that if he differed from other people in his judgment, that it was because of a deeper thinking about the possible consequences of our action than the other people had. I believe that it is an expression of loyalty—of particular loyalty—if a person tries to go beyond the obvious and tries to make available his deeper insight, even in making unpopular suggestions, even in making suggestions which are not the obvious ones to make, are not those which a normal intellect might be led to make.

I have absolutely no question that he has served this country very long and very well. I think everybody agrees that his service in Los Alamos was one of the greatest services that were given to this country. I believe he has served equally well in the GAC in reestablishing the strength of our atomic weapons program in 1947. I have faith in him quite generally.

Q. You and he are good friends?

A. Yes.

Q. Would you expect him to place his loyalty to his country even above his loyalty to a friend?

A. I suppose so.

Mr. Marks. That is all.

c. George F. Kennan.

Q. As a result of your experience with Dr. Oppenheimer in the cases that you have reference to, what convictions, if any, did you form about him?

A. I formed the conviction that he was an immensely useful person in the councils of our Government, and I felt a great sense of gratitude that we had his help. I am able to say that in the course of all these contacts and deliberations within the Government I never observed anything in his conduct or his words that could possibly, it seemed to me, have indicated that he was animated by any other motives than a devotion to the interests of this country.

Q. Did you ever observe anything that would possibly have suggested to you that he was taking positions that the Russians would have liked?

A. No. I cannot say that I did in any way. After all, the whole purpose of these exercises was to do things which were in the interest of this country, not in the interests of the Soviet Union, at least not in the interests of the Soviet Union as their leaders saw it at that time. Anyone who collaborated sincerely and enthusiastically in the attempt to reach our objectives, which Dr. Oppenheimer did, obviously was not serving Soviet purposes in any way.

Q. Have you said that he contributed significantly to the results?

A. I have, sir.

Q. Mr. Kennan, is there any possibility in your mind that he was dissembling?

A. There is in my mind no possibility that Dr. Oppenheimer was dissembling.

Q. How do you know that? How can anybody know that?

A. I realize that is not an assertion that one could make with confidence about everyone. If I make it with regard to Dr. Oppenheimer it is because I feel and believe that after years of seeing him in various ways, not only there in Government, but later as an associate and a neighbor, and a friend at Princeton, I know his intellectual makeup and something of his personal makeup and I consider it really out of the question that any man could have participated as he did in these discussions, could have bared his thoughts to us time after time in the way that he did, could have thought those thoughts, so to speak, in our presence and have been at the same time dissembling.

I realize that is still not wholly the answer. The reason I feel it is out of the question that could have happened is that I believed him to have an intellect of such a nature that it would be impossible for him to speak dishonestly about any subject to which he had given his deliberate and careful and professional attention.

That is the view I hold of him. I have the greatest respect for Dr. Oppenheimer's mind. I think it is one of the great minds of this generation of Americans. A mind like that is not without its implications.

Q. Without its what?

A. Implications for a man's general personality. I think it would be actually the one thing probably in life that Dr. Oppenheimer could never do, that is to speak dishonestly about a subject which had really engaged the responsible attention of his intellect. My whole impression of him is that he is a man who when he turns his mind to something in an orderly and responsible way, examines it with the most extraordinary scrupulousness and fastidiousness of intellectual process.

I must say that I cannot conceive that in these deliberations in Government he could have been speaking disingenuously to us about these matters. I would suppose that you might just as well have asked Leonardo da Vinci to distort an anatomical drawing as that you should ask Robert Oppenheimer to speak responsibly to the sort of questions we were talking about, and speak dishonestly.

Q. Mr. Kennan, in saying what you have just said, are you saying it with an awareness of the background that Dr. Oppenheimer has, the general nature of which is reflected in the letter which General Nichols addressed to him, which is the genesis of these proceedings, and his response?

A. I am, sir.

Q. How do you reconcile these two things?

A. I do not think that they are necessarily inconsistent one with the other. People advance in life for one thing. I saw Dr. Oppenheimer at a phase of his life in which most of these matters in General Nichols' letter did not apply. It seems to me also that I was concerned or associated with him in the examination of problems which both he and I had accepted as problems of governmental responsibility before us, and I do not suppose that was the case with all the things that were mentioned in General Nichols' letter about his early views about politics and his early activities and his early associations.

I also think it quite possible for a person to be himself profoundly honest and yet to have associates and friends who may be misguided and misled and for who either at the time or in retrospect he may feel intensely sorry and concerned. I think most of us have had the experience of having known people at one time in our lives of whom we felt that way.

d. Isador Rabi.

Q. Dr. Rabi, Mr. Robb asked you whether you had spoken to Chairman Strauss in behalf of Dr. Oppenheimer. Did you mean to suggest in your reply—in your reply to him you said you did among other things—did you mean to suggest that you had done that at Dr. Oppenheimer's instigation?

A. No; I had no communication from Dr. Oppenheimer before these charges were filed, or since, except that I called him once to just say that I believed in him, with no further discussion.

Another time I called on him and his attorney at the suggestion of Mr. Strauss. I never hid my opinion from Mr. Strauss that I thought that this whole proceeding was a most unfortunate one.

Dr. Evans: What was that?

The Witness: That the suspension of the clearance of Dr. Oppenheimer was a very unfortunate thing and should not have been done. In other words, there he was; he is a consultant, and if you don't want to consult the guy, you don't consult him, period. Why you have to then proceed to suspend clearance and go through all this sort of thing, he is only there when called, and that is all there was to it. So it didn't seem to me the sort of thing that called for this kind of proceeding at all against a man who had accomplished what Dr. Oppenheimer has accomplished. There is a real positive record, the way I expressed it to a friend of mine. We have an A-bomb and a whole series of it, and what more do you want, mermaids? This is just a tremendous achievement. If the end of that road is this kind of hearing, which can't help but be humiliating, I thought it was a pretty bad show. I still think so.

By Mr. Marks:

Q. Dr. Rabi, in response to a question of the Chairman, the substance of which I believe was, was Dr. Oppenheimer unalterably opposed to the H-bomb development at the time of the October 1949 GAC meeting, I think you said in substance no, and then you added by way of explanation immediately thereafter the two annexes or whatever they were—

A. During the discussion.

Q. During the discussion he said he would be willing to sign either or both. Can you explain what you meant by that rather paradoxical statement?

A. No, I was just reporting a recollection.

Q. What impression did you have?

A. What it means to me is that he was not unalterably opposed, but on sum, adding up everything, he thought it would have been a mistake at that time to proceed with a crash program with all that entailed with this object that we didn't understand, when we had an awfully good program on hand in the fission field, which we did not wish to jeopardize. At least we did not feel it should be jeopardized. It turned out in the events that both could be done. Los Alamos just simply rose to the occasion and worked miracles, absolute miracles.

Mr. Marks. That is all.

e. Edward Teller.

Q. Dr. Teller, you know Dr. Oppenheimer well; do you not?

A. I have known Dr. Oppenheimer for a long time. I first got closely associated with him in the summer of 1942 in connection with atomic energy work. Later in Los Alamos and after Los Alamos I knew him. I met him frequently, but I was not particularly closely associated with him, and I did not discuss with him very frequently or in very great detail matters outside of business matters.

Q. To simplify the issues here, perhaps, let me ask you this question: Is it your intention in anything that you are about to testify to, to suggest that Dr. Oppenheimer is disloyal to the United States?

A. I do not want to suggest anything of the kind. I know Oppenheimer as an intellectually most alert and very complicated person, and I think it would be presumptuous and wrong on my part if I would try in any way to analyze his motives. But I have always assumed, and I now assume that he is loyal to the United States. I believe this, and I shall believe it until I see very conclusive proof to the opposite.

Q. Now, a question which is the corollary of that. Do you or do you not believe that Dr. Oppenheimer is a security risk?

A. In a great number of cases I have seen Dr. Oppenheimer act—I understood that Dr. Oppenheimer acted—in a way which for me was exceedingly hard to understand. I thoroughly disagreed with him in numerous issues and his actions frankly appeared to me confused and complicated. To this extent I feel that I would like to see the vital interests of this country in hands which I understand better, and therefore trust more.

In this very limited sense I would like to express a feeling that I would feel personally more secure if public matters would rest in other hands.

38. Arthur Holly Compton to Gordon Gray, April 21, 1954

Park Hotel
Istanbul, Turkey

April 21, 1954

Dear President Gray:

I was in Pakistan when the news broke regarding the suspension of J. Robert Oppenheimer from the Advisory Board of the A.E.C., and it was not until yesterday that I received sufficiently full information about the case to make a statement that I knew would be relevant. Because of my close association with Oppenheimer, I believe that some facts from my first-hand knowledge will be useful to your committee in judging his loyalty. These facts I am stating here with the same responsibility for their veracity as I would accept in the case of sworn testimony before a court of law.

1. I was responsible for appointing Oppenheimer to the task of organizing whatever was necessary for designing the atomic bomb. This was about the end of April, 1942. (The dates given here are all approximate. I have here no notes regarding these matters, and all of my statements are from

memory.) It was a month or two later that responsibility for the atomic development was assigned to the U.S. Engineers (Army), and it was in September, 1942, that I recommended to General Groves that Oppenheimer be continued at this task, and that the responsibility be enlarged to include the construction of the bomb. This recommendation was in accord with that of others, in particular that of Dr. James Conant, and was accepted.

At the time of Oppenheimer's appointment the atomic development was, as you know, in civilian hands. By Dr. Conant, acting for the OSRD, I was given complete responsibility for such appointments, reporting to him, however, regarding actions taken. For my guidance, within our "Metallurgical Project," a personnel division had been established which investigated the loyalty background of employees. This investigation (and our other security measures) was carried on with the help of the Federal Bureau of Investigation. But we kept within our own hands the responsibility for the decisions in each case.

It was recognized that the task to which we were appointing Oppenheimer was not only of essential importance to the success of the total project, but also one in which a unique degree of discretion was required. It was my judgment that Oppenheimer's qualifications fitted him for such a post better than any other person who could be made available. This was after a diligent search for about a month which included consultation with many top-level scientists.

That Oppenheimer had had contacts with Communists was well known to me. These contacts, as I knew them, were essentially as described in his letter as published in *The New York Times* of April 13. It was my impression that, as one eager to find a solution to world problems, he had investigated Communism first-hand to see what it had to offer.

The much more significant fact was that he had already become disillusioned with Communism by what he had found. In a conversation with him about another matter, in the spring of 1940, I believe, he had given me his reasons for not associating himself with a professional organization that had some Communist ties. Later, in 1941, I believe, he had told me of his efforts to persuade his brother Frank to dissociate himself from Communist groups. As I recall it, it was during the early stages of our conversations about the atomic program, before I had approached him about accepting the responsibility for the design of the bomb, that Oppenheimer told me he was breaking completely every association with any organization that might be suspected of Communism, in order that he might be of maximum usefulness on war projects.

It was my judgment that a person who had known Communism and had found its faults was more to be relied upon than one who was innocent of such connections. This was the more important because it was evident that his task would necessarily draw in many men of foreign background, who were among those most competent in theoretical physics. Without the use of such men his task could hardly have been accomplished. A man with Oppenheimer's experience both in foreign countries and, in a limited way, with Communism, was in a most favorable position to recognize those whose loyalties might be directed elsewhere than to the United States and the free world.

The chief positive reason for selecting Oppenheimer for this post was that after working vigorously on rallying the help of theoretical physicists to the design of the bomb, Oppenheimer had shown the most eagerness and initiative, he was one of the very few American-born men who had the professional competence, and he had demonstrated a certain firmness of character. While his administrative competence remained to be demonstrated, it looked to me promising.

Looking back on this selection, I do not believe it would have been possible to find anywhere a man better suited to carry through this unique job in the nation's interest.

2. After the war was over I discussed with Oppenheimer several times the question of proceeding with the development of the H-bomb. Its possibility Oppenheimer had called to my attention in August, 1942. This possibility was kept under the utmost secrecy; but before the war was over its suggestion had arisen spontaneously from so many sources that it was evident that it would arise in any group working seriously on nuclear explosions. In particular, it was evident that the basic principle of the fusion bomb would be known in Russia.

Immediately after the war our great consideration was to reach a reliable agreement with Russia that would rule out atomic weapons but permit the development of atomic power. During this period there was no immediate occasion to work toward atomic fusion. The situation changed when Russia showed such reluctance that no atomic weapon agreement seemed feasible. We then recognized that Russia was working vigorously on her own atomic development. Presumably the H-bomb would eventually become part of her program.

I recall a conversation with Oppenheimer in 1947 or 48 in which I was advocating the initiation of a program of active research and development toward the H-bomb. My point was that this development, if it was physically possible, was sure to come before many years, and it was important that the availability of this "super" weapon should first be in our hands.

I found Oppenheimer reluctant. His chief reluctance was, I believe, on purely moral grounds. No nation should bring into being a power that would (or could) be so destructive of human lives. Even if another nation should do so, our morality should be higher than this. We should accept the military disadvantage in the interest of standing for a proper moral principle.

He had other reasons—the development of fear and antagonism among other nations, the substantial possibility that the effort to create an atomic explosion would fail, questions regarding the H-bomb's military value. He hoped that no urgent need for its development would arise.

In this and other conversations Oppenheimer brought up precisely those questions that needed to be considered. His thinking seemed to be aimed solely toward finding what was in the best interests of the United States. He took for granted, as did I, that the United States' interests are those of humanity. There was no shadow of a suspicion that his arguments were subtly working toward Russia's advantage. I am confident that no such thought was in his mind.

With the explosion of Russia's first atomic bomb in 1949 the situation was sharply changed. I do not recall any first-hand discussions with Oppenheimer on this matter after that date.

3. Having known Robert Oppenheimer since his days in Göttingen in 1927, having worked with him closely during the war years, and having kept in touch with him occasionally since, it is my judgment that he is completely loyal to the interests of the United States, and that any activity in the interest of a foreign power at the expense of the United States would be thoroughly repugnant to him. It is my judgment further that he is, and has been since 1941, just as thoroughly opposed to Communism.

Yours sincerely,
Arthur H. Compton

AHC:dbs
(Permanent address,
Washington University,
St. Louis, MO. U.S.A.)

VI

Nuclear Testing and the Test Ban

*I*n July 1946, just two weeks after Bernard Baruch had presented his plan for international control of atomic weapons, a B-29 dropped an atomic bomb on a fleet of captured Japanese naval vessels in the lagoon at Bikini Atoll in the Pacific Ocean. Several weeks later another weapon was exploded under the water. Operation Crossroads, as the two nuclear tests were named, were the first nuclear explosions in peacetime. Before the signing of the Limited Test Ban Treaty in 1963, the United States would announce 328 additional nuclear tests.

Coinciding as it did with the presentation of the Baruch Plan, Operation Crossroads would spark charges against the United States of conducting atomic diplomacy. The government, however, had discussed test plans since the end of the war. Like other parts of the armed services, the program in atomic weapons was also being demobilized, and the military wanted to test the effects of atomic bombs on ships and other equipment before the scientists returned to their universities.

Shortly after Crossroads, the civilian Atomic Energy Commission assumed responsibility for the nuclear program. The Commission learned that the Hiroshima-type bomb was too inefficient for the short supply of fissionable material. Moreover, the Nagasaki-type weapon was more a laboratory than a production model. Consequently, the Commission focused on rebuilding its laboratories, its plutonium production facilities, and its testing program. By the spring of 1948 the Commission conducted its first atomic tests, codenamed Operation Sandstone, on Enewetak, 180 miles west of Bikini. The primary purpose of Sandstone was to improve the efficiency of weapon design so that scarce fissionable material would be better utilized and bomb size and weight could be reduced.

To most Americans the rush of world events justified the Atomic Energy Commission's nuclear testing program. The Soviet rejection of the Baruch Plan, Soviet satellite governments in control in eastern Europe, the Greek Civil War, the Communist coup in Czechoslovakia, and the Berlin Blockade in 1948 heralded the realities of the Cold War. Within the next year Americans learned that Klaus Fuchs, a British scientist who had worked on the bomb at Los Alamos, had given atomic secrets to the Russians. In addition, Mao Tse-Tung's Communist armies had captured the Chinese mainland, and the Soviet Union had detonated its first Soviet atomic bomb. These events, especially the news of the Soviet atomic bomb, quickened a debate within the Atomic Energy Commission and between high government officials over the development of a thermonuclear weapon.

The North Korean invasion of South Korea in June 1950 appeared to justify President Truman's decision to proceed with a "super" bomb. The "police action" might become a full-scale war in which nuclear weapons would be the deciding factor.

The logistics of testing new weapons designs in the Pacific placed limitations on the efficiencies and costs of the testing program. Scientists at Los Alamos argued for a testing site within the United States. The threat that the Enewetak site might be cut off in a war brought a new urgency to open a continental test location. By late 1950 the government decided on a large area in Nevada, north of Las Vegas. Although the government located the site in a remote area, the Atomic Energy Commission now confronted the hazard of radioactive fallout within the United States. Because neither Greenhouse nor Sandstone had involved radiological hazards to nearby populations, precise information concerning the effect of radiation on people was less important than weapons data. Tests on the American continent, however, could pose real dangers to the general public.

Continental testing began in January 1951 with Operation Ranger, a series of four shots, followed by Operation Buster-Jangle in the fall. The weapons, mostly dropped from aircraft, did not produce significant amounts of fallout off the test site. However, fallout did occur in Rochester, New York, exposing film at the Eastman Kodak Company and causing the commission to establish a long-range radiation monitoring system throughout the United States.

The drive to produce a thermonuclear, or hydrogen, bomb led to further experimentation at the Nevada Test Site and in the Pacific. At Operation Ivy at Enewetak in November 1952, a prototype thermonuclear device named Mike yielded an explosive power of over ten megatons (ten million tons of TNT) and proved that a hydrogen bomb was possible. But Mike weighed over eight tons and was not a deliverable weapon. That was the next task.

To obtain precise measurements for designing and testing of hydrogen weapons in the Pacific (the Commission prohibited thermonuclear testing in Nevada), the continental tests carried out during Upshot-Knothole in the spring of 1953 were fired from 300-foot towers. As a result, for the first time the atomic fireball swept up great amounts of debris from the desert floor and deposited significant levels of radioactive fallout on Utah and Nevada. According to some estimates, 80 percent of offsite fallout from continental testing came from the Upshot-Knothole series.

The high levels of fallout caused the Atomic Energy Commission to re-examine continental testing and recommend tighter restrictions on test procedures, especially weather conditions, which largely determined where and how much fallout would occur. But two months after Upshot-Knothole concluded, the Soviet Union detonated its first thermonuclear device. The race for weapons of greater destruction intensified.

By 1954 the Atomic Energy Commission had established a second weapons laboratory in Livermore, California, and had made plans for a large thermonuclear test on Bikini, which had not been used since Crossroads. The Castle series, designed by the Los Alamos laboratory, would test a weapon so powerful that most civilian and military personnel were evacuated to ships thirty miles from ground zero.

The preparations were well taken but inadequate. Shot Bravo yielded fifteen megatons and spread radioactive fallout far beyond the thirty-mile perimeter. Ships received heavy radiation and sailed out to fifty miles. Later, when part of the task force returned to Bikini, it could not enter the area due to high radiation levels. Bravo also exposed natives on islands a hundred miles away. In addition, the fallout hit a Japanese fishing vessel, the Lucky Dragon, and brought worldwide attention to the dangers of thermonuclear weapons.

That American troops were exposed to fallout at Bravo was not a new situation. Since 1952 the Department of Defense had placed Army troops at the Nevada Test Site to test their reactions during a nuclear attack. Seven thousand soldiers had participated at Operation Tumbler-Snapper (1952) and more than 17,000 at Upshot-Knothole (1953). Psychiatrists charted troop reactions to the blast and their ability to perform simple tasks and brief maneuvers following the explosion. In addition, troops participated at Operation Teapot in 1955 and Operation Plumbbob in 1957.

Even as nuclear weapons development and testing proceeded, President Eisenhower in 1954 asked his aides to consider a moratorium on nuclear testing. Opposition from the Departments of Defense and State and the Atomic Energy Commission doomed this initial attempt. But Eisenhower appointed Harold Stassen as a special adviser on disarmament, and Stassen, with Henry Cabot Lodge, the ambassador to the United Nations, continued to explore Eisenhower's wishes for a test ban and disarmament.

In 1956 Adlai Stevenson, the Democratic candidate for president, spoke vigorously for a test ban agreement, but Eisenhower, angry that disarmament had become a campaign issue, refused to have a test ban without a general disarmament agreement. Americans, troubled by instability abroad—the Soviet invasion of Hungary and the British-French-Israeli invasion of Egypt—agreed with Eisenhower, who easily won reelection.

Still, Eisenhower hoped to end nuclear testing and announced in the summer of 1958 that the United States would voluntarily suspend weapons testing. There was a rush of testing in Operation Hardtack, first in the Pacific and then in Nevada, before the moratorium went into effect on October 31, 1958. Since Alamogordo, the United States had detonated 151 atomic devices.

On August 31, 1961, Communist Party Chairman Nikita Khrushchev announced that the Soviet Union would begin nuclear testing. Two weeks later the United States followed suit. Most tests in Nevada took place underground, except for cratering experiments done for the Plowshare program to analyze the peaceful uses of nuclear explosions (such as earth moving or digging a canal).

Two years later, in August 1963, President John F. Kennedy joined Great Britain and the Soviet Union in signing the Limited Test Ban Treaty prohibiting nuclear testing in the atmosphere, underwater, or in space. Since then, all American nuclear tests have been conducted underground.

On testing and events leading up to the 1963 test ban, see Robert A. Divine, Blowing on the Wind: The Nuclear Test Ban Debate, 1954–1960 (New York: Oxford University Press, 1978); Richard G. Hewlett and Jack M. Holl, Atoms for Peace and War, 1953–1961: Eisenhower and the Atomic Energy Commission (Berkeley: University of California Press, 1989); Bernard G. Bechoefer, Postwar Negotiations for Arms Control (Washington, D.C.: Brookings Institution, 1961); Joint Committee on Atomic Energy, Special Subcommittee on Radiation, Hearings on the Nature of Radioactive Fallout and Its Effects on Man, May 27–June 7, 1957 (Washington: Government Printing Office, 1957); Henry Kissinger, Nuclear Weapons and Foreign Policy (New York: Harper & Brothers, 1957); Harold K. Jacobson and Eric Stein, Diplomats, Scientists, and Politicians: The United States and the Nuclear Test Ban Negotiations (Ann Arbor: University of Michigan Press, 1966); and Glenn T. Seaborg, with Benjamin S. Loeb, Kennedy, Khrushchev, and the Test Ban (Berkeley: University of California Press, 1981). An excellent account of popular perceptions of weapons testing, fallout, and nuclear *energy is Spencer R. Weart's* Nuclear Fear: A History of Images *(Cambridge, Mass.: Harvard University Press, 1988).*

39. Ralph E. Lapp, "Civil Defense Faces New Perils," November 1954

[Abstract] Dr. Lapp assesses the radioactive hazard of "fall-out" and analyzes its impact upon civil defense. His data are gathered from Japanese sources, from independent calculations, and from the unclassified scientific literature. He urges that the federal government release classified data on fall-out to provide guidance to civil defense organizations.

On March 1, 1954 chalk-white dust fell on twenty-three Japanese fishermen 72 miles from Bikini. It took three hours for the pulverized coral to start falling-out from the air and coat the *Lucky Dragon* tuna ship with a mantle of radioactive debris. However, it was not until late this summer that the Federal Civil Defense Administration felt the impact of "fall-out."

Q-cleared officials in FCDA were briefed on the nature of the fall-out. They were shown colored charts with neat elliptical contours describing the range of lethality of the residual radioactivity from superweapon explosions. Up to this point FCDA had worked and thought mostly in terms of circles—the symmetric patterns of primary damage from superbombs which the *Bulletin* published last month. Now superimposed upon the great circles of H-bomb blast and heat, there were zeppelin-shaped ellipses which stretched far beyond the circles of primary damage.

These ellipses stunned civil defense planners. By a major shift of policy they had replaced previous "duck-and-cover" or "stay put" planning by a policy of pre-attack evacuation which dispersed metropolitan populations beyond the inner circles of near-total and heavy blast damage. Gov. Val Peterson, as head of FCDA, made the policy switch once he took a good hard look at the blast and heat effects of our MIKE explosion of November 1,

This essay appears in *Bulletin of the Atomic Scientists* 10 (November 1954): 349–51. Copyright © 1954 by the Educational Foundation for Nuclear Science. Reprinted by permission.

1952. What Gov. Peterson realized very clearly was the MIKE was not the last word in superweapon development. Indeed a weapon twice its power was tested in the March-April CASTLE series of tests in the Pacific in 1954. Moreover, this weapon was a bomb, not a "device"—meaning that the United States now possesses a droppable bomb in the range of 20 megatons.

Faced with the prospect of corresponding weapons in the Soviet arsenal, Gov. Peterson recognized that too much of America's metropolitan population resided inside the 14 mile radius of the superbomb's punch. It was sheer suicide, he reasoned, to put 35 million Americans in the sitting duck category. This is the simple background for the policy of evacuation which is now being implemented.

It was at this point in the evolution of C.D. policy that "fall-out" descended upon civil defense planners.

• • • •

Now radioactivity was not a new wrinkle to the planners. The Baker shot of a 20 kiloton bomb in the Bikini lagoon in July of 1946 saturated the mushroom cloud with awesome quantities of radioactivity. Millions of tons of salt water were erupted into the air and a misty radioactive steam surged across the lagoon surface. An egg-shaped ellipse about 3 miles in length constituted the lethal area of this radioactivity. This area was not comparable to that of a metropolis, so as time went by radioactivity as a menace shrunk to less formidable proportions in civil defense planning.

The dimensions of the superbomb "fall-out" greatly exceeded those for a 1946 A-bomb explosion. Unfortunately, the exact or even rough forecasting of these dimensions is subject to a number of uncertainties. These can be appreciated by describing the fall-out from a superbomb detonation. When the bomb explodes there is a flash of penetrating radiation consisting of neutrons and gamma rays. Fortunately, the area subjected to lethal bombardment by this primary flash is small compared with that affected by the heat and blast of the explosion. Although altitude-dependent, the area is much less than that of total destruction from the blast wave.

Once the bomb explodes and the heat-blast waves have run their course, we have to consider the fate of the bomb cloud. If the bomb is burst high in the air so that there is no significant cratering effect the bomb cloud will contain only such surface debris as is sucked up into the ascending column. In this case, the cloud radioactivity consists predominantly of split uranium or plutonium atoms (technically called fission products) which are intensely radioactive. Two factors take their toll in reducing the menace of this activity. One—the high velocity upper air winds disperse the fine (usually invisible) particles in the bomb cloud so that they are dispersed over a very large area before they finally settle out and come to earth. Two—radioactive decay sharply reduces the activity on the following time schedule:

Time (after burst)	Radioactivity (arbitrary units)
1 minute	1,000,000
1 hour	7,300
1 day	162
1 week	16
1 month	3

Assuming that there were no mass deposits of radioactive debris in the early history of the bomb cloud, then we would have to relegate the problem of cloud activity to the controversial area of global atmospheric contamination and human genetics.

Superbombs which burst close to the surface present quite another problem. In this case, the 3 1/4 mile wide fire-ball of a 10-megaton bomb introduces a radically new factor into the fall-out equation. Much of the substratum below the exploding bomb is dislodged and volatilized into particles impregnated with radioactivity. In addition, some of the elements in the substratum may become radioactive by the primary penetrating radiation from the bomb. Sodium in sea water, for example, is easily activated (made radioactive) and can become a hazard. The pulverized substratum is funneled upward much in the manner of a cyclone. In this way, the coral and sand of a low-lying Bikini atoll island were sucked up into the bomb cloud.

As is well known the cauliflower cloud mushrooms upward to a height of over 10 miles in about 12 minutes. The characteristic 40 to 60 knot winds of the stratosphere then distort the mushroom in a downwind direction. At this point the difference between a high air burst and a low one becomes significant. The subsurface particles tend to all out from the bomb cloud early in its history because of their massiveness. Fall-out of up to 50 per cent of the radioactive debris occurs in the first 24 hours with the maximum activity being deposited in the 1 to 3 hour period.

Estimation of the area of serious-to-lethal fall-out involves many unknowns. First, there is the power of the bomb. Second, the architecture of the weapon, involving principally the contribution of fission to the energy release in the bomb. Third, there is the height of the bomb burst. Fourth, one is confronted with the nature of the substratum—both as a carrier of the radioactivity and also as a producer of radioactivity. (In general, it would seem that induced activities cannot compete with the radioactivity produced in the bomb itself. However, a bomb burst close to sea water would produce vast quantities of radiosodium and radiochlorine. For example, a bomb burst in the Pacific Ocean off the coast of Los Angeles would probably cloak the city in a lethal fog even though no blast shook the city.) Finally, the meteorology at the time of the burst would determine the location and extent of the fall-out.

It is like seeking the Holy Grail to quest for hard data on the fall-out as applied to an American city. The unknowns enumerated above make prediction of the fall-out a highly tentative business. About all one can hope to do is to define what *might* happen. Assuming a 15 megaton superbomb burst close to the ground the author has made the following estimates for the fall-out ellipses:

Time (after burst)	Area	Average Intensity (gamma radiation)
1 hr	250 sq. mi.	2500 roentgen/hr
3 hr	1200 sq. mi.	200 roentgen/hr
6 hr	4000 sq. mi.	30 roentgen/hr

It must be remembered that his fall-out will occur downwind. Upwind for 15 miles and sideways for 20 miles the fall-out should not be lethal. Integration of the dosage rate for the 4,000 square mile area leads to the conclusion that people will receive a serious to lethal dose in the first day. The area and hazard represents a conservative calculation. The radioactive hazard is truly immense. The explosion of 50 superbombs could blanket the entire N.E. USA in a serious to lethal radioactive fog.

A schedule for the effects of external radiation upon a man in the open is as follows:

Roentgen Dose*	Effect on Man
50 to 100 r	Few per cent casualities
150 to 200 r	50 percent casualties
200 to 300 r	100% casualties plus some mortality
400 to 500 r	50% mortality
700 r	close to 100% mortality

Applying the 500 roentgen criterion to the fall-out pattern at 1 hour after the bomb burst it is clear that an individual within the 2500 r/hr area would

*Delivered in a period of less than 1 hour. If delivered over several days the required effect dosage is doubled.

accumulate this dose in 12 minutes. At 3 hours a person much farther downwind might be exposed to a very serious dose. Here we must consider the relation between the intensity of the radioactivity and the time of exposure. Fortunately, the process of radioactive decay reduces the intensity quite sharply as time goes by. However, one must seek shelter as a protection against the radiation if caught in a near-lethal area. That is, unless some means is provided to direct downwind populations out of the path of the fall-out.

Just what kind of shelter protection is required? The answer depends upon how close to ground zero one locates the shelter since blast then becomes the criterion. Assuming, however, that the shelter is located beyond the range of primary blast, the radiation shielding requirements are as follows:

Reduction Factor	10	50	100	1000
Inches of concrete	6	11	13	19
or Inches of packed soil	11	18	21	30

The relatively small thicknesses of concrete or earth shielding needed to reduce the incident radiation to one-hundredth of its topside value may surprise the layman. Exponential absorption of gamma radiation accounts for the fact that a foot and a half of hard packed soil can reduce an intensity of 2500 r/hr to 50 r/hr. Thirty inches of soil cuts this intensity down to 2.5 r/hr which can be regarded as acceptable for survival in a shelter.

• • • •

In addition to the external radiation hazard there is the enigma of ingested or inhaled radioactive debris. Judging from the brief reports issued by the Atomic Energy Commission on the medical histories of the Marshall Islanders exposed to fall-out, the internal radiation hazard may be less serious than generally believed. However, data on the Japanese survivors does not make for complacency on the significance of radioactive material taken into the body.

From the foregoing description it can be readily appreciated that fall-out presents civil defense with potentially greater perils than those of heat and blast. Blast can be readily felt as can heat and they both come in a flash. Radioactivity, on the other hand, cannot be felt and possesses all the terror of the unknown. It is something which evokes revulsion and helplessness—like a bubonic plague.

Having been already weighted down by the incubus of the immense primary effects of the superbomb, civil defense planners may well feel that they are condemned to a labor of a Sisyphus. Unless the challenge of the fall-out is met head-on, the very burden may kill civil defense in this country. Those who oppose the FCDA policy of evacuation may use fall-out as the excuse for rejecting the recommendation. They can (and with partial justification) point out that the mass removal of metropolitan population to the suburbs or open country may be like jumping from the frying pan into the fire.

Civil defense must reckon with the hazards of fall-out, but it would be utterly disastrous if it abandoned its policy of evacuation at this time. No one is going to come up with a perfect civil defense plan. As long as we have such huge agglomerations of people on a few bits of territory there can be no perfect civil defense. There will always have to be the element of the calculated risk to civil defense just as there is for the soldier at the front line. What civil defense must do is to acknowledge certain risks in its planning.

First things must come first in any good C.D. plan. This means that one must deal with certain primary effects of superweapons. The great circles of heat and blast effects for weapons in the 10 to 40 megaton range are simply too large to neglect. They are predictable whereas fall-out is not. Furthermore, *it makes no sense whatsoever to plan for a secondary hazard if you fail to survive the primary effects of the bomb.*

The Federal Civil Defense Administration has assessed the true significance of the circles of damage from H-bombs and has boldly put forth an evacuation policy to deploy urbanites to the suburbs beyond the 12 to 14 mile radius of the city. This policy has yet to be implemented but cities are going ahead with their plans. Milwaukee, Washington, Seattle, and Detroit are a few that have reached an advanced state of planning.

Consideration of the fall-out ellipses must be made and the C.D. plan of each city modified accordingly. One thing is perfectly clear—evacuation even on present plans does provide a large measure of protection for people since *distance is still the best defense against the bomb.* Those who are deployed upwind or laterally will escape the lethal fall-out. In general, viewed as a problem in geometry, it is the area of the evacuation circle which is intersected by the fall-out ellipse which requires special planning. It will be argued that since we have so many unknowns it is futile to plan. For example, the wind may be in a contrary direction. However, local data on upper air winds often show a pattern of prevailing winds which may indicate preferential evacuation in upwind directions. Moreover, when an alert comes civil defense directors can make an immediate prognostication of the probable fallout area based on current weather data.

The alert is the great and vexing problem in evacuation. It is often pointed out that there may be no warning whatsoever. This is possible but it certainly does not apply to all cities. Moreover, it flies in the face of probability. There may, in fact, be a long period of alert—a strategical alert of days, weeks, or even months.

Dr. Vannevar Bush, in his testimony before the Riehlman subcommittee, pointed out that conventional warfare might precede nuclear warfare due to the initial respect which each nation might have for the other's nuclear capability. Such a period of tension would make possible gradual decentralization of urban populations in anticipation of a nuclear attack. Time would also be given, plus the incentive of imminent attack, to prepare peripheral shelters to accommodate evacuees.

Besides the strategical alert period accompanying Dr. Bush's concept, there is a shorter term alert which would be concomitant with Soviet aggression into zones to which we guaranteed military support. Should the Soviets strike in such an area they might well limit themselves to conventional weapons, and it would then be up to the President to issue the fateful ultimatum for "retaliation in advance." Any such ultimatum would be, if sanity prevailed, accompanied by a nation-wide evacuation order.

The announcement that a Distant Early Warning line of radar posts is to be established in the far north provides hope that before long a 4 hour tactical alert will be assured for continental cities. Four hours is sufficient for most cities to deploy their populations to relatively safe sites provided evacuation plans are perfected and test drills are inaugurated.

Progress is being made in civil defense despite the vertiginous, almost exponential, rise in the hazards faced. The year 1954 may well mark the turning point in our C.D. activities. One very favorable index is that more and more top advisors in the government are becoming serious about civil defense. More and more, it is becoming clear that the security of the home base is of paramount importance. In this security, civil defense must assume a high priority.

The Federal Civil Defense Administration must be permitted access to classified data about fall-out. Furthermore, the agency must be able to translate these data or "sanitize" them so that a realistic picture of the radioactive hazard can be given to the American people. Local communities should not be left to plan in the dark nor should they be put in the position of planning on the basis of newspaper reports.

In the final analysis the solution to civil defense problems will not come from Washington. It is in the local communities that they will find a solution. The federal government can help with solid planning data, with technical assistance, and with financial support. But the real burden of the work must be shouldered by the people who reside in the cities. However, the new peril from radioactive fall-out is more than just a threat to civil defense—it is a peril to humanity.

40. Atomic Energy Commission Meeting 1377, May 28, 1958

10:15 A.M., Wednesday, May 28, 1958, Room A-410,
Germantown, Maryland

1. Weapons Test Limitations

Mr. Graham reviewed with the representatives of the AEC weapons laboratories a number of events which had led the Commission to request a meeting with them to discuss the question of weapons test limitations. He pointed out that the subject of weapons tests had been considered both by the General Advisory Committee and the Advisory Committee on Biology and Medicine at their recent meetings. In addition, the current Congressional authorization hearings were concerned, in part, with funds to be authorized for the weapons test program. Mr. Graham said that all of these events had made it evident that the question of future weapons testing was still far from being resolved, and therefore he and other Commissioners believed it would be helpful if a full discussion of the subject could be held with representatives from the weapons laboratories. General Starbird added that the Laboratory Directors had in particular been requested to consider what technical problems were involved and what limitations would result from a decision to test underground only.

Mr. Teller began the discussion by stating that scientists at Livermore Laboratory had concluded that nearly all the information needed by the U.S. about nuclear explosions could be obtained from underground tests, and that underground weapons testing is more easily carried out than testing above

This document appears in the United States Atomic Energy Commission Records, United States Department of Energy, Germantown, Maryland.

ground. He said that he would be in favor of conducting future weapons tests underground regardless of whether an international test limitation should be agreed upon. He added, however, that before it could be known with certainty that weapons in the megaton range could be successfully tested underground, there would have to be a series of tests gradually increasing in size to the megaton range. In addition, if all tests were conducted underground, there would be no opportunity to test weapons effects or to proof-test anti-ICBM missile systems. Therefore, he recommended that it would be desirable to have some above-ground testing each year but to limit each country to tests placing a maximum DELETED of fissionable material in the atmosphere per year. This amount of above-ground testing, he said, would permit all the diagnostic weapons experimentation necessary for the weapons program. If the maximum amount of fissionable material which could be put into the atmosphere each year by each country were limited DELETED it would still be possible to conduct valuable above-ground tests, although the amount of information gained would be less and the results would be limited.

Mr. Brown discussed with the Commissioners the three major types of measurements for a weapons test and how accurate these measurements would be if applied to an underground detonation. Yield measurement for the Rainier shot, he said, had a margin of error DELETED DELETED but with additional experimentation, this margin could be brought down DELETED. Mr. Teller remarked that weapons yields could not be quite as accurately determined underground as above-ground. However, diagnostic details from direct measurements can be obtained more easily and more accurately underground than above-ground, he said.

Mr. Brown said that, in his opinion, radio-chemistry measurement techniques for underground tests, although not as accurate as yield measurements, are adequate. Reaction history measurements, he said, can actually be more accurate underground than above-ground. Mr. Brown said he agreed with Mr. Teller that all necessary above-ground experiments could be conducted within the limitation DELETED DELETED. . . . to keep within this limitation, he said it would probably be necessary to substitute clean weapons for normal weapons shots. Mr. Brown said special weapons types would have to be developed for effects testing and for anti-missile missile warheads to be detonated underground.

Mr. Brown said the best information available to him indicated that weapons DELETED could be tested safely under-ground but, as Mr. Teller had stated, the size of the tests would have to be detonated deeper and in much harder rock, and there is uncertainty about the effect of such detonations since the transmission of shock would be more direct.

Mr. Sewell reviewed with the Commissioners the advantages of an underground test program over the present series of yearly above-ground test programs. Greater flexibility of scheduling tests and therefore more rapid progress in developing new weapons would be possible if the laboratories were not restricted to waiting for only one test series each year. With an underground test program, tests could be conducted periodically throughout the year whenever a weapon under development reached the point where certain test experiments needed to be conducted. More radical weapons designs could be tested because a laboratory would know that if a particular test failed, a year would not have to lapse before it could conduct another test experiment. This fact could also lead to more rapid progress in weapons development. Costly, full-scale test operations such as are conducted in alternate years at the Pacific Proving Grounds could be eliminated. Tests would no longer be dependent upon weather conditions, thus saving additional time and money now consumed in conducting above-ground tests. The cost of digging the tunnel for the Rainier shot cost no more than a five-hundred foot tower for a test device, and the cost of digging new tunnels out from a main one is about one-fourth the cost of the original tunnel Mr. Sewell said. In addition, public opposition to the tests because of the fallout danger could be eliminated by under-ground testing.

Mr. Teller stated that conducting all future U.S. weapons tests underground would interfere somewhat with Project PLOWSHARE, the program for developing peaceful uses of nuclear explosive devices.

In response to a question by Mr. Fields, Mr. Sewell said as many DELETED bombs could be detonated in a single mountain.

Mr. Bradbury commented that although the Livermore scientists may be correct in believing that nearly all necessary weapons information could be obtained from underground testing, if actual experience proves this to be incorrect, the U.S. would be placed in a very difficult position. He pointed out that much of this work on megaton weapons now involved detailed improvements of the weapons and that with only underground testing there would be a greater risk of being misled about what the tests actually demonstrated. Mr. Bradbury said he took the position that a diversified testing program using balloons, towers, and underground tunnels would be preferable to a complete underground test program.

Mr. Teller remarked that data on a weapons test is most needed when that device misfires, and that the diagnostics of such a test could be done as well or better if the shot were underground rather than above-ground.

General Starbird pointed out that estimates regarding a major underground test program are being based to a large extent upon extrapolations from the Rainier shot. He said he concurred with Messrs. Bradbury and

Graves in their view that the U.S. would be faced with a serious military problem if it agreed to underground testing for all weapons, and then discovered that the larger weapons could not be satisfactorily tested underground. He added that the public could not be expected to raise questions about the safety of even underground weapons tests, and the danger that the radio-activity created by the detonation might sometime contaminate water and plant life.

Mr. Graves said he favored the use of balloons for testing weapons devices and that this valuable testing technique would be lost if all aboveground tests were halted.

Mr. Teller stated the belief that there would be no chance that radioactivity from underground tests could contaminate animals or humans, and said he believed one could be overly cautious about questioning the feasibility of underground testing.

Mr. Vance referred to a statement by the GAC at their 58th meeting to the effect that the Commission must not wait too long in proposing an intermediate position between unlimited testing and a ban on all weapons testing. Alternatives suggested by Mr. Vance included: (1) A limitation on the amount of radioactivity released into the atmosphere; (2) Supervision by the U.N. of all weapons tests; and (3) Complete underground testing of all weapons. Mr. Vance said he favored the underground testing position because it would be the simplest to establish and because it would end completely any radioactive contamination of the atmosphere from weapons detonations. He said he considered that public concern with the question makes it an important consideration, even though the Commission recognized that the danger at present is negligible. Mr. Vance said the question he wished the representatives of the laboratories to discuss is what intermediate position would enable the U.S. to continue its weapons development program with the greatest freedom. Mr. Bradbury said it would be difficult to select the best alternative since each one would interfere with the development which is possible under unlimited testing.

Mr. Teller said he believed the most desirable alternative would be to restrict above-ground testing by each country to DELETED fission products released into the atmosphere each year, and to require that all other tests be conducted underground. Such a position would have a number of advantages, he said. It would permit the development of anti-missile missiles, which probably would be impossible if all testing had to be conducted underground. It would also make it quite difficult for nations which do not now have weapons capability to develop weapons. They would be forced to develop both very small weapons for underground testing and clean devices for above-ground testing. Finally, an agreement to limit the amount of fissionable material released into the atmosphere would necessitate establishing a mutual inspection system, thus giving U.S. observers an opportunity to gain first-hand information about the Soviet weapons program.

Mr. Teller said he would be reluctant to have the U.S. accept more stringent test limitations that the DELETED fission yield per year above ground, and unlimited underground testing.

Mr. Vance observed that Mr. Teller's arguments were quite logical; however, the question of limiting nuclear tests would not necessarily be resolved on a logical or strictly scientific basis. He cited the instance of Mr. Hans Bethe's recent statements to the effect that all nuclear testing should be halted as a first step in ending the arms race and reducing world tensions. These, he said, are political judgments, not scientific ones, although they are made by a scientist.

Mr. Floberg asked Mr. Teller how possible violations of limitations such as he recommended would be handled. Mr. Teller replied that he would not be too concerned if the violation were in the range DELETED of fission yield put into the atmosphere in a single year, DELETED. Violations substantially greater than this, however, would be a cause for serious concern, he said. He pointed out that all countries carrying out an atmospheric sampling program would know when serious violations had occurred. General Starbird remarked that continued violations of such a limitation might cause neutralist countries to conclude that the limitation was not effective and to demand a complete cessation of all types of weapons testing.

Mr. Strauss raised the question of how complete weapons systems could be proof-tested underground. Mr. Bradbury said he was not convinced that a final proof-test of a missile and warhead was absolutely necessary if the two could be adequately tested separately. Mr. Teller said a great amount of experimentation was still required for an anti-ICBM missile warhead and that some above-ground testing would be necessary.

Mr. Strauss then inquired whether fallout measurement techniques are accurate enough to determine precisely whether DELETED DELETED fission yield had been placed in the atmosphere by a particular country. Mr. Teller replied that this would be possible only if there were mutual inspection within the countries where the tests occurred.

The question of identifying from which country particular fallout originated was then discussed by Mr. Strauss and Mr. Teller. Mr. Strauss postulated that a country might detonate a number of weapons at extremely high altitudes, with the fallout not being detected until a year or more later. He asked whether it would be possible then to accurately determine which country was responsible for those tests. Mr. Teller said that with satellite counters it would be possible to detect almost immediately when high altitude shots have been detonated, but that the country carrying out the test could not be positively identified in such cases. He added, however, that scientists could probably make an intelligent guess about the origin of the shots.

Mr. Vance left the meeting briefly during the above discussion. The Commissioners returned to the question of the necessity of proof-testing complete weapons systems, and Mr. Graham inquired whether it would be satisfactory to test missile systems using chemical rather than nuclear warheads. Mr. Teller said it would be possible to test a missile system without the nuclear warhead and determining the performance of such things as the timing of the detonation device, but of course, such a test would not provide any information on the effects of a nuclear explosion. Mr. Floberg said he believed there have been many examples in the past of the importance of conducting tests of complete weapons systems, such as submarine torpedos. He said he was unconvinced that it would be unnecessary to carry out proof-tests of the complete missile system. The Psychological factor of working with missiles equipped with nuclear warheads was important in itself, he said.

Mr. Fields then discussed with Mr. Teller and Mr. Graves when it would be possible to carry out the first diagnostic tests of an anti-ICBM missile. Mr. Brown said the Air Force had mentioned 1960 as the date when such tests might be held; however, the weapons laboratories do not yet know what device might be used. Mr. Teller said the DELETED may be a basis for developing such a weapon but that much more experimentation is required before it could be perfected. Following the detonation DELETED at Operation HARDTACK, it might be possible to conduct further test of it in 1959. Mr. Graves said he believed that it probably would be 1960 before an anti-ICBM device would be ready for diagnostic testing.

Mr. Strauss concluded the discussion by expressing the Commissioners' appreciation to Messrs. Teller, Bradbury, Graves, Brown, and Sewell for an enlightening discussion of the test limitations questions.

•　•　•　•

W. B. McCool
Secretary

41. W. K. Wyant, Jr., "50,000 Baby Teeth," June 13, 1959

Ordinarily, a group that called itself the Greater St. Louis Citizens Committee for Nuclear Information would not be expected to last for any great period of time. Mortality among earnest and well-meaning organizations has been as great here as elsewhere, and the waters of the Mississippi have rolled over many such. Yet the committee now is striding vigorously into its second year. It has turned out to be an unusually happy union of scientific knowledge and civic leadership. Lay members are not afraid of Ph.D.s and M.D.s, the good doctors are neither fearful nor contemptuous of the laymen, and both doctors and laymen are unafraid of the United States Government.

At the outset, CNI—as the committee is called locally—decided it would not be an "action" group—that is to say, it would take no position for or against testing of nuclear weapons, even though the sentiment of the organizing spirits was clearly and outspokenly against. The view prevailed that what really was needed was information. It was felt that too many people—the politicians, the military and the oracles speaking "ex cathedra" from the Atomic Energy Commission—were taking decisive attitudes on the basis of indecisive information, or none.

This stand, useful in itself, also served the practical purpose of broadening the base of support, with the consequence that the CNI leadership now reflects a wide spectrum of opinion.

But while persisting in its "non-action" policy, the committee during its first year waged such a valiant fight on the information front that strontium-90 is now a household word in St. Louis. On the average of every other day, a member of the speakers' stable makes a talk on the subject. The CNI's monthly bulletin's circulation has jumped from 500 to 2,500. And the committee's drive to collect 50,000 baby teeth, to be analyzed for strontium-90, got headlines throughout the world press.

A few months ago, CNI's first anniversary meeting took place at the Union Avenue Christian Church, where the organization had been founded. Present was a cross-section of the medical and scientific elite of the city, along with lawyers, ministers and other civic leaders. Morale was high. There was evident a fine rapport, an atmosphere of comradeship and understanding, between scientists and laymen. One reason for this was that much had been accomplished with little.

Dr. Alfred S. Schwartz, assistant professor of clinical pediatrics at Washington University School of Medicine, the committee's vice president and treasurer, reported $5,077 had been spent during the year, leaving a balance of $469. Membership was more than 500. Tribute was paid to the women volunteer workers, many of them the wives of physicians and scientists.

Specifically, what had the committee done in its first year? There was a good deal besides the fact that it now had an office on the second floor of an old, converted red-brick house on West Pine Boulevard, with a number in the telephone book, a constantly ringing telephone, and a dedicated staff secretary to answer phone and mail—Mrs. Edward C. Roberts, who started as a part-time paid worker and now is a full-time volunteer.

1. A Speaker's Bureau, consisting of seventeen men and three women, has been organized to offer scientists as speakers to organizations in the St. Louis area. The speakers are all either M.D.s or Ph.D.s, capable of dealing with technical aspects of the fallout controversy. Since last October, some seventy church, parent, fraternal and business groups have heard addresses. Three organizers of the committee—Barry Commoner, professor of botany at Washington University, Dr. Walter C. Bauer, instructor in surgical pathology at the university's School of Medicine, and John M. Fowler, Washington University physicist—have made the knife-and-fork circuit a way of life. Fowler estimated they spend about a third of their time at it.

2. *Nuclear Information*, CNI's monthly publication, was started last fall to report new scientific facts in lay language. The March issue was entitled "Strontium-90 and Common Foods." It served to point out that the public had heard much about radioactivity in milk, but little about radioactivity in other foods. An article told of the three-year survey of wheat samples from Minnesota and the Dakotas.

3. CNI has sponsored two series of five seminars each on radiation for the scientific and medical community, and three public meetings at which qualified people discussed the consequences of radiation and fallout.

4. At the request of the St. Louis Dairy Council, CNI scientists last January issued a statement discussing the potential hazard from strontium-90 in milk (the local milk supply has been showing the highest strontium-90 concentration of 10 cities surveyed monthly by the United States Public Health Service—and nobody knows why). The statement said, in effect, that the harm that might result to children from milk containing radioactive substances could not yet be assayed with certainty. It emphasized that milk, as an essential food, must not be eliminated from the diet, and called for research on ways to lower its strontium-90 content.

5. The Baby Tooth Survey was started last December as a ten-year scientific project. Directed by Dr. Louise Reiss, an internist, the campaign has the cooperation of the School of Dentistry at both Washington and St. Louis universities. It has also served as an excellent device for calling attention to the more dreadful implications of nuclear fallout.

There was deep concern in St. Louis about nuclear-weapons testing for several years before CNI was organized. One of Washington University's most widely known scientists, former Chancellor Arthur H. Compton, Nobel Prize-winner and key figure in developing the A-bomb in World War II, has maintained a friendly attitude toward government atomic policies. But younger men in the physics department, headed by Edward U. Condon, former U.S. Bureau of Standards, chief, were openly hostile. In the spring of 1957, Linus C. Pauling of the California Institute of Technology, also a Nobel Prize-winner, attacked atom bomb testing in an eloquent address before Washington University's faculty and student body, warning of the price that might have to be paid by generations yet unborn. Later, in Condon's office, Pauling drew up a petition that called for immediate action to halt testing by international agreement.

Condon, Commoner and three other faculty members were among the twenty-seven signers of this petition; by the time Pauling presented it to the U.N. Secretary-General in January, 1958, more than 9,000 scientists from forty-three nations had subscribed. The scientific community at St. Louis University, a Catholic institution, had a share in the effort.

The formation of CNI followed several years of public needling of the AEC by Fowler, Commoner, Dr. Bauer, T. Alexander Pond of Washington University's physics department, and others. Some were horrified at the attacks made on Adlai Stevenson when he tried to debate the fallout question in the 1956 Presidential campaign.

St. Louis women, an enlightened and militant breed, got into the conflict early. A pioneer group calling itself "Eves Against Atoms," headed by Mrs. Thomas B. Sherman, wife of a *Post-Dispatch* editor, took a stand against testing. However, the CNI traces its origin directly to a meeting which took place in March of last year at the apartment of Mrs. George Gellhorn, a tireless and effective worker for civic causes in city, state and nation for sixty years. Now eighty, a widow since 1936 and the mother of writer Martha Gellhorn, Mrs. Gellhorn is a woman possessed of beauty, charm and sharp political savvy. Her assistance in any campaign to get something done in St. Louis is the rough equivalent of six Marine battalions.

From the start, CNI has had strong moral and religious overtones, sharpened by a kind of cheerful "let's look at the facts" iconoclasm. Like Mrs. Gellhorn, the Commoners—Barry and his wife, Gloria, committee vice president—are members of the St. Louis Ethical Society. So is Alexander S. Langsdorf, dean emeritus of Washington University's School of Engineering and CNI president. Fowler and Dr. Bauer and Quakers. At the initial meeting in the Gellhorn apartment, in addition to Commoner and Fowler, was the Rev. Ralph C. Abele, then head of the Metropolitan Church Federation. They decided to form a permanent body.

The next step was a meeting of about thirty people which took place several weeks later at the home of Mrs. Ernest W. Stix, another civic leader. Attending were university people, members of social-action groups, priests, ministers, housewives. The sequel was the founders' meeting, April 21, 1958.

While not all the CNI prime movers took a position against testing, the major figures did; and it is quite apparent that the group as a whole is motivated by deep misgivings as to the biological and moral implications of the weapons race. Nevertheless, the decision was to take no official stand and to concentrate instead on information.

Scientists with CNI have served as a Greek chorus to amplify, comment on, question, explain and criticize the thunderous pronunciamentos of the protagonists in Washington. Each new gobbet of official information is analyzed and put into context. Research papers are seized on, dissected and translated into lay language. The press is kept alerted to its duty. Any inaccuracies, omissions or glossings-over in official statements are promptly challenged.

The Baby Tooth Survey has given the organization an objective that should hold it together for some time. The idea was suggested last summer by Dr. Schwartz, who cited an article in the British scientific publication *Nature* written by Dr. Herman M. Kalckar, biochemist at Johns Hopkins University. CNI became the first group anywhere, apparently to initiate a large collection of "milk" teeth to be analyzed for radioactivity.

It was characteristic of the committee that it planned the project carefully. Before any announcement was made, its scope and purpose had the approval of the deans of both the local schools of dentistry; and Dr. John T. Bird, assistant dean of Washington University's school had agreed to head a group of dentists to examine and classify the teeth. Dr. E. S. Khalifah, editor of the *Journal* of the Missouri State Dental Association, gave his enthusiastic support. With the help of Martin Quigley, a public-relations adviser who is on the CNI board, the announcement, when it came, was deliberately designed to softpedal the "human interest" angle of the story and to stress sober scientific objectives. It said in part:

The importance of an immediate collection of deciduous, or baby, teeth lies in the fact that teeth now being shed by children represent an irreplaceable source of scientific information about the absorption of strontium-90 in the human body. Beginning about ten years ago, strontium-90 from nuclear-test fallout began to reach the earth and to contaminate human food.

Deciduous teeth now being shed were formed from the minerals present in food eaten by mothers and infants during the period 1948 to 1953—the first few years of the fallout era—and therefore represent invaluable baseline information with which analysis of later teeth and bones can be compared. Unless a collection of deciduous teeth is started immediately, scientists will lose the chance to learn how much strontium-90 human beings absorbed during the first years of the atomic age.

Strontium-90 present in food accumulates in bones and teeth; milk is the main food source of strontium-90. In sufficient amounts the radiation from strontium-90 may cause harmful effects, including bone tumors and other forms of cancer. . . .

The reaction to the announcement was instantaneous and world-wide: There were letters from New York, Hawaii, Calcutta, California, Spain. Some mail from afar included teeth, although the committee wants specimens only from the St. Louis area.

Dr. Louise Reiss and her assistants have had extraordinary success in getting local schools—public, private and parochial—to help in the teeth collecting. Some 250,000 forms have been distributed to reach all lower-grade students. The Council of Catholic Women, the public libraries, the dental societies and city dental clinics have been of great assistance. At present, baby teeth are reaching the little office on West Pine Boulevard at the rate of about fifty a day. The stockpile is still short of 10,000, although the estimated annual "fallout" of baby teeth in the St. Louis area is half-a-million. When enough have been accumulated, they will be classified, ground up and analyzed at the Washington University School of Dentistry. Mothers who request an individual report on their children's teeth are doomed to disappointment.

Summing up the first year, the CNI, adopting a "conservative" approach, has set a pattern already being followed by other communities. What official bodies will not do for them, citizens are seeking to do for themselves.

42. Treaty Banning Nuclear Weapon Tests in the Atmosphere, in Outer Space, and Under Water, August 5, 1963

Signed at Moscow August 5, 1963
Ratification advised by U.S. Senate September 24, 1963
Ratified by U.S. President October 7, 1963
U.S. ratification deposited at Washington, London, and Moscow
    October 10, 1963
Proclaimed by U.S. President October 10, 1963
Entered into force October 10, 1963

The Governments of the United States of America, the United Kingdom of Great Britain and Northern Ireland, and the Union of Soviet Socialist Republics, hereinafter referred to as the "Original Parties,"

Proclaiming as their principal aim the speediest possible achievement of an agreement on general and complete disarmament under strict international control in accordance with the objectives of the United Nations which would put an end to the armaments race and eliminate the incentive to the production and testing of all kinds of weapons, including nuclear weapons.

Seeking to achieve the discontinuance of all test explosions of nuclear weapons for all time, determined to continue negotiations to this end, and desiring to put an end to the contamination of man's environment by radioactive substances.

Have agreed as follows:

Article I

1. Each of the Parties to this Treaty undertakes to prohibit, to prevent, and not to carry out any nuclear weapon test explosion, or any other nuclear explosion, at any place under its jurisdiction or control:

(a) in the atmosphere; beyond its limits, including outer space; or under water, including territorial waters or high seas; or

This document appears in United States Arms Control and Disarmament Agency, *Arms Control and Disarmament Agreements* (Washington, D.C.: U.S. Government Printing Office, 1982), pp. 41–43.

(b) in any other environment if such explosion causes radioactive debris to be present outside the territorial limits of the State under whose jurisdiction or control such explosion is conducted. It is understood in this connection that the provisions of this subparagraph are without prejudice to the conclusion of a treaty resulting in the permanent banning of all nuclear test explosions, including all such explosions underground, the conclusion of which, as the Parties have stated in the Preamble to this Treaty, they seek to achieve.

2. Each of the Parties to this Treaty undertakes furthermore to refrain from causing, encouraging, or in any way participating in, the carrying out of any nuclear weapon test explosion, or any other nuclear explosion anywhere which would take place in any of the environments described, or have the effect referred to, in paragraph 1 of this Article.

Article II

1. Any Party may propose amendments to this Treaty. The text of any proposed amendment shall be submitted to the Depositary Governments which shall circulate it to all Parties to this Treaty. Thereafter, if requested to do so by one-third or more of the Parties, the Depositary Governments shall convene a conference, to which they shall invite all the Parties, to consider such amendment.

2. Any amendment to this Treaty must be approved by a majority of the votes of all the Parties to this Treaty, including the votes of all of the Original Parties. The amendment shall enter into force for all Parties upon the deposit of instruments of ratification by a majority of all the Parties, including the instruments of ratification of all of the Original Parties.

Article III

1. This Treaty shall be open to all States for signature. Any State which does not sign this Treaty before its entry into force in accordance with paragraph 3 of this Article may accede to it at any time.

2. This Treaty shall be subject to ratification by signatory States. Instruments of ratification and instruments of accession shall be deposited with the Governments of the Original Parties—the United States of America, the United Kingdom of Great Britain and Northern Ireland, and the Union of

Soviet Socialist Republics—which are hereby designated the Depositary Governments.

3. This Treaty shall enter into force after its ratification by all the Original Parties and the deposit of their instrument of ratification.

4. For States whose instruments of ratification or accession are deposited subsequent to the entry into force of this Treaty, it shall enter into force on the date of the deposit of their instruments of ratification or accession.

5. The Depositary Governments shall promptly inform all signatory and acceding States of the date of each signature, the date of deposit of each instrument of ratification of and accession to this Treaty, the date of its entry into force, and the date of receipt of any requests for conferences or other notices.

6. This Treaty shall be registered by the Depositary Governments pursuant to Article 102 of the Charter of the United Nations.

Article IV

This Treaty shall be of unlimited duration.

Each party shall in exercising its national sovereignty have the right to withdraw from the Treaty if it decides that extraordinary events, related to the subject matter of this Treaty, have jeopardized the supreme interests of its country. It shall give notice of such withdrawal to all other Parties to the Treaty three months in advance.

Article V

This Treaty, of which the English and Russian texts are equally authentic, shall be deposited in the archives of the Depositary Governments. Duly certified copies of this Treaty shall be transmitted by the Depositary Governments to the Governments of the signatory and acceding States.

IN WITNESS WHEREOF the undersigned, duly authorized, have signed this Treaty.

DONE in triplicate at the city of Moscow the fifth day of August, one thousand nine hundred and sixty-three.

LIMITED TEST BAN TREATY

| For the Government of the United States of America | For the Government of the United Kingdom of Great Britain and Northern Ireland | For the Government of the Union of Soviet Socialist Republics |

43. Treaty Between the United States of America and the Union of Soviet Socialist Republics on the Limitation of Underground Nuclear Weapon Tests, July 3, 1974

The United States of America and the Union of Soviet Socialist Republics, hereinafter referred to as the Parties,

Declaring their intention to achieve at the earliest possible date the cessation of the nuclear arms race and to take effective measures toward reductions in strategic arms, nuclear disarmament, and general and complete disarmament under strict and effective international control,

Recalling the determination expressed by the Parties to the 1963 Treaty Banning Nuclear Weapon Tests in the Atmosphere, in Outer Space and Under Water in its Preamble to seek to achieve the discontinuance of all test explosions of nuclear weapons for all time, and to continue negotiations to this end,

Noting that adoption of measures for the further limitation of underground nuclear weapon tests would contribute to the achievement of these objectives and would meet the interests of strengthening peace and the further relaxation of international tension,

Reaffirming their adherence to the objectives and principles of the Treaty Banning Nuclear Weapon Tests in the Atmosphere, in Outer Space and Under Water and of the Treaty on the Non-Proliferation of Nuclear Weapons,

Have agreed as follows:

Article I

1. Each Party undertakes to prohibit, to prevent, and not to carry out any underground nuclear weapon test having a yield exceeding 150 kilotons at any place under its jurisdiction or control, beginning March 31, 1976.

2. Each Party shall continue their negotiations with a view toward achieving a solution to the problem of the cessation of all underground nuclear weapon tests.

This document appears in United States Arms Control and Disarmament Agency, *Arms Control and Disarmament Agreements* (Washington, D.C.: U.S. Government Printing Office, 1982), pp. 167–68.

Article II

1. For the purpose of providing assurance of compliance with the provisions of this Treaty, each Party shall use national technical means of verification at its disposal in a manner consistent with the generally recognized principles of international law.

2. Each Party undertakes not to interfere with the national technical means of verification of the other Party operating in accordance with paragraph 1 of this Article.

To promote the objectives and implementation of the provisions of this Treaty the Parties shall, as necessary, consult with each other, make inquiries and furnish information in response to such inquiries.

TESTS OF ARMS CONTROL AGREEMENTS

Article III

The provisions of this Treaty do not extend to underground nuclear explosions carried out by the Parties for peaceful purposes. Underground nuclear explosions for peaceful purposes shall be governed by an agreement which is to be negotiated and concluded by the Parties at the earliest possible time.

Article IV

This Treaty shall be subject to ratification in accordance with the constitutional procedures of each Party. This Treaty shall enter into force on the day of exchange of instruments of ratification.

Article V

1. This Treaty shall remain in force for a period of five years. Unless replaced earlier by an agreement in implementation of the objectives specified in paragraph 3 of Article I of this Treaty, it shall be extended for successive five-year periods unless either Party notifies the other of its termination no later than six months prior to the expiration of the Treaty. Before the expiration of this period the Parties may, as necessary, hold consultations to

consider the situation relevant to the substance of this Treaty and to introduce possible amendments to the text of the Treaty.

2. Each Party shall, in exercising its national sovereignty, have the right
to withdraw from this Treaty if it decides that extraordinary events related
to the subject matter of this Treaty have jeopardized its supreme interests.
It shall give notice of its decision to the other Party six months prior to
withdrawal from this Treaty. Such notice shall include a statement of the
extraordinary events the notifying Party regards as having jeopardized its
supreme interests.

3. This Treaty shall be registered pursuant to Article 102 of the Charter
of the United Nations.

Done at Moscow on July 3, 1974, in duplicate, in the English and Russian languages, both texts being equally authentic.

For the United States of America:
    RICHARD NIXON
        The President of the United States of America

For the Union of Soviet Socialist Republics:
    L. I. BREZHNEV
        General Secretary of the Central Committee of the CPSU

---

44. Protocol to the Treaty Between the United States of America and the Union of Soviet Socialist Republics on the Limitation of Underground Nuclear Weapon Tests, July 3, 1974

The United States of America and the Union of Soviet Socialist Republics, hereinafter referred to as the Parties,
    Have agreed to limit underground nuclear weapon tests,
    Have agreed as follows:

This document appears in United States Arms Control and Disarmament Agency, *Arms
Control and Disarmament Agreements* (Washington, D.C.: U.S. Government Printing Office,
1982), pp. 169–70.

1. For the Purpose of ensuring verification of compliance with the obligations of the Parties under the Treaty by national technical means, the Parties shall on the basis of reciprocity, exchange the following data:

a. The geographic coordinates of the boundaries of each test site and of the boundaries of the geophysically distinct testing areas therein.

b. Information on the geology of the testing areas of the sites (the rock characteristics of geological formations and the basic physical properties of the rock, i.e., density, seismic velocity, water saturation, porosity and depth of water table).

c. The geographic coordinates of underground nuclear weapon tests, after they have been conducted.

d. Yield, date, time, depth and coordinates for two nuclear weapon tests for calibration purposes from each geophysically distinct testing area where underground nuclear weapon tests have been and are to be conducted. In this connection the yield of such explosions for calibration purposes should be as near as possible to the limit defined in Article I of the Treaty and not less than one-tenth of that limit. In the case of testing areas where data are not available on two tests for calibration purposes, the data pertaining to one such test shall be exchanged, if available, and the data pertaining to the second test shall be exchanged as soon as possible after the second test having a yield in the above-mentioned range. The provision of this Protocol shall not require the Parties to conduct tests solely for calibration purposes.

2. The Parties agree that the exchange of data pursuant to subparagraphs a, b, and d of paragraph 1 shall be carried out simultaneously with the exchange of instruments of ratification of the Treaty, as provided in Article VI of the Treaty, having in mind that the Parties shall, on the basis of reciprocity, afford each other the opportunity to familiarize themselves with these data before the exchange of instruments of ratification.

3. Should a Party specify a new test site or testing area after the entry into force of the Treaty, the data called for by subparagraphs a and b of paragraph 1 shall be transmitted to the other Party in advance of use of that site or area. The data called for be subparagraph d of paragraph 1 shall also be transmitted in advance of use of that site or area if they are available; if they are not available, they shall be transmitted as soon as possible after they have been obtained by the transmitting Party.

4. The Parties agree that the test sites of each Party shall be located at places under its jurisdiction or control and that all nuclear weapon tests shall be conducted solely within the testing areas specified in accordance with paragraph 1.

For the purposes of the Treaty, all underground nuclear explosions at the specified test sites shall be considered nuclear weapon tests and shall be subject to all the provisions of the Treaty relating to nuclear weapon tests. The provisions of Article III of the Treaty apply to all underground nuclear explosions conducted outside of the specified test sites, and only to such explosions.

This Protocol shall be considered an integral part of the Treaty.

Done at Moscow on July 3, 1974.[1]

For the United States of America:
RICHARD NIXON
The President of the United States of America.

For the Union of Soviet Socialist Republics:
L. I. BREZHNEV
General Secretary of the Central Committee of the CPSU

1. The Treaty and protocol were signed at a ceremony in St. Vladimir Hall of the Grand Kremlin Palace on Wednesday, July 3, 1974.

VII

Deterrence

*I*n 1945 the United States became the first nation to use a nuclear weapon in time of war. Even then many scientists and policy makers realized that the terrible power of the nuclear age implied a terrible paradox, namely, that nuclear weapons were both a problem and a solution. The problem was that their enormous destructive power made them ill suited to any rational military or foreign policy; two rival nations with nuclear weapons could threaten each other but offer no real defense against an attack. The solution was that the nuclear threat could avert nuclear war precisely by its existence, under threat of mutual annihilation, two nations, like the two scorpions in a bottle in J. Robert Oppenheimer's metaphor, could destroy the other only at the risk of destroying themselves.

As early as 1946, Bernard Brodie and other strategic thinkers associated with the RAND Corporation, an air force think tank in California, realized that with nuclear weapons full-scale wars could be deterred but not really won. "Fear of substantial retaliation," wrote Brodie, would be the key to nuclear policy, which would not be to win wars, but to avert them. Moreover, the only rational purpose in stockpiling more nuclear weapons would be to ensure that they need not be used in time of war. The notion of an arms race as a means to peace became the paradoxical cornerstone of the doctrine of deterrence.

After 1949 the American atomic monopoly gave way to a Soviet-American arms race centered on the construction of thermonuclear weapons. In 1954 Secretary of State John Foster Dulles established as the basis of deterrence (of both war and Soviet aggression) the doctrine of massive retaliation, whereby the United States would maintain the peace by maintaining its ability to respond to a nuclear attack or any other form of aggression with an all-out nuclear attack upon the Soviet Union. Massive retaliation would be the American response to a Soviet first-strike attack on the United States and also to an attack on Europe and its new North Atlantic Treaty Organization (NATO). The delivery vehicles would be the B-52 bombers of the Strategic Air Command, and the weapons would be nuclear and thermonuclear.

In the late 1950s the development of intercontinental ballistic missiles reduced the time of delivery of strategic nuclear weapons to minutes rather than hours. The winds of the Cold War blew vigorously, and Dulles had shifted the notion of deterrence from the deterrence of war to the deterrence of Soviet aggression. As the number of nuclear weapons multiplied, the danger of war by accident was compounded by an inflexible military doctrine that threatened nuclear response to nonnuclear but unspecified aggression. Thus was born the idea of limited nuclear war, laid down in quantitative and frightening terms by another RAND strategist, Herman Kahn, in his book On Thermonuclear War, published in 1960.

Kahn created a sensation by arguing that nuclear war might not mean the end of the world, as envisaged in On the Beach and other fictional scenarios of the 1950s. Rather, depending on how many weapons of what yield were used for how long a time, nuclear wars were distinguishable in their effects and postwar states; more than this, argued his critics, they were thinkable and thus winnable. Brodie had argued that nuclear war was irrational; Kahn made it rational.

Until the 1960s limited accuracy and numbers of weapons meant that deterrence was limited to a "second strike" and "countervalue" exchange. That is, a nation would not strike first if it knew that its enemy would still retain the ability to strike back at its population centers and cities. Targeting cities, it was argued, was a stable and rational arms control measure in that it constituted only a strategy of retaliation; the intent to strike first would be indicated by targeting military forces, not cities.

In fact, a counterforce doctrine was emerging to provide a whole new theory of extended, not limited, deterrence. Under the air force-dominated SIOP targeting program, an increasing number of missiles in the 1960s were assigned to Soviet military targets. This shift in targeting was articulated in 1962 by Secretary of Defense Robert McNamara, who declared that the most rational policy would be to target military forces, not cities. "In particular, relatively weak national nuclear forces with enemy cities as their targets are not likely to be sufficient to perform even the function of deterrence," argued McNamara, since they would be vulnerable to attack. Thus by the late 1960s the number of available weapons, targeting strategy, and missile technology made it possible to target many times over both the military and civilian centers of the United States and the Soviet Union.

In the 1970s deterrence became complicated further by the development

of another new technology, Multiple Independently Targeted Reentry Vehicles, or MIRVs. By placing as many as twelve warheads with independent
computer guidance systems on the nosecone of a single missile, the ratio of
available targets to each launcher was greatly increased. MIRV made the
dream of an antiballistic missile defense hopeless as well as expensive. It
also made deterrence increasingly unstable. Until MIRV the doctrine of
"mutual assured destruction," or MAD, was based on the fact that neither
side could hope to destroy the launchers of the other in a first strike, so that
a sufficient number of invulnerable second-strike launchers would always
remain. MIRV technology made a first strike thinkable by targeting enemy
forces many times over. The great success of the Strategic Arms Limitation
Talks (SALT) of the 1970s was the 1972 agreement to limit ABM systems;
the great failure was to allow the development of MIRVed missile launchers.

In 1979 extended deterrence was codified by the Carter administration
in Presidential Directive 59, a top secret and still unpublished statement of
American nuclear war-fighting policy. Coupled with Reagan administration
talk of winning, limiting, and surviving a nuclear war by decapitating Soviet
military command and control structures, PD 59 helped spawn a massive
antinuclear movement in both Europe and America. An entire generation of
young people joined concerned elders in questioning the entire basis of deterrence, a doctrine designed to avert war but widely viewed as merely a
justification for an accelerating arms race. Led by Catholic bishops in
America, the movement to question the entire philosophical and ethical justification for deterrence gathered momentum, especially as the December
1983 deadline for installing American cruise missiles in Western Europe
approached.

The new sense of alarm about prospects for nuclear was reflected most
dramatically in Jonathan Schell's apocalyptic essay entitled The Fate of the
Earth, published in 1982. Schell questioned the entire notion of deterrence
but offered little in its place except the dream of world government and
international control of atomic energy. The Soviet Union and the United
States continued to negotiate arms reduction agreements and to stay within
the rough limits of SALT II, an agreement limiting launchers that was signed
in 1979 by President Carter and Premier Brezhnev but not ratified by the
United States Senate. Yet deterrence remained the cornerstone of both
American and Soviet nuclear policy. The concept of deterrence has under-
gone many changes since 1946, but no alternative short of nuclear war or
unilateral disarmament has emerged. It remains the military theology of our
times and, until a nuclear war occurs, will have worked.

The best survey of the theory of deterrence is Lawrence Freedman, The
Evolution of Nuclear Strategy (New York: St. Martin's Press, 1981). On the

history of the RAND Corporation and its military intellectuals, see Fred Kaplan, The Wizards of Armageddon *(New York: Simon and Schuster, 1983). A good brief view of deterrence as a necessary future is Michael Mandelbaum,* The Nuclear Future *(Ithaca, N.Y. and London: Cornell University Press, 1983). Jonathan Schell's* The Fate of the Earth *(New York: Alfred A. Knopf, 1982) is a radical critique of deterrence that should be read in conjunction with the more recent book by the Harvard Nuclear Study Group,* Living with Nuclear Weapons *(New York: Bantam Books, 1983), and, especially on the European perspective, Solly Zuckerman,* Nuclear Illusion and Reality *(New York: Vintage, 1983).*

Also critical of deterrence is Robert Jay Lifton and Richard Falk, Indefensible Weapons *(New York: Basic Books, 1982), and the more scholarly analyses by Louis Rene Beres,* Apocalypse: Nuclear Catastrophe in World Politics *(Chicago: University of Chicago Press, 1980); Stephen J. Cimbala,* Nuclear War and Nuclear Strategy: Unfinished Business *(Westport, Conn.: Greenwood Press, 1987); and Robert E. Osgood,* The Nuclear Dilemma in American Strategic Thought *(Boulder, Col.: Westview Press, 1988). Michael Vlahos in* Strategic Defense and the American Ethos: Can the Nuclear World Be Changed? *(Boulder, Col.: Westview Press, 1986) explores the hypothesis that the failure of the concept of deterrence had led us to accept the precepts of the Strategic Defense Initiative (SDI). An historical survey of the development of post post-World War II nuclear strategy is Gregg Herkin,* Counsels of War *(New York: Knopf, 1985). See also Beres, "Embracing Omnicide: President Reagan and the Strategic Mythmakers,"* The Hudson Review *(Spring 1983). The best speculation concerning what the failure of deterrence would mean is Arthur M. Katz,* Life After Nuclear War *(Cambridge, Mass.: Ballinger, 1982).*

---

45. Bernard Brodie on the Absolute Weapon, 1946

There is happily little disposition to believe that the atomic bomb by its mere existence and by the horror implicit in it "makes war impossible." In

This material appears in Bernard Brodie, ed., *The Absolute Weapon: Atomic Power and World Order* (New York: Harcourt, Brace, 1946), pp. 75–77, 83–85, 106–7. Copyright 1946 by Yale Institute of International Studies, renewed 1974 by Yale Concilium on International and Area Studies. Reprinted by permission of Harcourt Brace Jovanovich, Inc.

the sense that war is something not to be endured if any reasonable alternative remains, it has long been "impossible." But for that very reason we cannot hope that the bomb makes war impossible in the narrower sense of the word. Even without it the conditions of modern war should have been a sufficient deterrent but proved not to be such. If the atomic bomb can be used without fear of substantial retaliation in kind, it will clearly encourage aggression. So much the more reason, therefore, to take all possible steps to assure that multilateral possession of the bomb, should that prove inevitable, be attended by arrangements to make as nearly certain as possible that the aggressor who uses the bomb will have it used against him.

If such arrangements are made, the bomb cannot but prove in the net a powerful inhibition to aggression. It would make relatively little difference if one power had more bombs and were better prepared to resist them than its opponent. It would in any case undergo incalculable destruction of life and property. It is clear that there existed in the thirties a deeper and probably more generalized revulsion against war than in any other era of history. Under those circumstances the breeding of a new war required a situation combining dictators of singular responsibility with a notion among them and their general staffs that aggression would be both successful and cheap. The possibility of irresponsible or desperate men again becoming rulers of powerful states cannot under the prevailing system of international politics be ruled out in the future. But it does seem possible to erase the idea—if not among madmen rulers then at least among their military supporters—this aggression will be cheap.

Thus, the first and most vital step in any American security program for the age of atomic bombs is to take measures to guarantee to ourselves in case of attack the possibility of retaliation in kind. The writer in making that statement is not for the moment concerned about who will *win* the next war in which atomic bombs are used. Thus far the chief purpose of our military establishment has been to win wars. From now on its chief purpose must be to avert them. It can have almost no other useful purpose.

Neither is the writer especially concerned with whether the guarantee of retaliation is based on national or international power. However, one cannot be unmindful of one obvious fact: for the period immediately ahead, we must evolve our plans with the knowledge that there is a vast difference between what a nation can do domestically of its own volition and on its own initiative and what it can do with respect to programs which depend on achieving agreement with other nations. Naturally, our domestic policies concerning the atomic bomb and the national defense generally should not be such as to prejudice real opportunities for achieving world security agreements of a worth-while sort. That is an important provision and may become a markedly restraining one.

Some means of international protection for those states which cannot protect themselves will remain as necessary in the future as it has been in the past.* Upon the security of such states our own security must ultimately depend. But only a great state which has taken the necessary steps to reduce its own direct vulnerability to atomic bomb attack is in a position to offer the necessary support. Reducing vulnerability is at least one way of reducing temptation to potential aggressors. And if the technological realities make reduction of vulnerability largely synonymous with reservation of striking power, that is a fact which must be faced. Under those circumstances any domestic measures which effectively guaranteed such preservation of striking power under attack would contribute to a more solid basis for the operation of an international security system.

Outlines of a Defense in the Atomic Age

What are the criteria by which we can appraise realistic military thinking in the age of atomic bombs? The burden of the answer will depend primarily on whether one accepts as true the several postulates presented and argued in the previous chapter. One might go further and say that since none of them is obviously untrue, no program of military preparedness which fails to consider the likelihood of their being true can be regarded as comprehensive or even reasonably adequate.

It is of course always possible that the world may see another major war in which the atomic bomb is not used. The awful menace to both parties of a reciprocal use of the bomb may prevent the resort to that weapon by either side, even if it does not prevent the outbreak of hostilities. But even so, the shadow of the atomic bomb would so govern the strategic and tactical dispositions of either side as to create a wholly novel form of war. The kind of

---

*The argument has been made that once the middle or small powers have atomic bombs they will have restored to them the ability to resist effectively the aggressions of their great-power neighbors—an ability which otherwise has well-nigh disappeared. This is of course an interesting speculation on which no final answer is forthcoming. It is true that a small power, while admitting that it could not win a war against a great neighbor, could nevertheless threaten to use the bomb as a penalizing instrument if it were invaded. But it is also true that the great-power aggressor could make counterthreats concerning its conduct while occupying the country which had used atomic bombs against it. It seems to this writer highly unlikely that a small power would dare threaten use of the bomb against a great neighbor which was sure to overrun it quickly once hostilities began. It seems, on the contrary, much more likely that Denmark's course in the second World War will be widely emulated if there is a third. The aggressor will not "atomize" a city occupied by its own troops, and the opposing belligerent will hesitate to destroy by such an unselective weapon the cities of an occupied friendly state.

spatial concentrations of force by which in the past great decisions have been achieved would be considered too risky. The whole economy of war would be affected, for even if the governments were willing to assume responsibility for keeping the urban populations in their homes, the spontaneous exodus of those populations from the cities might reach such proportions as to make it difficult to service the machines of war. The conclusion is inescapable that war will be vastly different because of the atomic bomb whether or not the bomb is actually used.

But let us now consider the degree of probability inherent in each of the three main situations which might follow from a failure to prevent a major war. These three situations may be listed as follows:

(a)   a war fought without atomic bombs or other forms of radioactive energy;
(b)   a war in which atomic bombs were introduced only considerably after the outbreak of hostilities;
(c)   a war in which atomic bombs were used at or near the very outset of hostilities.

We are assuming that this hypothetical conflict occurs at a time when each of the opposing sides possesses at least the "know-how" of bomb production, a situation which, as argued in the previous chapter, approximates the realities to be expected not more than five to ten years hence.

Under such conditions the situation described under (a) above could obtain only as a result of a mutual fear of retaliation, perhaps supported by international instruments outlawing the bomb as a weapon of war. It would *not* be likely to result from the operation of an international system for the suppression of bomb production, since such a system would almost certainly not survive the outbreak of a major war. If such a system were in fact effective at the opening of hostilities, the situation resulting would be far more likely to fall under (b) than under (a), unless the war were very short. For the race to get the bomb would not be an even one, and the side which got it first in quantity would be under enormous temptation to use it before the opponent had it. Of course, it is more reasonable to assume that an international situation which had so far deteriorated as to permit the outbreak of a major war would have long since seen the collapse of whatever arrangements for bomb production control had previously been imposed, unless the conflict were indeed precipitated by an exercise of sanctions for the violation of such a control system.

Thus we see that a war in which atomic bombs are not used is more likely to occur if both sides have the bombs in quantity from the beginning than if neither side has it at the outset or if only one side has it.* But how likely is it to occur? Since the prime motive in refraining from using it would be fear of retaliation, it is difficult to see why a fear of reciprocal use should be strong enough to prevent resort to the bomb without being strong enough to prevent the outbreak of war in the first place.

Of course, the bomb may act as a powerful deterrent to direct aggression against great powers without preventing the political crises out of which wars generally develop. In a world in which great wars become "inevitable" as a result of aggression by great powers upon weak neighbors, the bomb may easily have the contrary effect. Hitler made a good many bloodless gains by mere blackmail, in which he relied heavily on the too obvious horror of modern war among the great nations which might have opposed him earlier. A comparable kind of blackmail in the future may actually find its encouragement in the existence of the atomic bomb. Horror of its implications is not likely to be spared evenly, at least not in the form of overt expression. The result may be a series of *faits accomplis* eventuating in that final deterioration of international affairs in which war, however terrible, can no longer be avoided.

Those who have been predicting attacks of 15,000 atomic bombs and upward will no doubt look with jaundiced eye upon these speculations. For they will say that a country so struck will not merely be overwhelmed but for all practical purposes will vanish. Those areas not directly struck will be covered with clouds of radioactive dust under which all living beings will perish.

No doubt there is a possibility that initial attack can be so overwhelming as to void all opportunity of resistance or retaliation, regardless of the precautions taken in the target state. Not *all* eventualities can be provided against. But preparation to launch such an attack would have to be on so gigantic a scale as to eliminate all chances of surprise. Moreover, while there is perhaps little solace in the thought that the lethal effect of radioactivity is generally considerably delayed, the idea will not be lost on the aggressor. The more horrible the results of attack, the more he will be deterred by even a marginal chance of retaliation.

Finally, one can scarcely assume that the world will remain either long ignorant of or acquiescent in the accumulations of such vast stockpiles of atomic bombs. If existing international organization should prove inadequate

---

*One can almost rule out too the possibility that war would break out between two great powers where both knew that only one of them had the bombs in quantity. It is one of the old maxims of power politics that *c'est une crime de faire la guerre sans compter sur la superiorité*, and certainly a monopoly of atomic bombs would be a sufficiently clear definition of superiority to dissuade the other side from accepting the gage of war unless directly attacked.

to cope with the problem of controlling bomb production—and it would be premature to predict that it will prove inadequate, especially in view of the favorable official and public reception accorded the Board of Consultants' report of March 16, 1946—a runaway competition in such production would certainly bring new force into the picture. In this chapter and in the preceding one, the writer has been under no illusions concerning the adequacy of a purely military solution.

Concern with the efficiency of the national defenses is obviously inadequate in itself as an approach to the problem of the atomic bomb. In so far as such concern prevails over the more fundamental considerations of eliminating war or at least of reducing the chance of its recurrence, it clearly defeats its purpose. That has perhaps always been true, but it is a truth which is less escapable today than ever before. Nations can still save themselves by their own armed strength from subjugation, but not from a destruction so colossal as to involve complete ruin. Nevertheless, it also remains true that a nation which is as well girded for its own defense as is reasonably possible is not a tempting target to an aggressor. Such a nation is therefore better able to pursue actively that progressive improvement in world affairs by which alone it finds its true security.

---

46. John Foster Dulles on Massive Retaliation, 1954

The Need for Long-Range Policies

This "long time" factor is of critical importance.

The Soviet Communists are planning for what they call "an entire historical era," and we should do the same. They seek, through many types of maneuvers, gradually to divide and weaken the free nations by overextending them in efforts which, as Lenin put it, are "beyond their strength, so that they come to practical bankruptcy." Then, said Lenin, "our victory is assured." Then, said Stalin, will be "the moment for the decisive blow."

In the face of this strategy, measures cannot be judged adequate merely

This material appears in John Foster Dulles, "The Evolution of Foreign Policy," *Department of State Bulletin* 30 (January 25, 1962): 107–10.

because they ward off an immediate danger. It is essential to do this, but it is also essential to do so without exhausting ourselves.

When the Eisenhower administration applied this test, we felt that some transformations were needed.

It is not sound military strategy permanently to commit U.S. land forces to Asia to a degree that leaves us no strategic reserves.

It is not sound economics, or good foreign policy, to support permanently other countries; for in the long run, that creates as much ill will as good will.

Also, it is not sound to become permanently committed to military expenditures so vast that they lead to "practical bankruptcy."

Change was imperative to assure the stamina needed for permanent security. But it was equally imperative that change should be accompanied by understanding of our true purposes. Sudden and spectacular change had to be avoided. Otherwise, there might have been a panic among our friends and miscalculated aggression by our enemies. We can, I believe, make a good report in these respects.

We need allies and collective security. Our purpose is to make these relations more effective, less costly. This can be done by placing more reliance on deterrent power and less dependence on local defensive power.

This is accepted practice so far as local communities are concerned. We keep locks on our doors, but we do not have an armed guard in every home. We rely principally on a community security system so well equipped to punish any who break in and steal that, in fact, would-be aggressors are generally deterred. That is the modern way of getting maximum protection at a bearable cost.

What the Eisenhower administration seeks is a similar international security system. We want, for ourselves and the other free nations, a maximum deterrent at a bearable cost.

Local defense will always be important. But there is no local defense which alone will contain the mighty landpower of the Communist world. Local defenses must be reinforced by the further deterrent of massive retaliatory power. A potential aggressor must know that he cannot always prescribe battle conditions that suit him. Otherwise, for example, a potential aggressor, who is glutted with manpower, might be tempted to attack in confidence that resistance would be confined to manpower. He might be tempted to attack in places where his superiority was decisive.

The way to deter aggression is for the free community to be willing and able to respond vigorously at places and with means of its own choosing.

47. Herman Kahn on Thermonuclear War, 1960

Some decision makers who accept the Finite Deterrence view are willing to pay for insurance against unreliability for more than political or psychological reasons. Even those who hold that war means mutual annihilation are sometimes willing for us to act beyond their beliefs—or fears. While this is inconsistent, it is not necessarily irrational. They understand that paper calculations can be wrong and are willing to hedge against this possibility. Sometimes these decision makers are making a distinction that (rather surprisingly) is not usually made. They may distinguish, for example, between 100 million dead and 50 million dead, and argue that the latter state is better than the former. They may distinguish between war damage which sets the economy of a country back fifty years or only ten years. *Actually when one examines the possible effects of thermonuclear war carefully, one notices that there are indeed many postwar states that should be distinguished.* If most people do not or cannot distinguish among these states it is because the gradations occur as a result of a totally bizarre circumstance—a thermonuclear war. The mind recoils from thinking hard about that; one prefers to believe it will never happen. If asked, "How does a country look on the day of the war?" the only answer a reasonable person can give is "awful." It takes an act of iron will or an unpleasant degree of detachment or callousness to go about the task of distinguishing among the possible degrees of awfulness.

But surely one can ask a more specific question. For example, "How does a country look five or ten years after the close of war, as a function of three variables: (1) the preparations made before the war, (2) the way the war started, and (3) the course of military events?" Both very sensitive and very callous individuals should be able to distinguish (and choose, perhaps) between a country which survives a war with, say 150 million people and a gross national product (GNP) of $300 billion a year, and a nation which emerges with only 50 million people and a GNP of $10 billion. The former would be the richest and the fourth largest nation in the world, and one which would be able to restore a reasonable facsimile of the prewar society; the latter would be a pitiful remnant that would contain few traces of the prewar way of life. When one asks this kind of question and examines the circumstances and possible outcomes of a future war in some detail, it appears

This material appears in Herman Kahn, *On Thermonuclear War* (Princeton, N.J.: Princeton University Press, 1961), pp. 18–23. Copyright © 1960 by Princeton University Press. Reprinted by permission of Princeton University Press.

that it is useful and necessary to make many distinctions among the results of thermonuclear war. The figures in Table 3 illustrate some simple distinctions which one may wish to make at the outset of his deliberations in this field.

Here I have tried to make the point that if we have a posture which might result in 40 million dead in a general war, and as a result of poor planning, apathy, or other causes, our posture deteriorates and a war occurs with 80 million dead, we have suffered an additional disaster, and *unnecessary* additional disaster that is almost as bad as the original disaster. If on the contrary, by spending a few billion dollars, or by being more competent or lucky, we can cut the number of dead from 40 million to 20 million, we have done something vastly worth doing! The survivors will not dance in the streets or congratulate each other if there have been 20 million men, women, and children killed; yet it would have been a worthwhile achievement to limit casualties to this number. It is very difficult to get this point across to laymen or experts with enough intensity to move them to action. The average citizen has a dour attitude toward planners who say that if we do thus and so it will not be 40 million dead—it will be 20 million dead. Somehow the impression is left that the planner said that there will be *only* 20 million dead. To him is often attributed the idea that this will be a tolerable or even astonishingly enough, a desirable state!

The rate of economic recuperation, like the number of lives saved, is also of extreme importance. Very few Americans can get interested in spending

Table 3[1]
Tragic but Distinguishable Postwar States

Dead	Economic Recuperation
2,000,000	1 year
5,000,000	2 years
10,000,000	5 years
20,000,000	10 years
40,000,000	20 years
80,000,000	50 years
160,000,000	100 years

Will the survivors envy the dead?

1. Tables 1, 2 not reproduced.

money or energy on preparations which, even if they worked, would result in preindustrial living standards for the survivors of a war. As will be explained later, our analysis indicates that if a country is moderately well prepared to use the assets which survive there is unlikely to be a critical level of damage to production. A properly prepared country is not "killed" by the destruction of even a major fraction of its wealth; it is more likely to be set back a given number of years in its economic growth. While recuperation times may range all the way from one to a hundred years, even the latter is far different from the "end of history."

Perhaps the most important item on the table of distinguishable states is not the numbers of dead or the number of years it takes for economic recuperation; rather, it is the question at the bottom: "Will the survivors envy the dead?" It is in some sense true that one may never recuperate from a thermonuclear war. The world may be permanently (i.e., for perhaps 10,000 years) more hostile to human life as a result of such a war. Therefore, if the question, "Can we restore the prewar conditions of life?" is asked, the answer must be "No!" But there are other relevant questions to be asked. For example: "How much more hostile will the environment be? Will it be so hostile that we or our descendants would prefer being dead than alive?" Perhaps even more pertinent is this question, "How happy or normal a life can the survivors and their descendants hope to have?" *Despite a widespread belief to the contrary, objective studies indicate that even though the amount of human tragedy would be greatly increased in the postwar world, the increase would not preclude normal and happy lives for the majority of survivors and their descendants.*

My colleagues and I came to this conclusion reluctantly; not because we did not *want* to believe it, but because it is so *hard* to believe. Thermonuclear bombs are so destructive, and destructive in so many ways, that it is difficult to imagine that there would be anything left after their large-scale use. One of my tasks with The RAND Corporation was to serve as a project leader for a study of the possibilities for alleviating the consequences of a thermonuclear war. That study was made as quantitatively and objectively as we could make it with the resources, information, and intellectual tools available to us. *We concluded that for at least the next decade or so, any picture of total world annihilation appears to be wrong, irrespective of the military course of events.* * Equally important, the picture of total disaster is likely to be wrong even for the two antagonists. Barring an extraordinary course for the war, or that most of the technical uncertainties turn out to lie

*Report on a Study of Non-Military Defense, the RAND Corporation, Report R-322-RC, July 1, 1958.

at the disastrous end of the spectrum, one and maybe both of the antagonists should be able to restore a reasonable semblance of prewar conditions quite rapidly. Typical estimates run between one and ten years for a reasonably successful and well-prepared attacker and somewhat longer for the defender, depending mainly on the tactics of the attacker and the preparations of the defender. In the RAND study we tried to avoid using optimistic assumptions. With the exceptions to be noted, we used what were in our judgment the best values available, or we used slightly pessimistic ones. We believe that the situation is likely to be better than we indicate, rather than worse, though the latter possibility cannot be ruled out.

Exactly what is it that one must believe if he is to be convinced that it is worthwhile to buy Counterforce as Insurance? Listed below are eight phases of a thermonuclear war. If our decision makers are to justify the expense (and possible risk of strategic destabilization) that would be incurred in trying to acquire a capability for alleviating the consequences of a war, they must believe they can successfully negotiate each and every one of these phases, or that there is a reasonable chance that they can negotiate each of these phases.

I repeat: To survive a war it is necessary to negotiate *all eight* stages. If there is a catastrophic failure in any one of them, there will be little value in being able to cope with the other seven. Differences among exponents of the different strategic views can often be traced to the different estimates they make on the difficulty of negotiating one or more of these eight stages. While all of them present difficulties, most civilian military experts seem to

Table 4
A Complete Description of a Thermonuclear War

Includes the Analysis of:

1. Various time-phased programs for deterrence and defense and their possible impact on us, our allies, and others.
2. Wartime performance with different preattack and attack conditions.
3. Acute fallout problems.
4. Survival and patch-up.
5. Maintenance of economic momentum.
6. Long-term recuperation.
7. Postwar medical problems.
8. Genetic problems.

consider the *last six* the critical ones. Nevertheless, most discussions among "classical" military experts concentrate on the *first two*. To get a sober and balanced view of the problem, one must examine all *eight*.

48. Robert McNamara and Counterforce "No Cities" Doctrine, 1962

A central military issue facing NATO today is the role of nuclear strategy. Four facts seem to us to dominate consideration of that role. All of them point in the direction of increased integration to achieve our common defense. First, the alliance has overall nuclear strength adequate to any challenge confronting it. Second, this strength not only minimizes the likelihood of major nuclear war but makes possible a strategy designed to preserve the fabric of our societies if war should occur. Third, damage to the civil societies of the alliance resulting from nuclear warfare could be very grave. Fourth, improved non-nuclear forces, well within alliance resources, could enhance deterrence of any aggressive moves short of direct, all-out attack on Western Europe.

Let us look at the situation today. First, given the current balance of nuclear power, which we confidently expect to maintain in the years ahead, a surprise nuclear attack is simply not a rational act for any enemy. Nor would it be rational for an enemy to take the initiative in the use of nuclear weapons as an outgrowth of a limited engagement in Europe or elsewhere. I think we are entitled to conclude that either of these actions has been made highly unlikely.

Second, and equally important, the mere fact that no nation could rationally take steps leading to a nuclear war does not guarantee that a nuclear war cannot take place. Not only do nations sometimes act in ways that are hard to explain on a rational basis, but even when acting in a "rational" way they sometimes, indeed disturbingly often, act on the basis of misunderstandings of the true facts of a situation. They misjudge the way others will react and the way others will interpret what they are doing.

This material appears in Robert McNamara, "Defense Arrangements of the North Atlantic Community," *Department of State Bulletin* 47 (July 9, 1962): 64–70.

We must hope—indeed I think we have good reason to hope—that all sides will understand this danger and will refrain from steps that even raise the possibility of such a mutually disastrous misunderstanding. We have taken unilateral steps to reduce the likelihood of such an occurrence. We look forward to the prospect that through arms control the actual use of these terrible weapons may be completely avoided. It is a problem not just for us in the West but for all nations that are involved in this struggle we call the cold war.

For our part we feel we and our NATO allies must frame our strategy with this terrible contingency, however remote, in mind. Simply ignoring the problem is not going to make it go away.

The United States has come to the conclusion that, to the extent feasible, basic military strategy in a possible general nuclear war should be approached in much the same way that more conventional military operations have been regarded in the past. That is to say, principal military objectives, in the event of a nuclear war stemming from a major attack on the alliance, should be the destruction of the enemy's military forces, not of his civilian population.

The very strength and nature of the alliance forces make it possible for us to retain, even in the face of a massive surprise attack, sufficient reserve striking power to destroy an enemy society if driven to it. In other words, we are giving a possible opponent the strongest imaginable incentive to refrain from striking our own cities.

The strength that makes these contributions to deterrence and to the hope of deterring attack upon civil societies even in wartime does not come cheap. We are confident that our current nuclear programs are adequate and will continue to be adequate for as far into the future as we can reasonably foresee. During the coming fiscal year the United States plans to spend close to $15 billion on its nuclear weapons to assure their adequacy. For what this money buys, there is no substitute.

In particular, relatively weak national nuclear forces with enemy cities as their targets are not likely to be sufficient to perform even the function of deterrence. If they are small, and perhaps vulnerable on the ground or in the air, or inaccurate, a major antagonist can take a variety of measures to counter them. Indeed, if a major antagonist came to believe there was a substantial likelihood of its being used independently, this force would be inviting a preemptive first strike against it. In the event of war, the use of such a force against the cities of a major nuclear power would be tantamount to suicide, whereas its employment against significant military targets would have a negligible effect on the outcome of the conflict. Meanwhile the

creation of a single additional national nuclear force encourages the proliferation of nuclear power with all of its attendant dangers.

In short, then, limited nuclear capabilities, operating independently, are
dangerous, expensive, prone to obsolescence, and lacking in credibility as a
deterrent. Clearly, the United States nuclear contribution to the alliance is
neither obsolete nor dispensable.

At the same time, the general strategy I have summarized magnifies the
importance of unity of planning, concentration of executive authority, and
central direction. There must not be competing and conflicting strategies to
meet the contingency of nuclear war. We are convinced that a general nuclear war target system is indivisible and if, despite all our efforts, nuclear
war should occur, our best hope lies in conducting a centrally controlled
campaign against all of the enemy's vital nuclear capabilities, while retaining
reserve forces, all centrally controlled.

We know that the same forces which are targeted on ourselves are also
targeted on our allies. Our own strategic retaliatory forces are prepared to
respond against these forces, wherever they are and whatever their targets.
This mission is assigned not only in fulfillment of our treaty commitments
but also because the character of nuclear war compels it. More specifically,
the United States is as much concerned with that portion of Soviet nuclear
striking power that can reach Western Europe as with that portion that also
can reach the United States. In short, we have undertaken the nuclear defense of NATO on a global basis. This will continue to be our objective. In
the execution of this mission, the weapons in the European theater are only
one resource among many.

There is, for example, the Polaris force, which we have been substantially increasing and which, because of its specially invulnerable nature, is
peculiarly well suited to serve as a strategic reserve force. We have already
announced the commitment of five of these ships, fully operational to the
NATO Command.

This sort of commitment has a corollary for the alliance as a whole. We
want and need a greater degree of alliance participation in formulating nuclear weapons policy to the greatest extent possible. We would all find it
intolerable to contemplate having only a part of the strategic force launched
in isolation from our main striking power.

We shall continue to maintain powerful nuclear forces for the alliance as
a whole. As the President has said, "Only through such strength can we be
certain of deterring a nuclear strike, or an overwhelming ground attack,
upon our forces and allies."

But let us be quite clear about what we are saying and what we would have to face if the deterrent should fail. This is the almost certain prospect that, despite our nuclear strength, all of us would suffer deeply in the event of major nuclear war.

We accept our share of this responsibility within the alliance. And we believe that the combination of our nuclear strength and a strategy of controlled response gives us some hope of minimizing damage in the event that we have to fulfill our pledge. But I must point out that we do not regard this as a desirable prospect, nor do we believe that the alliance should depend solely on our nuclear power to deter actions not involving a massive commitment of any hostile force. Surely an alliance with the wealth, talent, and experience that we possess can find a better way than extreme reliance on nuclear weapons to meet our common threat. We do not believe that if the formula $E = MC^2$ had not been discovered, we should all be Communist slaves. On this question I can see no valid reason for a fundamental difference of view on the two sides of the Atlantic.

---

49. L. Hagen, Comments on Presidential Directive 59, 1980

---

1. The so-called Countervailing Strategy, embodied as American policy on Presidential Directive 59, and signed by President Carter on 25 July 1980, has been portrayed as a significant shift in US nuclear policy away from Mutual Assured Destruction (MAD) towards a doctrine based on the capability and intent to fight—and win—a limited nuclear war fought through the medium of counterforce/point-target exchanges. In this sense, critics have pointed to a return to the "unstable" years of the McNamara Doctrine of 1962, with its implications of arms race and deterrence instability.

2. It is clear, however, that PD-59 does not in itself represent a radical shift in American targetting policy. It is, rather, the result of several years of study within the Pentagon and the National Security Council. Indeed, it is a natural offspring of studies done, and concerns evidenced, from the beginning of the first Nixon Administration. This pre-occupation with the rigidity

This material appears in L. Hagen, "PD-59 and the Countervailing Strategy: Continuity or Change?" Department of National Defence, Canada; Project Report No. PR 170 (Ottawa, October 1981).

and inadequacy of American strategic policy was expressed as early as February 1970 by President Nixon in his "State of the World" message to Congress:

Should a President, in the event of a nuclear attack, be left with the single option of ordering the mass destruction of enemy civilians in the face of the certainty that it would be followed by the mass slaughter of Americans?

3. Studies underway at that time led to the signing of National Security Decision Memorandum 242 by President Nixon on 17 January 1974. This was supplemented by the promulgation of the *Policy Guidance for the Employment of Nuclear Weapons*, signed by Secretary of Defense Schlesinger on 4 April 1974, which in turn led to a new Single Integrated Operational Plan (SIOP-5) which took effect on 1 January 1976. The general purpose behind these revisions is indicated in the *Defense Department Annual Report* for FY 1975:

What we need is a series of measured responses to aggression which bear some relationship to provocation, have prospect of terminating hostilities before general nuclear war breaks out, and leave some possibility of restoring deterrence.

4. Following the Nuclear Targetting Planning Review of 1977–79, PD-59 continues these themes, arguing for greater targetting of military assets (soft and hard), war-fighting (as opposed to recovery) industry assets, and political and C^3 centres.[1] In this sense, there is a new emphasis on the first three general target sets of the SIOP—Soviet Nuclear forces, conventional forces, military and political leadership centres at the expense, *but not to the exclusion*, of the fourth target set—the Soviet industrial and economic base. It must be recalled that fully 50 percent of the 40,000 targets in the SIOP remain dedicated to non-nuclear force targets.

5. There is little doubt, moreover, that targetting plans prior to the Nixon Administration contained plans for the targetting of Soviet military installations—although on a relatively massive and undifferentiated scale. Indeed, explicit American policy in the early McNamara period emphasized nuclear bargaining through the medium of nuclear exchanges directed at opposing military forces. This policy was de-emphasized, and that of MAD became the strategic cornerstone of American policy, when it became clear that the growth in Soviet capability, and the costs—fiscal and political— involved in maintaining a capacity for nuclear victory were unsupportable.

1. C^3 is military terminology for command, control, and communication: three areas to target in war.

Soviet military targets remained in the SIOP, however, with a capacity for sole attack of these within the confines of existing technology.

6. Thus, while the Countervailing Strategy *may* involve the refinement and multiplication of limited nuclear options plans using extant computer capabilities and multiple targetting memories, there is nothing *inherent* in the policy which is radically at variance with prior evolutions in American doctrine. The *tone* and *content* of the announcement may, however, have significant impact.

7. PD-59 must also be seen as part of the response of the Western Alliance to growing concern for the credibility and flexibility of the American nuclear guarantee in Central Europe. In this connection, the philosophy behind PD-59 is similar to that embodied in NATO LRTNF (Long Range Theatre Nuclear Forces) decisions and refinements in the tactical nuclear weapons area, designed to strengthen the fabric of deterrence by increasing the number and strength of ladders in the escalation process.

8. In this sense, there is general support for such a strategy within Europe, which is to be contrasted with the antagonism which greeted efforts during the McNamara period to institute somewhat similar policies at the nuclear level. This reflects increased concern in Western Europe over perceived Soviet advances at the intercontinental and Eurostrategic levels, which has partially obfuscated traditional European fears that war may be waged in Europe while preserving Soviet and American homelands.

9. The Countervailing Strategy is a reflection of the following tenets of deterrence theory:

(i) Deterrence is best preserved by some measure of proportionality in response which maintains the credibility of the threatened use of nuclear forces by meeting each threat with roughly the same level of response. It is necessary, therefore, to develop a capability and policy which does not leave significant gaps in strategic forces or policies for their use, which can then be politically or militarily exploited by the enemy.

(ii) Plans to use nuclear weapons in a limited fashion not only increase pre-war deterrence, but indicate a willingness and capability to respond at less than all-out levels, thereby enhancing intra-war deterrence through denying the enemy his objectives, as opposed to merely punishing him through massive societal destruction. The escalation process is thereby seen to be more rational and controllable, with the extension of some sort of strategic bargaining possible into nuclear war.

(iii) A limited nuclear war policy enhances the credibility of *extended* deterrence in Europe and elsewhere by presenting the possibility of limited use of nuclear weapons which does not imply total destruction of the United States, and which will inflict damage against opposing military forces which will physically prevent the attainment of his military aims. In this sense, PD-59 is directed towards doctrinally minimizing the self-deterrence problems which might obtain in the event of a disaster in the theatre.

(iv)   Plans to use nuclear weapons in a limited and controlled manner do not increase the possibility of, or incentives to engage in, a nuclear war, but help prevent conflict through increased credibility, as outlined above, present the possibility—although not certainty—of military and political dialogue once conflict occurs, and by examining the question of limited use options, help ensure more responsible behaviour in a crisis situation than if no contingencies were available or considered in this area.

10. Although elements of PD-59 have characterized major concerns with, and modifications to, American nuclear doctrine since at least 1970, a series of recent developments in the strategic environment have provoked greater interest in the development of limited war strategies:

(i)   improvements in Soviet conventional and eurostrategic capabilities (such as the SS-20 and *Backfire* bomber) which simultaneously pose the perceived danger of a capability for military and political pressure in the area and heighten anxiety over the credibility of the American nuclear guarantee;

(ii)   peripheral conflict involving the Soviet Union, particularly in the Persian Gulf/South-West Asia region;

(iii)   growing Soviet strategic capabilities which, through a combination of numbers, accuracy, yield, and throw-weight, indicate at least a theoretical capability to destroy virtually the entire American land-based missile force by 1982–83. Particular developments of concern have been:

a)   the SS17/18/19, which are, in their warhead size and accuracy characteristics, ideal silo-busting weapons;

b)   the testing of SLBM forces in depressed-trajectory modes to reduce warning time, which can be perceived only as counterforce in their intent (primarily against SAC bases); and

c)   the capacity for cold launch—and hence re-load—on the SS18/19 systems.

11. These developments, never satisfactorily explained by the Soviet Union during the SALT process, and which generally occurred much sooner

than projected, have increased American doubts about the acceptance by the USSR of the Assured Destruction model of deterrence advocated by the United States, and led to a general tendency to take prudent measures to counter the possible political or military use of Soviet strategic forces, and to, in effect, the adoption by American of a war-fightable posture as a doctrinal response.

12. PD-59, while largely a response to these developments, may also have been intended to variously:

(i)   counter Republican criticism of Carter Administration defense policy;
(ii) serve as a signal to the Soviet Union that the US is willing doctrinally to meet Soviet capabilities on their own terms;
(iii) act as a general signal in the post-Afghanistan environment;
(iv) pave the way for significant upgrading of American limited nuclear option capabilities.

13. Technologies which could receive a push from PD-59, and which would be required for any radical improvement in American capabilities to conduct limited nuclear war include:

(i)   the Mk 12A re-entry vehicles, whose deployment on Minuteman III and possibly the M-X and/or Trident D-5 missiles would be necessary to pose a significant counter-silo threat against Soviet SS18/19 ICBMs;
(ii) the Trident D-5 SLBM which could be used against hardened command and control bunkers and Soviet ICBMs such as the SS17;
(iii) M-X development in sufficient numbers to ensure survivability (PD-59 means very little without M-X deployment);
(iv) a manned penetrating bomber which could be used for follow-up strikes against point targets;
(v)   improved US command and control facilities and procedures along the lines of the Presidential Directives 53 and 58 issued in this area;
(vi) developments in ballistic missile defense; and
(vii) increased production facilities for strategic nuclear materials and such special materials as tritium required for warhead production, which are currently in short supply; significant increases in production will be required for new weapons.

14. Only when action is taken on these various weapons programs will the real impact—intended or otherwise—of PD-59 be realized. However, it should be noted that the timing and tone of the announcement, coming during a hiatus in the detente process, and in the middle of the US

Presidential election campaign, may have served to generate some undesirable side-effects. These, when coupled with certain observations on the technological level, indicate some potential problems with the Countervailing Strategy.

15. The United States does not currently have a significant capacity for major counterforce strikes against the Soviet Union. Depending on the assumptions made, and targetting scenarios adopted, a launch by the US of its entire ICBM force against Soviet ICBMs would leave 60–72% of Soviet silos and 48–57% of RVs intact.

16. If the Soviet Union ignored SALT II limits on its ICBM force and deployed new launchers at maximum current rates, and the US deployed Mk 12A warheads (in maximum numbers) on the M-X and Minuteman III forces (assuming 200 M-X and 550 Minuteman III ICBMs), a total American strike against Soviet silos in 1995 would still leave about 4200 warheads (assuming 10 RVs per Soviet ICBM). This would represent a massive remaining capability.

17. Deployment of the Mk 12A warhead on M-X and Minuteman III ICBM (even in limited numbers: i.e., 3 warheads per 1 Minuteman III instead of its maximum capacity [7]) will allow the United States to credibly threaten remaining Soviet ICBMs following a first-strike against American ICBM forces, assuming SALT II limits. It would appear, then, that PD-59 is meant to provide doctrinal justification both for the development by the United States of a *Counter-third strike retaliatory counterforce capability* and of a first-strike *non-disarming* counterforce strike which could preserve escalation dominance and not, thereby, leave the Soviet Union in a favourable position vis à vis remaining US ICBM assets. No first-strike counterforce capability is likely to exist for the United States under any foreseeable scenario.

18. The combination of a credible Soviet counterforce capability against US ICBM fields, with a US strategic posture based on PD-59, may encourage Launch On Warning/Laungh Through Attack postures in the Soviet Union. This would be highly destabilizing in crisis situations. If, however, force deployments coupled with doctrinal adjustments are made in the US, LOW/LTA incentives may be reduced on the American side.

19. It has been mentioned in public statements concerning PD-59 that Soviet command and control facilities will be targeted in the future. It is, however, difficult to see how a controlled nuclear exchange can take place if the capacity for that control is eliminated. There are, moreover, two additional problems on the Soviet side. First, political C^3 facilities are hardened and dispersed such that their location is either unknown, or such that there are so many plausible locations for actual leadership location that there

would not be enough warheads to target them all. Second, such facilities as are known are co-located with major population centres; this would obviously hamper attempts at limiting the conflict.

20. There are equally serious problems with American C^3. Several points deserve emphasis here. Unless American C^3 assets are improved dramatically, the capacity for prolonged nuclear exchange is likely to be illusory. Even if improved, it is questionable whether C^3 can in any case be hardened, dispersed, or proliferated, so as to allow for the sort of intra-war control and prolonged use which more enthusiastic proponents of nuclear flexibility envisage. This relates not so much to C^3 at the systems level but at the National Command Authority-system interface. (In any case, the sort of C^3 improvements called for by a robust PD-59 will not be available until approximately 1989, if then.) The vulnerability of satellites, the "hot line," FBM SUBMARINE C^3, the NCA and airborne C^3 systems are such that it is difficult to imagine adequate defenses against electromagnetic pulse (EMP), jamming, and deception. This is not to question the need for better C^3; it is, rather, to emphasize the likely limitations which C^3 is likely to put on more ambitious strategies of flexibility and nuclear bargaining.

21. Significant casualties may be expected from any American attack against Soviet economic or military assets. For example, recent American studies demonstrate that a very limited strike against 10 Soviet oil refinery and storage centres will generate approximately 1.5 million fatalities. Many ICBM fields are co-located with large Soviet industrial conurbations and, depending on the nature of the attack and prevailing wind conditions, casualties would undoubtedly be high. Such damage levels, combined with inevitable difficulties in attack assessment might make *identification* of limited nuclear strikes difficult—perhaps provoking an all-out Soviet response—and provoke serious doubts as to how "limited" strikes of this sort would be perceived, *even assuming* accurate attack assessment.

By merely identifying the need for sufficient forces and flexibility for limited war prosecution, PD-59 does not indicate an upper limit for such efforts. As during the early McNamara years, this may be perceived as providing service rationale for increased defense spending beyond that intended, and may, under a Republican Administration, allow the movement from a counter-third strike capability towards a first-strike threat. This may be facilitated by the identification of the strategy with a previous administration; it could not then be identified as a Republican ploy.

23. An intended or unintended by-product of PD-59 may be to provide an opportunity to gauge Soviet strategic intentions. US officials are likely to watch closely to see if Soviet responses to the new strategy and associated force improvements are in the direction of:

(i)   going mobile (stabilizing if verifiable);

(ii) going to sea (stabilizing);

(iii) deploying a new ICBM generation with increased vigour (destabilizing);

(iv) ignoring SALT II limits and building more launchers with more warheads (destablizing); or

(v) returning to the SALT table with an increased eagerness to negotiate limitations or reductions (stabilizing).

24. Therefore, the impact of PD-59 on SALT and deterrence stability is unclear at this point. If Soviet responses remain temperate, and US force improvements observe the limits established by current programs, there is unlikely to be much effect. It *is* likely, in the absence of Soviet willingness to radically decrease its ICBM capability, to increase American reluctance to curb its own development programs. It may, however, also help to demonstrate to the USSR that doctrine and capability to the American side is such as to *mathematically* indicate the futility of increased ICBM deployments, which may in turn ease the negotiating process. And, as a response to the Soviet threat to US ICBMs, prudent force deployments in conjunction with the doctrinal emphasis of PD-59 may act to stabilize deterrence by enhancing Alliance cohesion and demonstrating the will and capacity to deny the USSR any first-strike political or military advantage. This virtue, however, depends for its realization on the sober recognition of some of the limitations and doubts which PD-59 evinces in the absence of concrete force planning decisions.

50. Ronald Reagan on Deterrence, November 23, 1982

What do we mean when we speak of nuclear deterrence? Certainly we do not want such weapons for their own sake. We do not desire excessive forces, or what some people have called overkill. Basically, it is a matter of

This essay appears in the *New York Times*, November 23, 1982, p. 4. Copyright © 1982 by the New York Times Company. Reprinted by permission.

others' knowing that starting a conflict would be more costly to them than anything they might hope to gain. And, yes, it is sadly ironic that in these modern times it still takes weapons to prevent war. I wish it did not.

We desire peace, but peace is a goal, not a policy. Lasting peace is what we hope for at the end of our journey; it does not describe the steps we must take, nor the paths we should follow to reach that goal. I intend to search for peace along two parallel paths—deterrence and arms reduction. I believe these are the only paths that offer any real hope for an enduring peace.

And let me say I believe that if we follow prudent policies, the risk of nuclear conflict will be reduced. Certainly the United States will never use its forces except in response to attack. Through the years, Soviet leaders have also expressed a sober view of nuclear war; and if we maintain a strong deterrent, they are exceedingly unlikely to launch an attack.

Now, while the policy of deterrence has stood the test of time, the things we must do in order to maintain deterrence have changed.

You often hear that the United States and the Soviet Union are in an arms race. The truth is that while the Soviet Union has raced, we have not. In constant dollars our defense spending in the 1960's went up because of Vietnam and then it went downward through much of the 1970's. Soviet spending has gone up and up and up. In spite of a stagnating Soviet economy, Soviet leaders invest 12 to 14 percent of their country's gross national product in military spending, two to three times the level we invest.

I might add that the defense share of our United States Federal budget has gone way down, too. In 1962, when John Kennedy was President, 46 percent, almost half of the Federal budget, went to our national defense. In recent years, about one-quarter of our budget has gone to defense, while the share for social programs has nearly doubled. And most of our defense budget is spent on people, not weapons.

The combination of the Soviets' spending more and the U.S. spending proportionately less changed the military balance and weakened our deterrent. Today, in virtually every measure of military power the Soviet Union enjoys a decided advantage.

This chart (pointing to chart, "Strategic Missiles and Bombers") shows the changes in the total number of intercontinental missiles and bombers. You will see that in 1962 and in 1972, the United States forces remained about the same, even dropping some by 1982. But take a look now at the Soviet side. In 1962, at the time of the Cuban missile crisis, the Soviets could not compare with us in terms of strength. In 1972, when we signed the SALT I Treaty, we were nearly equal. But in 1982, well, that red Soviet bar stretching above the blue American bar tells the story.

I could show you chart after chart where there is a great deal of red and a much lesser amount of U.S. blue. For example, the Soviet Union has deployed a third more land-based intercontinental ballistic missiles than we have. Believe it or not, we froze our number in 1965 and have deployed no additional missiles since then.

The Soviet Union put to sea 60 new ballistic missile submarines in the last 15 years. Until last year we had not commissioned one in that same period.

The Soviet Union has built over 200 modern Backfire bombers—and is building 30 more a year. For 20 years, the United States has deployed no new strategic bombers. Many of our B-52 bombers are now older than the pilots who fly them.

The Soviet Union now has 600 of the missiles considered most threatening by both sides—the intermediate-range missiles based on land. We have none. The U.S. withdrew its intermediate-range land-based missiles from Europe almost 20 years ago.

The world has also witnessed unprecedented growth in the area of Soviet conventional forces; the Soviets far exceed us in the number of tanks, artillery pieces, aircraft and ships they produce every year. What is more, when I arrived in this office I learned that in our own forces we had planes that could not fly and ships that could not leave port, mainly for lack of spare parts and crew members.

The Soviet military buildup must not be ignored. We have recognized the problem and together with our allies we have begun to correct the imbalance. Look at this chart (accompanying chart, "Projected Defense Spending") of projected real defense spending for the next several years. Here's the Soviet line. Let us assume the Soviets' rate of spending remains at the level they have followed since the 1960's.

The blue line is the United States. If my defense proposals are passed, it will still take five years before we come close to Soviet level. Yet the modernization of our strategic and conventional forces will assure that deterrence works and peace prevails.

Our deployed nuclear forces were built before the age of microcircuits. It is not right to ask our young men and women in uniform to maintain and operate such antiques. Many have already given their lives in missile explosions and aircraft accidents caused by the old age of their equipment. We must replace and modernize our forces, and that is why I have decided to proceed with production and deployment of the new ICBM known as the MX.

Three earlier Presidents worked to develop this missile. Based on the best advice I could get, I concluded that the MX is the right missile at the right time. On the other hand, when I arrived in office, I felt the proposal on where and how to base the missile simply cost too much in terms of money, and the impact on our citizens' lives.

I have concluded, however, it is absolutely essential that we proceed to produce this missile, and that we base it in a series of closely based silos at Warren Air Force Base near Cheyenne, Wyo.

This plan requires only half as many missiles as the earlier plan and will fit in an area of only 20 square miles. It is the product of around-the-clock research that has been under way since I directed a search for a better, cheaper way. I urge the members of Congress who must pass this plan to listen and examine the facts, before they come to their own conclusion.

Some may question what modernizing our military has to do with peace. Well, as I explained earlier, a secure force keeps others from threatening us and that keeps the peace. And just as important, it also increased the products of reaching significant arms reductions with the Soviets, and that is what we really want. The United States wants deep cuts in the world's arsenal of weapons.

But unless we demonstraste the will to rebuild our strength and restore the military balance, the Soviets, since they are so far ahead, have little incentive to negotiate with us. If we had not begun to modernize, the Soviet negotiators would know we had nothing to bargain with except talk. They would know we were bluffing without a good hand because they know what cards we hold—just as we know what is in their hand.

51. Caspar Weinberger on United States Nuclear Deterrence Policy, December 14, 1982

I. *Introduction*

I welcome this opportunity to brief the Committee on U.S. nuclear policy. Over the past several months few subjects have been treated so extensively—and so incorrectly—in the press. Our meeting today provides an

Testimony of the Honorable Caspar Weinberger, Secretary of Defense, before the United States Senate Foreign Relations Committee, December 14, 1982.

opportunity to dispel the misconceptions about our nuclear strategy which these recent articles may have produced.

We live in a nuclear age. As of 1945, nuclear weapons became a fact of life. We can neither wish them away nor pretend that they do not exist. A review of the ways in which the United States has dealt with these realities for the past 37 years contributes significantly to understanding U.S. nuclear policy today. With that goal in mind I will this morning:

- discuss our concept of deterrence;
- trace the evolution of U.S. strategic policy; and
- describe the systems we deploy to support the policy.

II. *Deterrence—The Concept*

In the wake of World War II, the United States and the Western democracies developed a policy intended to prevent any future recurrence of the tremendous carnage and devastation which the war had caused. The answer lay in addressing the root cause of the problem rather than the symptoms it produced. Thus the course of action chosen aimed at preventing wars from occurring in the first place. To that end, the United States made clear that it would use its atomic weapons not for conquest or coercion, but for discouraging—for *deterring*—aggression and attack against ourselves and our allies.

Today, deterrence remains—as it has for the past 37 years—the cornerstone of our strategic nuclear policy, and, indeed, of our entire national security posture. Our strategy is a defensive one, designed to prevent attack—particularly nuclear attack—against us or our allies. To deter successfully, we must be able—and must be seen to be able—to retaliate against any potential aggressor in such a manner that the costs we will exact will substantially exceed any gains he might hope to achieve through aggression. We, for our part, are under no illusions about the consequences of a nuclear war: we believe there would be no winners in such a war. But this recognition on *our* part is not sufficient to ensure effective deterrence or to prevent the outbreak of war: it is essential that the Soviet leadership understands this as well. We must make sure that the *Soviet* leadership, in calculating the risks of aggression, recognizes that *because of our retaliatory capability, there can be no circumstance where the initiation of a nuclear war at any level or of any duration would make sense.* If they recognize that our forces can deny them their objectives at whatever level of conflict they contemplate, and in addition that such a conflict could lead to the destructions of those political, military, and economic assets which they value most highly, then deterrence is enhanced and the risk of war diminished. It is this outcome which we seek to achieve.

III. *The Evolution of U.S. Strategic Policy*

During the late 1940s and early 1950s, America's virtual monopoly of intercontinental nuclear systems meant that our requirements for conventional deterrence were relatively small. The Soviet Union understood that, under our policy of "massive retaliation," we might respond to a Soviet attack, however limited, on the U.S. or our allies, with an atomic attack on the USSR. As the fifties ended and the sixties began, however, the Soviets began developing and acquiring long-range nuclear capabilities. As their capacity for nuclear and conventional aggression continued to grow, the U.S. threat to respond to a conventional or even limited nuclear attack with a massive nuclear retaliation became less and less credible. Accordingly, in the 1960s the U.S.—and later the NATO Allies—adopted a policy of "flexible response." Under this concept, the United States and NATO planned to strengthen general purpose warfare forces in order to better equip them to deal with a Soviet conventional attack; at the same time, U.S. nuclear capabilities were increased in order to provide the President with the option of using nuclear forces both to support our general purpose forces and to respond selectively (on less than an all-out basis) to a limited Soviet nuclear attack. The option of retaliation on a more massive scale was retained in order to deter the possibility of a major Soviet nuclear attack. This concept of flexible response remains as a central principle of our strategy today.

Of paramount importance to the flexible response strategy is the requirement for *flexibility*—for our nuclear forces and plans for their use to be designed and developed in such a way that our response is appropriate to the circumstances of the aggression against us. This means that they should be capable of being used on a very limited basis as well as more massively. This does not imply that through flexible response we seek to fight a limited nuclear war, or, for that matter, to fight a nuclear war under any conditions. Our basic strategy, in direct support of our policy of deterrence has been, and remains, the prevention of any aggression, nuclear or conventional. But it would be irresponsible—indeed immoral—to reject the possibility that the terrible consequences of a nuclear conflict might be limited if deterrence should fail. To be sure, there is no guarantee that we would be successful in creating such limits. But there is every guarantee that restrictions cannot be achieved if we do *not* attempt to do so.

While we work toward insuring deterrence, we need to think about the failures of deterrence (for whatever reason). If that were to occur we cannot predict the nature of a Soviet nuclear strike, nor assure with any certainty that what may have started out as a limited Soviet attack would remain confined at that level. Nevertheless, we must plan for flexibility in our forces and in our response options so that there is a possibility of re-establishing deterrence at the lowest possible level of violence, and avoiding further escalation. I assure you it is not pleasant to think in these terms, but it would certainly be the gravest irresponsibility for those of us who are charged with the nation's defense *not* to do so.

Of course, this concept of seeking to contain the level of destruction by having flexible and enduring forces is not new. It has been squarely in the mainstream of American strategic thinking for over two decades. A brief review of the record illustrates this point well. In the early 1960s, then Secretary of Defense Robert McNamara told the Congress:

> The major mission of the Strategic Retaliatory Forces is to deter war by their capability to destroy the enemy's war-making potential, including not only his nuclear strike forces and military installations, but also his urban society, if necessary. . . . What we are proposing is a capability to strike back after absorbing the first blow. This means we have to build and maintain a second strike force. Such a force should have sufficient flexibility to permit a choice of strategies, particularly an ability to: (1) strike back decisively at the entire Soviet target system simultaneously or (2) strike back first at the Soviet bomber bases, missile sites, and other military installations associated with their long-range nuclear forces to reduce the power of any follow-on attack—and then if necessary, strike back at the Soviet urban and industrial complex in a controlled and deliberate way.*

Almost a decade later, then Secretary of Defense James Schlesinger reaffirmed the importance of strategic force flexibility when he told the Congress that, to remain credible, a deterrent strategy must be consistent with the threat it seeks to deter.

> If anything, the need for options other than suicide or surrender, and other than escalation to all out nuclerar war, is more important for us today than it was in 1960, because of the growth of the capabilities possessed by other powers . . . The Soviet Union now has the capability in its missile forces to undertake selective attacks against targets other than cities. This poses for us an obligation, if we are to ensure the credibility of our strategic deterrent, to be certain that we have a comparable

---

*Statement of Secretary of Defense Robert S. McNamara before the House Armed Services Committee, *The Fiscal Year 1964–68 Defense Program and 1964 Defense Budget*, January 30, 1963, pp. 28–30.

capability in our strategic systems and in our targeting doctrine, and to be certain that the USSR has no misunderstanding on this point. . . .*

As the 1980s began Harold Brown, in explaining the policy of deterrence to the Congress, emphasized the importance of flexibility in much the same way as did his predecessors:

To the Soviet Union, our strategy makes clear that no course of aggression by them that led to the use of nuclear weapons, on any scale of attack and at any stage of conflict, could lead to victory, however they define victory. Besides our power to devastate the full target system of the USSR, the United States would have the option for more selective, lesser retaliatory attacks that would exact a prohibitively high price from the things the Soviet leadership prizes most—political and military control, nuclear and conventional military force, and the economic base needed to sustain a war. . . . Our planning must provide a continuum of options, ranging from small numbers of strategic and/or theater nuclear weapons aimed at narrowly defined targets, to employment of large portions of our nuclear forces against a broad spectrum of targets.†

Thus, the past three and a half decades have taught us two central lessons with regard to implementing our policy . . . lessons which we must continue to take into account in the years ahead:

• first, in order for our retaliatory threat to be seen as credible, we must be able—and be seen to have the means—to respond appropriately to a wide range of aggressive actions; if our threatened response is perceived as inadequate or inappropriate, it will be seen as a bluff and ignored;

• secondly, deterrence is a dynamic effort, not a static one. In order to continue to deter successfully, our capabilities must change as the threat changes, and as our knowledge of what is necessary to deter improves.

IV. *The Forces for Nuclear Deterrence*

A policy of deterrence through flexible response would be hollow and not credible—to us, and to our allies and adversaries alike—if we did not possess the military forces and capabilities necessary to enforce it. As a result, we have maintained—and will continue to maintain—strategic nuclear weapons systems which we have procured and deployed for the express purpose of avoiding their use in anger. This seeming paradox lies at the heart of our strategic deterrent posture.

*Department of Defense Annual Report, Fiscal Year 1975, March 4, 1974, p. 4.
†Department of Defense Annual Report, Fiscal Year 1982, January, 1981, pp. 40–41.

In the 1960s the United States' nuclear forces evolved from one composed primarily of manned bombers to the current balanced force structure of manned bombers, intercontinental land-based ballistic missiles, and ballistic missiles fired from our fleet submarine forces. This force configuration, commonly known as the "Triad", has been retained because it complicates an enemy's ability to attack our strategic nuclear forces, provides us with maximum flexibility, and hedges against a catastrophic failure (either through technical difficulties or hostile action) of one leg of the Triad. The Triad has served us well in the past and will continue to do so in the future.

By the early 1960s, the U.S. had over 7,000 strategic nuclear weapons, most of which were carried by B-47s, and the then-new B-52s. The Soviet Union had fewer than 500 strategic warheads. Throughout the 1960s our strategic posture presented the Soviet planner with a dilemma if he decided to pre-empt against the United States: due to the relatively small number of weapons the USSR possessed and their ineffectiveness against any U.S. strategic forces, such an attack was impossible to execute successfully. If the Soviet planner targeted our hardened missile silos and alert bomber bases, he found that he would deplete his arsenal while not significantly reducing U.S. retaliatory forces. In other words, his ability to limit the certain massive destruction to his own forces and society was rather small. If on the other hand, the Soviet planner targeted U.S. cities he would feel the full brunt of a U.S. retaliatory strike against his own cities, a U.S. arsenal quite larger and much more capable than his own, by any measure. Again, he was deterred.

During the course of the 1970s the Soviet arsenal began growing both in quantity and in quality (although the U.S. qualitative edge remained). The Soviets expanded their land-based missile force and hardened their protective silos. At the same time, the U.S. made a conscious choice not to upgrade the yield/accuracy combination of its own missile forces or to build force levels of sufficient size to threaten the Soviet Union with a sudden disarming first strike. The net result of this was to allow the Soviets a "sanctuary" for its ICBM force, since U.S. forces could not attack it effectively. The Soviets, however, did not follow our lead and developed a new generation of ICBMs specifically designed to attack U.S. missile silos. By the late 1970s, this combination of vulnerable U.S. missiles and a Soviet sanctuary has eased the earlier dilemma of the Soviet planner. Now, he potentially could envision nuclear confrontation in which he probed U.S. resolve to retaliate by attacking a smaller and smaller subset of the U.S. military forces—while U.S. options for retaliation were limited. If the Soviet leadership came to accept this as plausible, our deterrent policy, and, as a result, global stability, would be severely threatened. As the "imbalance of imbalances" continued to grow in the late 1970s, that is, as the Soviets began tipping the theater nuclear balance in their favor while maintaining their superiority in conventional forces, the risk became greater that this type of limited attack would appear attractive to the Soviet military.

The strategic modernization program which President Reagan set forth in October 1981 is designed to address in part this adverse and imbalanced situation. It restores the margin of safety we require in order to continue to deter successfully Soviet strategic aggression. In essence, the program is designed to accomplish two general goals: first, to improve the survivability of our present and planned forces in order that they do not serve to destabilize potential crises by offering lucrative targets for Soviet preemption; and, secondly, to sustain the credibility of our deterrent policy by developing the capability to threaten, and destroy if necessary, the full spectrum of potential Soviet targets. This combination of improved survivability and military capability is intended to assure that Soviet leadership will continue to recognize clearly and unambiguously that they can realize no conceivable benefit from initiating nuclear aggression.

Let me review briefly our modernization program to highlight how the specific elements contribute to these goals. With regard to survivability, we are taking the following steps:

—In the area of command, control and communications we are upgrading the survivability of command centers and communications links upon which the effective use of our forces depends; we are also improving the survivability, performance and coverage of our warning and attack assessment systems in order that we may provide the President—and our forces—with the maximum amount of control and timely information to respond to nuclear attack.

—With regard to our bomber forces, we have begun production of a variant of the B-1 bomber which—because of its superior base escape ability—is far less vulnerable to destruction on the ground than the B-52; by equipping the B-52 with cruise missiles we are extending the in-flight survivability of these aircraft, which because of their age are increasingly less able to survive penetration missions in the face of the large and sophisticated Soviet air defense net.

—We have taken advantage of the high survivability currently enjoyed by our sea-based systems by developing and deploying two new systems, the sea-launched cruise missile and the Trident II; deployment of the cruise missiles on selected attack submarines, as part of our secure reserve force, will ensure continued long-term survivability and the extended range potential of the Trident II will provide a hedge against any unanticipated advances in Soviet anti-submarine warfare.

—Perhaps the most vulnerable of our retaliatory forces are the Minuteman and Titan ICBMs which were designed and housed in the late 1950s and 1960s. Designed to deter a Soviet threat of the past, these missiles have served us well: The Peacekeeper is designed to meet both the current and future Soviet threat, both in capability and survivability. Peacekeeper in Closely Spaced Basing will, at long last, provide a feasible, affordable solution to the ICBM vulnerability problem.

—Finally, by upgrading our air defense and warning capability, we will remove any Soviet misperceptions they might have about being able to use small scale raids by bomber or cruise missile forces to disable our warning and control systems.

With regard to developing the military capabilities necessary to make our retaliatory deterrent posture stronger and more credible, we are pursuing the following measures:

—By deploying the B-1 variant as soon as possible and by equipping less obsolescent model B-52s with air-launched cruise missiles, we will be able to continue to pose both a penetrating bomber and a stand-off cruise missile threat to the USSR. This combination places the greatest amount of stress on Soviet air defenses and thereby ensures that a high number of our bomber-delivered weapons will reach their targets; once the advanced technology, or "Stealth" bomber enters the inventory, it can assume a major share of the penetration mission, and, in combination with B-1s in a mixed air-launched cruise missile carrier and penetration role, will continue to give us high confidence that our retaliatory strike can be succesful.

—The Trident II missile, with its substantial increase in accuracy over previous SLBMs, will provide a hard target kill capability which will contribute significantly to our ability to hold the full Soviet target base at risk. A sea-based capability of this kind increases the flexibility of our strategic posture; provides an important hedge to the loss, through unexpected system failure, of similar land-based missiles; reduces Soviet incentives to attempt a pre-emptive strike against our ICBM force; and complicates severely Soviet offensive and defensive planning.

—The Peacekeeper has had a highly successful research and development program and will be ready for deployment in 1986; we must be able to draw on its prompt hard target kill capability as soon as possible so that the Soviet Union would have to assume that even as it attempted to evaluate the consequences of its first strike it would be losing military capability it would need to implement its strategy. This will offset the current Soviet monopoly in this area.

In sum, the strategic modernization program is tailored precisely to provide the capabilities we need in order to ensure that our deterrent continues to be credible and effective well into the next century, as well as to support our arms control efforts to reduce significantly the nuclear aresenals of both sides.

•   •   •   •

Let me conclude by returning to the point at which I began. There is nothing new about our policy. Since the era of nuclear weapons began, the United States has sought to prevent a nuclear war through a policy of deterrence. That policy has worked successfully for almost four decades. We are dedicated to ensuring that it continues to do so.

---

52. Ronald Reagan on the Strategic Defense Initiative, December 28, 1984

Since the advent of nuclear weapons, every President has sought to minimize the risk of nuclear destruction by maintaining effective forces to deter aggression and by pursuing complementary arms control agreements. This approach has worked. We and our allies have succeeded in preventing nuclear war while protecting Western security for nearly four decades.

Originally, we relied on balanced defensive and offensive forces to deter. But over the last twenty years, the United States has nearly abandoned efforts to develop and deploy defenses against nuclear weapons, relying instead almost exclusively on the threat of nuclear retaliation. We accepted the notion that if both we and the Soviet Union were able to retaliate with devastating power even after absorbing a first strike, that stable deterrence would endure. That rather novel concept seemed at the time to be sensible for two reasons. First, the Soviets stated that they believed that both sides should have roughly equal forces and neither side should seek to alter the balance to gain unilateral advantage. Second, there did not seem to be any

This text appears as "Foreword Written for a Report on the Initiative," December 28, 1984, *Weekly Compilation of Presidential Documents* 21, 1, January 7, 1985, pp. 8–9.

alternative. The state of the art in defensive systems did not permit an effective defensive system.

Today both of these basic assumptions are being called into question. The pace of the Soviet offensive and defensive buildup has upset the balance in the areas of greatest importance during crises. Furthermore, new technologies are now at hand which may make possible a truly effective nonnuclear defense.

For these reasons and because of the awesome destructive potential of nuclear weapons, we must seek another means of deterring war. It is both militarily and morally necessary. Certainly, there should be a better way to strengthen peace and stability, a way to move away from a future that relies so heavily on the prospect of rapid and massive nuclear retaliation and toward greater reliance on defensive systems which threaten no one.

On March 23, 1983, I announced my decision to take an important first step toward this goal by directing the establishment of a comprehensive and intensive research program, the Strategic Defense Initiative, aimed at eventually eliminating the threat posed by nuclear armed ballistic missiles.

The Strategic Defense Initiative (SDI) is a program of vigorous research focused on advanced defensive technologies with the aim of finding ways to provide a better basis for deterring aggression, strengthening stability, and increasing the security of the United States and our allies. The SDI research program will provide to a future President and a future Congress the technical knowledge required to support a decision on whether to develop and later deploy advanced defensive systems.

At the same time, the United States is committed to the negotiation of equal and verifiable agreements which bring real reductions in the power of the nuclear arsenals of both sides. To this end, my Administration has proposed to the Soviet Union a comprehensive set of arms control proposals. We are working tirelessly for the success of these efforts, but we can and must go further in trying to strengthen the peace.

Our research under the Strategic Defense Initiative complements our arms reduction efforts and helps to pave the way for creating a more stable and secure world. The research that we are undertaking is consistent with all of our treaty obligations, including the 1972 Anti-Ballistic Missile Treaty.

In the near term, the SDI research program also responds to the ongoing and extensive Soviet anti-ballistic missile (ABM) effort, which includes actual deployments. It provides a powerful deterrent to any Soviet decision to expand its ballistic missile defense capability beyond that permitted by the ABM Treaty. And, in the long-term, we have confidence that SDI will be a crucial means by which both the United States and the Soviet Union can safely agree to very deep reductions, and eventually, even the elimination of ballistic missiles and the nuclear weapons they carry.

Our vital interests and those of our allies are inextricably linked. Their safety and ours are one. They, too, rely upon our nuclear forces to deter attack against them. Therefore, as we pursue the promise offered by the Strategic Defense Initiative, we will continue to work closely with our friends and allies. We will ensure that, in the event of a future decision to develop and deploy defensive systems—a decision in which consultation with our allies will play an important part—allied, as well as U.S. security against aggression would be enhanced.

Through the SDI research program, I have called upon the great scientific talents of our country to turn to the cause of strengthening world peace by rendering ballistic missiles impotent and obsolete. In short, I propose to channel our technological prowess toward building a more secure and stable world. And I want to emphasize that in carrying out this research program, the United States seeks neither military superiority nor political advantage. Our only purpose is to search for ways to reduce the danger of nuclear war.

As you review the following pages, I would ask you to remember that the quality of our future is at stake and to reflect on what we are trying to achieve—the strengthening of our ability to preserve the peace while shifting away from our current dependence upon the threat of nuclear retaliation. I would also ask you to consider the SDI research program in light of both the Soviet Union's extensive, ongoing efforts in this area and our own government's constitutional responsibility to provide for the common defense. I hope that you will conclude by lending your own strong and continuing support to this research effort—an effort which could prove to be critical to our nation's future.

Ronald Reagan

December 28, 1984.

VIII

Arms Control

T*he term "arms control" first came into use in the 1950s in the United States among military planners and specialists interested in avoiding accidental nuclear war through miscalculation and aware of the pitfalls involved in Soviet proposals for "general and complete disarmament." Since the failure of the Baruch Plan in 1946, international control of nuclear weapons and disarmament had become utopian dreams amid the realities of a Soviet-American arms race. The devastating power of thermonuclear weapons, combined after 1957 with the rapid new delivery systems spawned by intercontinental ballistic missiles (ICBMs), made arms control imperative. Talks between Soviet and American scientists at businessman Cyrus Eaton's Nova Scotia estate, Pugwash, initiated a new tradition of bilateral talks between arms control experts on both sides of the Iron Curtain.*

Neither the Soviet Union nor China accepted the concept of arms control in public, on the grounds that it was a mere bourgeois ploy to continue the arms race and that, in wartime, men and morale were of greater importance than technology. Then, in October 1962, the Soviet Union installed medium-range missiles in Cuba and precipitated a crisis that brought the powers to the brink of nuclear war. Two arms control measures that quickly resulted from the Cuban missile crisis were the partial ban on atmospheric testing of 1963 and the establishment of a "hot line" communications link between Moscow and Washington. In this way both powers sought to provide room for weapons development while avoiding misunderstandings that could lead to war by accident.

Further arms control agreements followed in the 1970s, including non-nuclear zones in Antarctica and Latin America. The key agreements were bilateral between the Soviet Union and the United States: the strategic arms limitation talks (SALT) process led, in 1972, to an agreement to limit each side's antiballistic missile sites (ABM) and to set ceilings on strategic launch vehicles. The more comprehensive SALT II agreement of 1979 was signed by Premier Brezhnev and President Carter but never ratified by the United States Senate.

Soviet-American arms control agreements helped widen the Sino-Soviet split, especially after the Soviet Union refused to provide the Chinese with a sample nuclear weapon in 1959. The Chinese were sharply critical of any "nuclear club" arrangement that would exclude the acquisition of nuclear weapons by Third World nations such as their own. In 1964 China successfully tested her first nuclear weapon and in the 1970s attained both thermonuclear and ICBM capability.

The crucial breakthroughs in the 1970s and early 1980s in arms control were technological, as well as political. New techniques of satellite photography and seismology made the violation of national security by inspection teams unnecessary and provided both sides with "national technical means of verification" for any major arms control measures. As technology spurred advances in arms control in the first four decades after World War II, political changes in the United States, the Soviet Union, and Eastern Europe fueled change after 1985. President Ronald Reagan inaugurated a massive strategic rearmament program as a precondition for negotiating arms reduction agreements with the Soviet Union. In the arms talks, the U.S. also insisted upon verification by technological surveillance as well as on site inspections. Many believed that the Reagan administration's policy of total disarmament, the so-called "zero-zero option," would be rejected out of hand by the Soviets and allow the U.S. more latitude in rearming.

However, when Mikhail Gorbachev became General Secretary of the Soviet Union, he initiated perestroika, or a restructuring of the Soviet economy, and accepted western technology, capital, and some of the trappings of capitalism. In arms control, Gorbachev accepted dramatic arms reduction in a series of treaties and ongoing discussions such as the Intermediate-range Nuclear Forces Treaty (INF), the Conventional Forces Europe Treaty (CFE), the Strategic Arms Reduction Talks (START), the Chemical Weapons Convention (CW), the Peaceful Nuclear Explosions Treaty (PNET), and the Threshold Test Ban Treaty (TTBT).

By the end of the 1980s political upheavals in eastern Europe marked the end of the Cold War. East and West Germany moved dramatically toward economic and political integration. The Eastern Bloc, solidly under Soviet domination for forty years, began to crumble. Across the European continent, from the Atlantic to the Urals, NATO and Warsaw Pact nations reduced their military bases, conventional and nuclear weapons, and armed forces. Not since the 1920s had the industrialized nations of the world pursued so vigorously and ambitiously the goal of arms control.

On the 1963 test ban, see H. K. Jacobson and Eric Stein, Diplomats, Scientists, and Politicians: The United States and the Nuclear Test Ban Negotiations *(Ann Arbor: University of Michigan Press, 1966). Milton S. Katz,* Ban the Bomb: A History of SANE, the Committee for a Sane Nuclear Policy, 1957–1985 *(Westport, Conn.: Greenwood Press, 1986) is a narrative history of the organization. A study of the nuclear freeze movement is Douglas C. Waller,* Congress and the Nuclear Freeze: An Inside Look at the Politics of a Mass Movement *(Amherst: University of Massachusetts Press, 1987). The SALT negotiations are chronicled in Thomas Wolfe,* The Salt Experience *(Cambridge, Mass.: Ballinger, 1979) and in Strobe Talbott,* Endgame: The Inside Story of Salt II *(New York: Harper & Row, 1979). A valuable personal account of the test ban negotiations is by President Eisenhower's science adviser, George B. Kistiakowsky,* A Scientist at the White House *(Cambridge, Mass.: Harvard University Press, 1976).*

---

53. Memorandum of Understanding Between the United States of America and the Union of Soviet Socialist Republics Regarding the Establishment of a Direct Communications Link, June 20, 1963 (with Annex)

Signed at Geneva June 20, 1963
Entered into force June 20, 1963

For use in time of emergency the Government of the United States of America and the Government of the Union of Soviet Socialist Republics have agreed to establish as soon as technically feasible a direct communications link between the two Governments.

Each Government shall be responsible for the arrangements for the link on its own territory. Each Government shall take the necessary steps to ensure continuous functioning of the link and prompt delivery to its head of

This document appears in United States Arms Control and Disarmament Agency, *Arms Control and Disarmament Agreements* (Washington, D.C.: U.S. Government Printing Office, 1982), pp. 31–33.

government of any communications received by means of the link from the head of government of the other party.

Arrangements for establishing and operating the link are set forth in the Annex which is attached hereto and forms an integral part hereof.

Done in duplicate in the English and Russian languages at Geneva, Switzerland, this 20th day of June, 1963.

FOR THE GOVERNMENT OF
THE UNITED STATES OF
AMERICA:

Acting Representative of the United States of America to the Eighteen-Nation Committee on Disarmament

FOR THE GOVERNMENT OF
THE UNION OF SOVIET
SOCIALIST REPUBLICS:

Acting Representative of the Union of Soviet Socialist Republics to the Eighteen-Nation Committee on Disarmament

(SEAL)

Annex

To the Memorandum of Understanding Between the United States of America and the Union of Soviet Socialist Republics Regarding the Establishment of a Direct Communications Link

The direct communications link between Washington and Moscow established in accordance with the Memorandum, and the operation of such link, shall be governed by the following provisions.

1. The direct communications link shall consist of:

a. Two terminal points with telegraph-teleprinter equipment between which communications shall be directly exchanged.

b. One full-time duplex wire telegraph circuit, routed Washington-London-Copenhagen-Stockholm-Helsinki-Moscow, which shall be used for the transmission of messages;

c. One full-time duplex radiotelegraph circuit, routed Washington-Tangier-Moscow, which shall be used for service communications and for coordination of operations between the two terminal points.

If experience in operating the direct communications link should demonstrate that the establishment of an additional wire telegraph circuit is advisable, such circuit may be established by mutual agreement between authorized representatives of both Governments.

2. In case of interruption of the wire circuit, transmission of messages shall be effected via the radio circuit, and for this purpose provision shall be made at the terminal points for the capability of prompt switching of all necessary equipment from one circuit to another.

3. The terminal points of the link shall be so equipped as to provide for the transmission and reception of messages from Moscow to Washington in the Russian language and from Washington to Moscow in the English language. In this connection, the USSR shall furnish the United States four sets of telegraph terminal equipment, including page printers, transmitters, and reperforators, with one year's supply of spare parts and all necessary special tools, test equipment, operating instructions, and other technical literature, to provide for transmission and reception of messages in the Russian language.

The United States shall furnish the Soviet Union four sets of telegraph terminal equipment, including page printers, transmitters, and reperforators, with one year's supply of spare parts and all necessary special tools, test equipment, operating instructions and other technical literature, to provide for transmission and reception of messages in the English language.

Equipment described in this paragraph shall be exchanged directly between the parties without any payment being required therefor.

4. The terminal points of the direct communications link shall be provided with encoding equipment. For the terminal point in the USSR, four sets of such equipment (each capable of simplex operation), with one year's supply of spare parts, with all necessary special tools, test equipment, operating instructions and other technical literature, and with all necessary blank tape, shall be furnished by the United States to the USSR against payment of the cost thereof by the USSR.

The USSR shall provide for preparation and delivery of keying tapes to the terminal point of the link in the United States for reception of messages from the USSR. The United States shall provide for the preparation and delivery of keying tapes to the terminal point of the link in the USSR for reception of messages from the United States. Delivery of prepared keying tapes to the terminal points of the link shall be effected through the Embassy of the USSR in Washington (for the terminal of the link in the USSR) and through the Embassy of the United States in Moscow (for the terminal of the link in the United States).

5. The United States and USSR shall designate the agencies responsible for the arrangements regarding the direct communications link, for its

technical maintenance, continuity and reliability, and for the timely transmission of messages.

Such agencies may, by mutual agreement, decide matters and develop instructions relating to the technical maintenance and operation of the direct communications link and effect arrangements to improve the operation of the link.

6. The technical parameters of the telegraph circuits of the link and of the terminal equipment, as well as the maintenance of such circuits and equipment shall be in accordance with CCITT and CCIR recommendations.

Transmission and reception of messages over the direct communications link shall be effected in accordance with applicable recommendations of international telegraph and radio communications regulations, as well as with mutually agreed instructions.

7. The costs of the direct communications link shall be borne as follows:

a. The USSR shall pay the full cost of leasing the portion of the telegraph circuit from Moscow to Helsinki and 50% of the cost of leasing the portion of the telegraph circuit from Helsinki to London. The United States shall pay the full cost of leasing the portion of the telegraph circuit from London to Helsinki.

Payment of the cost of leasing the radio telegraph circuit between Washington and Moscow shall be effected without any transfer of payments between the parties. The USSR shall bear the expenses relating to the transmission of messages from Moscow to Washington. The United States shall bear the expenses relating to the transmission of messages from Washington to Moscow.

54. Treaty on the Nonproliferation of Nuclear Weapons, March 5, 1970

Signed at Washington, London, and Moscow July 1, 1968
Ratification advised by U.S. Senate March 13, 1969
Ratified by U.S. President November 24, 1969

This treaty is F. R. Document 76-3382; filed February 2, 1976.

U.S. ratification deposited at Washington, London,
and Moscow March 5, 1970
Proclaimed by U.S. President March 5, 1970
Entered into force March 5, 1970

The States concluding this Treaty, hereinafter referred to as the "Parties to the Treaty,"

Considering the devastation that would be visited upon all mankind by a nuclear war and the consequent need to make every effort to avert the danger of such a war and to take measures to safeguard the security of peoples,

Believing that the proliferation of nuclear weapons would seriously enhance the danger of nuclear war,

In conformity with resolutions of the United Nations General Assembly calling for the conclusion of an agreement on the prevention of wider dissemination of nuclear weapons,

Undertaking to cooperate in facilitating the application of International Atomic Energy Agency safeguards on peaceful nuclear activities,

Expressing their support for research, development and other efforts to further the application, within the framework of the International Atomic Energy Agency safeguards system, of the principle of safeguarding effectively the flow of source and special fissionable materials by the use of instruments and other techniques at certain strategic points,

Affirming the principle that the benefits of peaceful applications of nuclear technology, including any technological by-products which may be derived by nuclear-weapon States from the development of nuclear explosive devices, should be available for peaceful purposes to all Parties of the Treaty, whether nuclear-weapon or non-nuclear weapon States,

Convinced that, in furtherance of this principle, all Parties to the Treaty are entitled to participate in the fullest possible exchange of scientific information for, and to contribute alone or in cooperation with other States to, the further development of the applications of atomic energy for peaceful purposes,

Declaring their intention to achieve at the earliest possible date the cessation of the nuclear arms race and to undertake effective measures in the direction of nuclear disarmament,

Urging the cooperation of all States in the attainment of this objective,

Recalling the determination expressed by the Parties to the 1963 Treaty banning nuclear weapon tests in the atmosphere, in outer space and under water in its Preamble to seek to achieve the discontinuance of all test explosions of nuclear weapons for all time and to continue negotiations to this end,

Desiring to further the easing of international tension and the strengthening of trust between States in order to facilitate the cessation of the manufacture of nuclear weapons, the liquidation of all their existing stockpiles, and the elimination from national arsenals of nuclear weapons and the means of their delivery pursuant to a treaty on general and complete disarmament under strict and effective international control,

Recalling that, in accordance with the Charter of the United Nations States must refrain in their international relations from the threat of use of force against the territorial integrity or political independence of any State, or in any other manner inconsistent with the Purposes of the United Nations, and that the establishment and maintenance of international peace and security are to be promoted with the least diversion for armaments of the world's human and economic resources,

Have agreed as follows:

Article I

Each nuclear-weapon State Party to the Treaty undertakes not to transfer to any recipient whatsoever nuclear weapons or other nuclear explosive devices or control over such weapons or explosive devices directly, or indirectly; and not in any way to assist, encourage, or induce any non-nuclear-weapon State to manufacture or otherwise acquire nuclear weapons or other nuclear explosive devices, or control over such weapons or explosive devices.

Article II

Each non-nuclear-weapon State Party to the Treaty undertakes not to receive the transfer from any transfer or whatsoever of nuclear weapons or other nuclear explosive devices or of control over such weapons or explosive devices directly, or indirectly; not to manufacture or otherwise acquire nuclear weapons or other nuclear explosive devices; and not to seek or receive any assistance in the manufacture of nuclear weapons or other nuclear explosive devices.

Article III

1. Each non-nuclear-weapon State Party to the Treaty undertakes to accept safe-guards, as set forth in an agreement to be negotiated and concluded with the International Atomic Energy Agency in accordance with the

Statute of the International Atomic Energy Agency and the Agency's safe-guards system, for the exclusive purpose of verification of the fulfillment of its obligations assumed under this Treaty with a view to preventing diversion on nuclear energy from peaceful uses to nuclear weapons or other nuclear explosive devices. Procedures for the safeguards required by this article shall be followed with respect to source or fissionable material whether it is being produced, processed or used in any principle nuclear facility or is outside any such facility. The safeguards required by this article shall be applied to all source or special fissionable material in all peaceful nuclear activities within the territory of such State, under its jurisdiction, or carried out under its control anywhere.

2. Each State Party to the Treaty undertakes not to provide: (a) source or special fissionable material, or (b) equipment or material especially de-signed or prepared for the processing, use or production of special fission-able material, to any non-nuclear-weapon State for peaceful purposes, unless the source or special fissionable material shall be subject to the safeguards required by this article.

3. The safeguards required by this article shall be implemented in a manner designed to comply with Article IV of this Treaty, and to avoid ham-pering the economic or technological development of the Parties or inter-national cooperation in the field of peaceful nuclear activities, including the international exchange of nuclear material and equipment for the processing, use or production of nuclear material for peaceful purposes in accordance with the provisions of this article and the principle of safeguarding set forth in the Preamble of the Treaty.

4. Non-nuclear-weapon States Party to the Treaty shall conclude agree-ments with the International Atomic Energy Agency to meet the re-quirements of this article either individually or together with other States in accordance with the Statute of the International Atomic Energy Agency. Negotiation of such agreements shall commence within 180 days from the original entry into force of this Treaty. For States depositing their in-struments of ratification or accession after the 180-day period, negotiation of such agreements shall commence not later than the date of such deposit. Such agreements shall enter into force not later than eighteen months after the date of initiation of negotiations.

Article IV

1. Nothing in the Treaty shall be interpreted as affecting the inalienable right of all the Parties to the Treaty to develop research, production and use

of nuclear energy for peaceful purposes without discrimination and in conformity with Articles I and II of this Treaty.

2. All the Parties to the Treaty undertake to facilitate, and have the right to participate in, the fullest possible exchange of equipment, materials and scientific and technological information for the peaceful uses of nuclear energy. Parties to the Treaty in a position to do so shall also cooperate in contributing alone or together with other States or international organizations to the further development of the applications of nuclear energy for peaceful purposes, especially in the territories of non-nuclear-weapon States Party to the Treaty, with due consideration for the needs of the developing areas of the world.

Article V

Each Party to the Treaty undertakes to take appropriate measures to ensure that, in accordance with this Treaty, under appropriate international observation and through appropriate international procedures, potential benefits from any peaceful applications of nuclear explosions will be made available to non-nuclear-weapon States Party to the Treaty on a nondiscriminatory basis and that the charge to such Parties for the explosive devices used will be as low as possible and exclude any charge for research and development. Non-nuclear-weapon States Party to the Treaty shall be able to obtain such benefits, pursuant to a special international agreement or agreements, through an appropriate international body with adequate representation of non-nuclear-weapon States. Negotiations on this subject shall commence as soon as possible after the Treaty enters into force. Non-nuclear-weapon States Party to the Treaty so desiring may also obtain such benefits pursuant to bilateral agreements.

Article VI

Each of the Parties to the Treaty undertakes to pursue negotiations in good faith on effective measures relating to cessation of the nuclear arms race at an early date and to nuclear disarmament, and on a treaty on general and complete disarmament under strict and effective international control.

Article VII

Nothing in this Treaty affects the right of any group of States to conclude regional treaties in order to assure the total absence of nuclear weapons in their respective territories.

Article VIII

1. Any Party to the Treaty may propose amendments to this Treaty. The text of any proposed amendment shall be submitted to the Depositary Governments which shall circulate it to all Parties to the Treaty. Thereupon, if requested to do so by one-third or more of the Parties to the Treaty, the Depositary Governments shall convene a conference to which they shall invite all the Parties to the Treaty, to consider such an amendment.

2. Any amendment to this Treaty must be approved by a majority of the votes of all the Parties to the Treaty, including the votes of all nuclear-weapon States Party to the Treaty and all other Parties which, on the date the amendment is circulated, are members of the Board of Governors of the International Atomic Energy Agency. The amendment shall enter into force for each Party that deposits its instrument of ratification of the amendment upon the deposit of such instruments of ratification by a majority of all the Parties, including the instruments of ratification of all nuclear-weapon States Party to the Treaty and all other Parties which, on the date the amendment is circulated, are members of the Board of Governors of the International Atomic Energy Agency. Thereafter, it shall enter into force for any other Party upon the deposit of its instrument of ratification of the amendment.

3. Five years after the entry into force of this Treaty, a conference of Parties to the Treaty shall be held in Geneva, Switzerland, in order to review the operation of this Treaty with a view to assuring that the purposes of the Preamble and the provisions of the Treaty are being realized. At intervals of five years thereafter, a majority of the Parties to the Treaty may obtain, by submitting a proposal to this effect to the Depositary Governments, the convening of further conferences with the same objective of reviewing the operation of the Treaty.

Article IX

1. This Treaty shall be open to all States for signature. Any State which does not sign the Treaty before its entry into force in accordance with paragraph 3 of this Article may accede to it at any time.

2. This Treaty shall be subject to ratification by signatory States. Instruments of ratification and instruments of accession shall be deposited with the Governments of the United States of America, the United Kingdom of Great Britain and Northern Ireland and the Union of Soviet Socialist Republics, which are hereby designated the Depositary Governments.

3. This Treaty shall enter into force after its ratification by the States, the Governments of which are designated Depositaries of the Treaty, and

forty other States signatory to this Treaty and the deposit of their instruments of ratification. For the purposes of this Treaty, a nuclear-weapon State is one which has manufactured and exploded a nuclear weapon or other nuclear explosive device prior to January 1, 1967.

4. For States whose instruments of ratification or accession are deposited subsequent to the entry into force of this Treaty, it shall enter into force on the date of the deposit of their instruments of ratification or accession.

5. The Depositary Governments shall promptly inform all signatory and acceding States of the date of each signature, the date of deposit of each instrument of ratification or of accession, the date of the entry into force of this Treaty, and the date of receipt of any requests for convening a conference or other notices.

6. This Treaty shall be registered by the Depositary Governments pursuant to article 102 of the Charter of the United Nations.

Article X

1. Each Party shall in exercising its national sovereignty have the right to withdraw from the Treaty if it decides that extraordinary events, related to the subject matter of this Treaty, have jeopardized the supreme interests of its country. It shall give notice of such withdrawals of all other Parties to the Treaty and to the United Nations Security Council three months in advance. Such notice shall include a statement of the extraordinary events it regards as having jeopardized its supreme interests.

2. Twenty-five years after the entry into force of the Treaty, a conference shall be convened to decide whether the Treaty shall continue in force indefinitely, or shall be extended for an additional fixed period or periods. The decision shall be taken by a majority of the Parties to the Treaty.

Article XI

This Treaty, the English, Russian, French, Spanish and Chinese texts of which are equally authentic, shall be deposited in the archives of the Depositary Governments. Duly certified copies of this Treaty shall be transmitted by the Depositary Governments to the Governments of the signatory and acceding States.

55. Treaty Between the United States of America and the Union of Soviet Socialist Republics on the Limitation of Anti-Ballistic Missile Systems, May 26, 1972

The United States of America and the Union of Soviet Socialist Republics, hereinafter referred to as the Parties,

Proceeding from the premise that nuclear war would have devastating consequences for all mankind,

Considering that effective measures to limit anti-ballistic missile systems would be a substantial factor in curbing the race in strategic offensive arms and would lead to a decrease in the risk of outbreak of war involving nuclear weapons,

Proceeding from the premise that the limitation of anti-ballistic missile systems, as well as certain agreed measures with respect to the limitation of strategic offensive arms,[1] would contribute to the creation of more favorable conditions for further negotiations on limiting strategic arms,

Mindful of their obligations under Article VI of the Treaty on the Non-Proliferation of Nuclear Weapons,[2]

Declaring their intention to achieve at the earliest possible date the cessation of the nuclear arms race and to take effective measures toward reductions in strategic arms, nuclear disarmament, and general and complete disarmament,

Desiring to contribute to the relaxation of international tension and the strengthening of trust between States,

Have agreed as follows:

Article I

1. Each Party undertakes to limit anti-ballistic missile (ABM) systems and to adopt other measures in accordance with the provisions of this Treaty.

This document appears in United States Arms Control and Disarmament Agency, *Arms Control and Disarmament Agreements* (Washington, D.C.: U.S. Government Printing Office, 1982), pp. 139–42.

1. For interim agreement and protocol between the United States and the Soviet Union, signed May 26, 1972, see TIAS 7504; 23 UST.

2. TIAS 6839; 21 UST 490.

2. Each Party undertakes not to deploy ABM systems for a defense of the territory of its country and not to provide a base for such a defense, and not to deploy ABM systems for defense of an individual region except as provided for in Article III of this Treaty.

Article II

1. For the purposes of this Treaty an ABM system is a system to counter strategic ballistic missiles or their elements in flight trajectory, currently consisting of:

(a) ABM interceptor missiles, which are interceptor missiles constructed and deployed for an ABM role, or of a type tested in an ABM mode;

(b) ABM launchers, which are launchers constructed and deployed for launching ABM interceptor missiles; and

(c) ABM radars, which are radars constructed and deployed for an ABM role, or of a type tested in an ABM mode.

2. The ABM system components listed in paragraph 1 of this Article include those which are:

(a) operational;

(b) under construction;

(c) undergoing testing;

(d) undergoing overhaul, repair or conversion; or

(e) mothballed.

Article III

Each Party undertakes not to deploy ABM systems or their components except that:

(a) within one ABM system deployment area having a radius of one hundred and fifty kilometers and centered on the Party's national capital, a Party may deploy: (1) no more than one hundred ABM launchers and no more than one hundred ABM interceptor missiles at launch sites, and (2) ABM radars within no more than six ABM radar complexes, the area of each complex being circular and having a diameter of no more than three kilometers; and

(b) within one ABM system deployment area having a radius of one hundred and fifty kilometers and containing ICBM silo launchers, a Party may deploy: (1) no more than one hundred ABM launchers and no more than one hundred ABM interceptor missiles at launch sites, (2) two large phased-array

ABM radars comparable in potential to corresponding ABM radars operational or under construction on the date of signature of the Treaty in an ABM system deployment area containing ICBM silo launchers, and (3) no more than eighteen ABM radars each having a potential less than the potential of the smaller of the above-mentioned two large phased-array ABM radars.

Article IV

The limitations provided for in Article III shall not apply to ABM systems or their components used for development or testing, and located within current or additionally agreed test ranges. Each Party may have no more than a total of fifteen ABM launchers at test ranges.

Article V

1. Each Party undertakes not to develop, test, or deploy ABM systems or components which are sea-based, air-based, space-based, or mobile land-based.

2. Each Party undertakes not to develop, test, or deploy ABM launchers for launching more than one ABM interceptor missile at a time from each launcher, nor to modify deployed launchers to provide them with such a capability, nor to develop, test, or deploy automatic or semi-automatic or other similar systems for rapid reload of ABM launchers.

Article VI

To enhance assurance of the effectiveness of the limitations on ABM systems and their components provided by this Treaty, each Party undertakes:

(a) not to give missiles, launchers, or radars, other than ABM interceptor missiles, ABM launchers, or ABM radars, capabilities to counter strategic ballistic missiles or their elements in flight trajectory, and not to test them in an ABM mode; and

(b) not to deploy in the future radars for early warning of strategic ballistic missiles attack except at locations along the periphery of its national territory and oriented outward.

Article VII

Subject to the provisions of this Treaty, modernization and replacement of ABM systems or their components may be carried out.

Article VIII

ABM systems or their components in excess of the numbers or outside the areas specified in this Treaty, as well as ABM systems or their components prohibited by this Treaty, shall be destroyed or dismantled under agreed procedures within the shortest possible agreed period of time.

Article IX

To assure the viability and effectiveness of this Treaty, each Party undertakes not to transfer to other States, and not to deploy outside its national territory, ABM systems or their components limited by this Treaty.

Article X

Each Party undertakes not to assume any international obligations which would conflict with this Treaty.

Article XI

The Parties undertake to continue active negotiations for limitations on strategic offensive arms.

Article XII

1. For the purpose of providing assurance of compliance with the provisions of this Treaty, each Party shall use national technical means of verification at its disposal in a manner consistent with generally recognized principles of international law.

2. Each Party undertakes not to interfere with the national technical

means of verification of the other Party operating in accordance with paragraph 1 of this Article.

3. Each Party undertakes not to use deliberate concealment measures which impede verification by national technical means of compliance with the provisions of this Treaty. This obligation shall not require changes in current construction, assembly, conversion, or overhaul practices.

Article XIII

1. To promote the objectives and implementation of the provisions of this Treaty, the Parties shall establish promptly a Standing Consultative Commission, within the framework of which they will:

(a) consider questions concerning compliance with the obligations assumed and related situations which may be considered ambiguous;

(b) provide on a voluntary basis such information as either Party considers necessary to assure confidence in compliance with the obligations assumed;

(c) consider questions involving unintended interference with national technical means of verification;

(d) consider possible changes in the strategic situation which have a bearing on the provisions of this Treaty;

(e) agree upon procedures and dates for destruction or dismantling of ABM systems or their components in cases provided for by the provisions of this Treaty;

(f) consider, as appropriate, possible proposals for further increasing the viability of this Treaty, including proposals for amendments in accordance with the provisions of this Treaty;

(g) consider, as appropriate, proposals for further measures aimed at limiting strategic arms.

2. The Parties through consultation shall establish, and may amend as appropriate, Regulations for the Standing Consultative Commission governing procedures, composition and other relevant matters.

Article XIV

1. Each Party may propose amendments to this Treaty. Agreed amendments shall enter into force in accordance with the procedures governing the entry into force of this Treaty.

2. Five years after entry into force of this Treaty, and at five year intervals thereafter, the Parties shall together conduct a review of this Treaty.

Article XV

1. This Treaty shall be of unlimited duration.
2. Each Party shall, in exercising its national sovereignty, have the right to withdraw from this Treaty if it decides that extraordinary events related to the subject matter of this Treaty have jeopardized its supreme interests. It shall give notice of its decision to the other Party six months prior to withdrawal from the Treaty. Such notice shall include a statement of the extraordinary events the notifying Party regards as having jeopardized its supreme interests.

Article XVI

1. This Treaty shall be subject to ratification in accordance with the constitutional procedures of each Party. The Treaty shall enter into force on the day of the exchange of instruments of ratification.
2. This Treaty shall be registered pursuant to Article 102 of the Charter of the United Nations.[3]

Done at Moscow on May 26, 1972, in two copies, each in the English and Russian languages, both texts being equally authentic.

FOR THE UNITED STATES
OF AMERICA:[4]

FOR THE UNION OF SOVIET
SOCIALIST REPUBLICS:[5]

President of the United States
of America

General Secretary of the
Central Committee of the CPSU

3. TS 993; 59 Stat. 1052.
4. Richard Nixon.
5. L. I. Brezhnev.

56. Interim Agreement Between the United States of America and the Union of Soviet Socialist Republics on Certain Measures with Respect to the Limitation of Strategic Offensive Arms, May 26, 1972 (with Protocol)

The United States of America and the Union of Soviet Socialist Republics, hereinafter referred to as the Parties,

Convinced that the Treaty on the Limitation of Anti-Ballistic Missile Systems[1] and this Interim Agreement on Certain Measures with Respect to the Limitation of Strategic Offensive Arms will contribute to the creation of more favorable conditions for active negotiations on limiting strategic arms as well as to the relaxation of international tension and the strengthening of trust between States,

Taking into account the relationship between strategic offensive and defensive arms,

Mindful of their obligations under Article VI of the Treaty on the Non-Proliferation of Nuclear Weapons,[2]

Have agreed as follows:

Article I

The Parties undertake not to start construction of additional fixed land-based intercontinental ballistic missile (ICBM) launchers after July 1, 1972.

Article II

The Parties undertake not to convert land-based launchers for light ICBM's, or for ICBM's of older types deployed prior to 1964, into land-based launchers for heavy ICBM's of types deployed after that time.

This document appears in United States Arms Control and Disarmament Agency, *Arms Control and Disarmament Agreements* (Washington, D.C.: U.S. Government Printing Office, 1982), pp. 150–53.

1. TIAS 7503; 23 UST.
2. TIAS 6839; 21 UST 490.

Article III

The Parties undertake to limit submarine-launched ballistic missile (SLBM) launchers and modern ballistic missile submarines to the numbers operational and under construction on the date of signature of this Interim Agreement, and in addition to launchers and submarines constructed under procedures established by the Parties as replacements for an equal number of ICBM launchers of older types deployed prior to 1964 or for launchers on older submarines.

Article IV

Subject to the provisions of this Interim Agreement, modernization and replacement of strategic offensive ballistic missiles and launchers covered by this Interim Agreement may be undertaken.

Article V

1. For the purpose of providing assurance of compliance with the provisions of this Interim Agreement, each Party shall use national technical means of verification at its disposal in a manner consistent with generally recognized principles of international law.

2. Each Party undertakes not to interfere with the national technical means of verification of the other Party operating in accordance with paragraph 1 of this Article.

3. Each Party undertakes not to use deliberate concealment measures which impede verification by national technical means of compliance with the provisions of this Interim Agreement. This obligation shall not require changes in current construction, assembly, conversion, or overhaul practices.

Article VI

To promote the objectives and implementation of the provisions of this Interim Agreement, the Parties shall use the Standing Consultative Commission established under Article XIII of the Treaty on the Limitation of

Anti-Ballistic Missile Systems in accordance with the provisions of that Article.

Article VII

The Parties undertake to continue active negotiations for limitations on strategic offensive arms. The obligations provided for in this Interim Agreement shall not prejudice the scope or terms of the limitations on strategic offensive arms which may be worked out in the course of further negotiations.

Article VIII

1. This Interim Agreement shall enter into force[3] upon exchange of written notices of acceptance by each Party, which exchange shall take place simultaneously with the exchange of instruments of ratification of the Treaty on the Limitation of Anti-Ballistic Missile Systems.

2. This Interim Agreement shall remain in force for a period of five years unless replaced earlier by an agreement on more complete measures limiting strategic offensive arms. It is the objective of the Parties to conduct active follow-on negotiations with the aim of concluding such an agreement as soon as possible.

3. Each Party shall, in exercising its national sovereignty, have the right to withdraw from this Interim Agreement if it decides that extraordinary events related to the subject matter of this Interim Agreement have jeopardized its supreme interests. It shall give notice of its decision to the other Party six months prior to withdrawal from this Interim Agreement. Such notice shall include a statement of the extraordinary events the notifying Party regards as having jeopardized its supreme interests.

Done at Moscow on May 26, 1972, in two copies, each in the English and Russian languages, both texts being equally authentic.

FOR THE UNITED STATES
OF AMERICA:[4]

President of the United States
of America

FOR THE UNION OF SOVIET
SOCIALIST REPUBLICS:[5]

General Secretary of the
Central Committee of the CPSU

3. October 3, 1972.
4. Richard Nixon.
5. L. I. Brezhnev.

Protocol

To the Interim Agreement Between
the United States of America and the Union of Soviet
Socialist Republics on Certain Measures with Respect
to the Limitation of Strategic Offensive Arms

The United States of America and the Union of Soviet Socialist Republics, hereinafter referred to as the Parties,

Having agreed on certain limitations relating to submarine-launched ballistic missile launchers and modern ballistic missile submarines, and to replacement procedures, in the Interim Agreement,

Have agreed as follows:

The Parties understand that, under Article III of the Interim Agreement, for the period during which that Agreement remains in force:

The US may have no more than 710 ballistic missile launchers on submarines (SLBM's) and no more than 44 modern ballistic missile submarines. The Soviet Union may have no more than 950 ballistic missile launchers on submarines and no more than 62 modern ballistic missile submarines.

Additional ballistic missile launchers on submarines up to the above-mentioned levels, in the U.S.—over 656 ballistic missile launchers on nuclear-powered submarines, and in the U.S.S.R.—over 740 ballistic missile launchers on nuclear-powered submarines, operational and under construction, may become operational as replacements for equal numbers of ballistic missile launchers of older types deployed prior to 1964 or of ballistic missile launchers on older submarines.

The deployment of modern SLBM's on any submarine, regardless of type, will be counted against the total level of SLBM's permitted for the U.S. and the U.S.S.R.

This Protocol shall be considered an integral part of the Interim Agreement.

Done at Moscow this 26th day of May, 1972.

FOR THE UNITED STATES
OF AMERICA:

President of the United States
of America

FOR THE UNION OF SOVIET
SOCIALIST REPUBLICS:

General Secretary of the
Central Committee of the CPSU

57. USA-USSR Agreement to Reduce the Risk of Nuclear War Outbreak, September 30, 1972

The United States of America and the Union of Soviet Socialist Republics, hereinafter referred to as the Parties:

Taking into account the devastating consequences that nuclear war would have for all mankind, and recognizing the need to exert every effort to avert the risk of outbreak of such a war, including measures to guard against accidental or unauthorized use of nuclear weapons,

Believing that agreement on measures for reducing the risk of outbreak of nuclear war serves the interests of strengthening international peace and security, and is in no way contrary to the interests of any other country,

Bearing in mind that continued efforts are also needed in the future to seek ways of reducing the risk of outbreak of nuclear war,

Have agreed as follows:

Article 1

Each Party undertakes to maintain and to improve, as it deems necessary, its existing organizational and technical arrangements to guard against the accidental or unauthorized use of nuclear weapons under its control.

Article 2

The Parties undertake to notify each other immediately in the event of an accidental, unauthorized or any other unexplained incident involving a possible detonation of a nuclear weapon which could create a risk of outbreak of nuclear war. In the event of such an incident, the Party whose nuclear weapon is involved will immediately make every effort to take necessary measures to render harmless or destroy such weapon without its causing damage.

This document appears in United States Arms Control and Disarmament Agency, *Arms Control and Disarmament Agreements* (Washington, D.C.: U.S. Government Printing Office, 1982), pp. 111–12.

Article 3

The Parties undertake to notify each other immediately in the event of detection by missile warning systems of unidentified objects, or in the event of signs of interference with these systems or with related communications facilities, if such occurrences could create a risk of outbreak of nuclear war between the two countries.

Article 4

Each Party undertakes to notify the other Party in advance of any planned missile launches if such launches will extend beyond its national territory in the direction of the other Party.

Article 5

Each Party, in other situations involving unexplained nuclear incidents, undertakes to act in such a manner as to reduce the possibility of its actions being misinterpreted by the other Party. In any such situation, each Party may inform the other Party or request information when, in its view, this is warranted by the interests of averting the risk of outbreak of nuclear war.

Article 6

For transmission of urgent information, notifications and requests for information in situations requiring prompt clarification, the Parties shall make primary use of the Direct Communications Link between the Governments of the United States of America and the Union of Soviet Socialist Republics.[1]

For transmission of other information, notifications and requests for information, the Parties, at their own discretion, may use any communications facilities, including diplomatic channels, depending on the degree of urgency.

Article 7

The Parties undertake to hold consultations, as mutually agreed, to consider questions relating to implementation of the provisions of this

1. See TIAS 5362, 14 UST 825; 22 UST.

Agreement, as well as to discuss possible amendments thereto aimed at further implementation of the purposes of this Agreement.

Article 8

This Agreement shall be of unlimited duration.

Article 9

This Agreement shall enter into force upon signature.

Done at Washington on September 30, 1971, in two copies, each in the English and Russian languages, both texts being equally authentic.

FOR THE UNITED STATES   FOR THE UNION OF SOVIET
OF AMERICA:[2]   SOCIALIST REPUBLICS:[3]

---

58. Protocol to the Treaty Between the United States of America and the Union of Soviet Socialist Republics on the Limitation of Anti-Ballistic Missile Systems, July 3, 1974

The United States of America and the Union of Soviet Socialist Republics, hereinafter referred to as the Parties,

Proceeding from the Basic Principles of Relations between the United States of America and the Union of Soviet Socialist Republics signed on May 29, 1972,

Desiring to further the objectives of the Treaty between the United States of America and the Union of Soviet Socialist Republics on the

2. William P. Rogers.

3. A. Gromyko.

[Selection 58] This document appears in United States Arms Control and Disarmament Agency, *Arms Control and Disarmament Agreements* (Washington, D.C.: U.S. Government Printing Office, 1982), pp. 162–63.

Limitation of Anti-Ballistic Missile Systems signed on May 26, 1972, hereinafter referred to as the Treaty,

Reaffirming their conviction that the adoption of further measures for the limitation of strategic arms would contribute to strengthening international peace and security,

Proceeding from the premise that further limitation of anti-ballistic missile systems will create more favorable conditions for the completion of work on a permanent agreement on more complete measures for the limitation of strategic offensive arms,

Have agreed as follows:

Article I

1. Each Party shall be limited at any one time to a single area out of the two provided in Article III of the Treaty for deployment of anti-ballistic missile (ABM) systems or their components and accordingly shall not exercise its right to deploy an ABM system or its components in the second of the two ABM system deployment areas permitted by Article III of the Treaty, except as an exchange of one permitted area for the other in accordance with Article II of this Protocol.

2. Accordingly, except as permitted by Article II of this Protocol: The United States of America shall not deploy an ABM system or its components in the area centered on its capital, as permitted by Article III(s) of the Treaty, and the Soviet Union shall not deploy an ABM system or its components in the deployment area of intercontinental ballistic missile (ICBM) silo launchers permitted by Article III of the Treaty.

Article II

1. Each Party shall have the right to dismantle or destroy its ABM system and the components thereof in the area where they are presently deployed and to deploy an ABM system or its components in the alternative area permitted by Article III of the Treaty provided that prior to initiation of construction, notification is given in accord with the procedure agreed to by the Standing Consultative Commission, during the year beginning October 3, 1977, and ending October 2, 1978, or during any year which commences at five year intervals thereafter, those being the years for periodic review of the Treaty, as provided in Article XIV of the Treaty. This right may be exercised only once.

2. Accordingly, in the event of such notice, the United States would have the right to dismantle or destroy the ABM system and its components in the deployment area of ICBM silo launchers and to deploy an ABM system or its components in an area centered on its capital, as permitted by Article III(a) of the Treaty, and the Soviet Union would have the right to dismantle or destroy the ABM system and its components in the area centered on its capital and to deploy an ABM system or its components in an area containing ICBM silo launchers, as permitted by Article III(b) of the Treaty.

3. Dismantling or destruction and deployment of ABM systems or their components and the notification thereof shall be carried out in accordance with Article VIII of the ABM Treaty and procedures agreed to in the Standing Consultative Commission.

Article III

The rights and obligations established by the Treaty remain in force and shall be complied with by the Parties except to the extent modified by this Protocol. In particular, the deployment of an ABM system or its components within the area selected shall remain limited by the levels and other requirements established by the Treaty.

Article IV

This Protocol shall be subject to ratification in accordance with the constitutional procedures of each Party. It shall enter into force on the day of the exchange of instruments of ratification and shall thereafter be considered an integral part of the Treaty.

Done at Moscow on July 3, 1974, in duplicate, in the English and Russian languages, both texts being equally authentic.

For the United States of America:
  Richard Nixon
    President of the United States of America

For the Union of Soviet Socialist Republics:
  L. I. Brezhnev
    General Secretary of the Central Committee of the CPSU

59. SALT II Treaty Between the United States of America and the Union of Soviet Socialist Republics on the Limitation of Strategic Offensive Arms, June 18, 1979 (including Protocol and Agreed Understandings)

The United States of America and the Union of Soviet Socialist Republics, hereinafter referred to as the Parties,

Conscious that nuclear war would have devastating consequences for all mankind,

Proceeding from the Basic Principles of Relations Between the United States of America and the Union of Soviet Socialist Republics of May 29, 1972,

Attaching particular significance to the limitation of strategic arms and determined to continue their efforts begun with the Treaty on the Limitation of Anti-Ballistic Missile Systems and the Interim Agreement on Certain Measures with Respect to the Limitation of Strategic Offensive Arms, of May 26, 1972,

Convinced that the additional measures limiting strategic offensive arms provided for in this Treaty will contribute to the improvement of relations between the Parties, help to reduce the risk of outbreak of nuclear war, and strengthen international peace and security,

Mindful of their obligations under Article VI of the Treaty on the Non-Proliferation of Nuclear Weapons,

Guided by the principle of equality and equal security,

Recognizing that the strengthening of strategic stability meets the interests of the Parties and the interests of international security,

Reaffirming their desire to take measures for the further limitation and for the further reduction of strategic arms, having in mind the goal of achieving general and complete disarmament,

Declaring their intention to undertake in the near future negotiations further to limit and further to reduce strategic offensive arms,

Have agreed as follows:

This document appears in United States Arms Control and Disarmament Agency, *Arms Control and Disarmament Agreements* (Washington, D.C.: U.S. Government Printing Office, 1982), pp. 246–69.

Article I

Each Party undertakes, in accordance with the provisions of this Treaty, to limit strategic offensive arms quantitatively and qualitatively, to exercise restraint in the development of new types of strategic offensive arms, and to adopt other measures provided for in this Treaty.

Article II

For the purposes of this Treaty:

1. Intercontinental ballistic missile (ICBM) launchers are land-based launchers of ballistic missiles capable of a range in excess of the shortest distance between the northeastern border of the continental part of the territory of the United States of America and the northwestern border of the continental part of the territory of the Union of Soviet Socialist Republics, that is, a range in excess of 5,500 kilometers.

2. Submarine-launched ballistic missile (SLBM) launchers are launchers of ballistic missiles installed on any nuclear-powered submarine or launchers of modern ballistic missiles installed on any submarine, regardless of its type.

3. Heavy bombers are considered to be:

(a) currently, for the United States of America, bombers of the B-52 and B-1 types, and for the Union of Soviet Socialist Republics, bombers of the Tupolev-95 and Myasishchev types;

(b) in the future, types of bombers which can carry out the mission of a heavy bomber in a manner similar or superior to that of bombers listed in subparagraph (a) above;

(c) types of bombers equipped for cruise missiles capable of a range in excess of 600 kilometers; and

(d) types of bombers equipped for ASBMs.

4. Air-to-surface ballistic missiles (ASBMs) are any such missiles capable of a range in excess of 600 kilometers and installed in an aircraft or on its external mountings.

5. Launchers of ICBMs and SLBMs equipped with multiple independently targetable reentry vehicles (MIRVs) are launchers of the types developed and tested for launching ICBMs or SLBMs equipped with MIRVs.

6. ASBMs equipped with MIRVs are ASBMs of the types which have been flight-tested with MIRVs.

7. Heavy ICBMs are ICBMs which have a launch-weight greater or a throw-weight greater than that of the heaviest, in terms of either launch-weight or throw-weight, respectively, of the light ICBMs deployed by either Party as of the date of signature of this Treaty.

8. Cruise missiles are unmanned, self-propelled, guided, weapon-delivery vehicles which sustain flight through the use of aerodynamic lift over most of their flight path and which are flight-tested from or deployed on aircraft, that is, air-launched cruise missiles, or such vehicles which are referred to as cruise missiles in subparagraph I(b) of Article IX.

Article III

1. Upon entry into force of this Treaty, each Party undertakes to limit ICBM launchers, SLBM launchers, heavy bombers, and ASBMs to an aggregate number not to exceed 2,400.

2. Each Party undertakes to limit, from January 1, 1982, strategic offensive arms referred to in paragraph 1 of this Article to an aggregate number not to exceed 2,250 and to initiate reductions of those arms which as of that date would be in excess of this aggregate number.

3. Within the aggregate numbers provided for in paragraphs 1 and 2 of this Article and subject to the provisions of this Treaty, each Party has the right to determine the composition of these aggregates.

4. For each bomber of a type equipped for ASBMs, the aggregate number provided for in paragraphs 1 and 2 of this Article shall include the maximum number of such missiles for which a bomber of that type is equipped for one operational mission.

5. A heavy bomber equipped only for ASBMs shall not itself be included in the aggregate numbers provided for in paragraphs 1 and 2 of this Article.

6. Reductions of the numbers of strategic offensive arms required to comply with the provisions of paragraphs 1 and 2 of this Article shall be carried out as provided for in Article XI.

Article IV

1. Each Party undertakes not to start construction of additional fixed ICBM launchers.

2. Each Party undertakes not to relocate fixed ICBM launchers.

3. Each Party undertakes not to convert launchers of light ICBMs, or of ICBMs of older types deployed prior to 1964, into launchers of heavy ICBMs of types deployed after that time.

4. Each Party undertakes in the process of modernization and replacement of ICBM silo launchers not to increase the original internal volume of an ICBM silo launcher by more than thirty-two percent. Within this limit each Party has the right to determine whether such an increase will be made through an increase in the original diameter or in the original depth of an ICBM silo launcher, or in both of these dimensions.

5. Each Party undertakes:

(a) not to supply ICBM launcher deployment areas with intercontinental ballistic missiles in excess of a number consistent with normal deployment, maintenance, training, and replacement requirements;

(b) not to provide storage facilities for or to store ICBMs in excess of normal deployment requirements at launch sites of ICBM launchers;

(c) not to develop, test, or deploy systems for rapid reload of ICBM launchers.

6. Subject to the provisions of this Treaty, each Party undertakes not to have under construction at any time strategic offensive arms referred to in paragraph 1 of Article III in excess of numbers consistent with a normal construction schedule.

7. Each Party undertakes not to develop, test, or deploy ICBMs which have a launch-weight greater or a throw-weight greater than that of the heaviest, in terms of either launch-weight or throw-weight, respectively, of the heavy ICBMs deployed by either Party as of the date of signature of this Treaty.

8. Each Party undertakes not to convert land-based launchers of ballistic missiles which are not ICBMs into launchers for launching ICBMs, and not to test them for this purpose.

9. Each Party undertakes not to flight-test or deploy new types of ICBMs, that is, types of ICBMs not flight-tested as of May 1, 1979, except that each Party may flight-test and deploy one new type of light ICBM.

10. Each Party undertakes not to flight-test or deploy ICBMs of a type flight-tested as of May 1, 1979 with a number of reentry vehicles greater than the maximum number of reentry vehicles with which an ICBM of that type has been flight-tested as of that date.

11. Each Party undertakes not to flight-test or deploy ICBMs of the one

new type permitted pursuant to paragraph 9 of this Article with a number of reentry vehicles greater than the maximum number of reentry vehicles with which an ICBM of either Party has been flight-tested as of May 1, 1979, that is ten.

12. Each Party undertakes not to flight-test or deploy SLBMs with a number of reentry vehicles greater than the maximum number of reentry vehicles with which an SLBM of either Party has been flight-tested as of May 1, 1979, that is, fourteen.

13. Each Party undertakes not to flight-test or deploy ASBMs with a number of reentry vehicles greater than the maximum number of reentry vehicles with which an ICBM of either Party has been flight-tested as of May 1, 1979, that is, ten.

14. Each Party undertakes not to deploy at any one time on heavy bombers equipped for cruise missiles capable of a range in excess of 600 kilometers a number of such cruise missiles which exceeds the product of 28 and the number of such heavy bombers. . . .

Soviet Backfire Statement

On June 16, 1979, President Brezhnev handed President Carter the following written statement:

"The Soviet side informs the U.S. side that the Soviet 'Tu-22M' [alternatively designated Tu-29] airplane, called 'Backfire' in the U.S.A., is a medium-range bomber and that it does not intend to give this airplane the capability of operating at intercontinental distances. In this connection, the Soviet side states that it will not increase the radius of action of this airplane in such a way as to enable it to strike targets on the territory of the U.S.A. Nor does it intend to give it such a capability in any other manner, including by in-flight refueling. At the same time, the Soviet side states that it will not increase the production rate of this airplane as compared to the present rate."

President Brezhnev confirmed that the Soviet Backfire production rate would not exceed 30 per year.

President Carter stated that the United States enters into the SALT II agreement on the basis of the commitments contained in the Soviet statement and that it considers the carrying out of these commitments to be essential to the obligation assumed under the treaty.

Memorandum of Understanding Between the United States of America and the Union of Soviet Socialist Republics Regarding the Establishment of a Data Base on the Numbers of Strategic Offensive Arms

For the purposes of the Treaty between the United States of America and the Union of Soviet Socialist Republics on the Limitation of Strategic Offensive Arms, the Parties have considered data on numbers of strategic offensive arms and agree that as of November 1, 1978 there existed the following numbers of strategic offensive arms subject to the limitations provided for in the Treaty which is being signed today.

	U.S.A.	U.S.S.R.
Launchers of ICBMs	1,054	1,398
Fixed launchers of ICBMs	1,054	1,398
Launchers of ICBMs equipped with MIRVs	550	576
Launchers of SLBMs	656	950
Launchers of SLBMs equipped with MIRVs	496	128
Heavy bombers	574	156
Heavy bombers equipped for cruise missiles capable of a range in excess of 600 kilometers	0	0
Heavy bombers equipped only for ASBMs	0	0
ASBMs	0	0
ASBMs equipped with MIRVs	0	0

At the time of entry into force of the Treaty the Parties will update the above agreed data in the categories listed in this Memorandum.

Done at Vienna on June 18, 1979, in two copies, each in the English and Russian languages, both texts being equally authentic.

FOR THE UNITED STATES
OF AMERICA

FOR THE UNION OF SOVIET
SOCIALIST REPUBLICS

CHIEF OF THE UNITED
STATES DELEGATION TO
THE STRATEGIC ARMS
LIMITATION TALKS

CHIEF OF THE U.S.S.R.
DELEGATION TO THE
STRATEGIC ARMS
LIMITATION TALKS

Statement of Data on the Numbers of Strategic Offensive Arms of the Date of Signature of the Treaty

The United States of America declares that as of June 18, 1979 it possesses the following numbers of strategic offensive arms subject to the limitations provided for in the Treaty which is being signed today.

Launchers of ICBMs	1,054
Fixed launchers of ICBMs	1,054
Launchers of ICBMs equipped with MIRVs	550
Launchers of SLBMs	656
Launchers of SLBMs equipped with MIRVs	496
Heavy bombers	573
Heavy bombers equipped for cruise missiles capable of a range in excess of 600 kilometers	3
Heavy bombers equipped only for ASBMs	0
ASBMs	0
ASBMs equipped with MIRVs	0

June 18, 1979

CHIEF OF THE
UNITED STATES DELEGATION
TO THE STRATEGIC ARMS
LIMITATION TALKS

Statement of Data on the Numbers of Strategic Offensive Arms of the Date of Signature of the Treaty

The Union of Soviet Socialist Republics declares that as of June 18, 1979 it possesses the following numbers of strategic offensive arms subject to the limitations provided for in the Treaty which is being signed today:

Launchers of ICBMs	1,398
Fixed launchers of ICBMs	1,398
Launchers of ICBMs equipped with MIRVs	608
Launchers of SLBMs	950
Launchers of SLBMs equipped with MIRVs	144

Heavy bombers	156
Heavy bombers equipped for cruise missiles capable of a range in excess of 600 kilometers	0
Heavy bombers equipped only for ASBMs	0
ASBMs	0
ASBMs equipped with MIRVs	0

June 18, 1979

CHIEF OF THE
U.S.S.R. DELEGATION
TO THE STRATEGIC ARMS
LIMITATION TALKS

Joint Statement of Principles and Basic Guidelines for Subsequent Negotiations on the Limitation of Strategic Arms

The United States of America and the Union of Soviet Socialist Republics, hereinafter referred to as the Parties,

Having concluded the Treaty on the Limitation of Strategic Offensive Arms,

Reaffirming that the strengthening of strategic stability meets the interests of the Parties and the interests of international security,

Convinced that early agreement on the further limitation and further reduction of strategic arms would serve to strengthen international peace and security and to reduce the risk of outbreak of nuclear war,

Have agreed as follows:

First. The Parties will continue to pursue negotiations, in accordance with the principle of equality and equal security, on measures for the further limitation and reduction in the numbers of strategic arms, as well as for their further qualitative limitation.

In furtherance of existing agreements between the Parties on the limitation and reduction of strategic arms, the Parties will continue, for the purposes of reducing and averting the risk of outbreak of nuclear war, to seek measures to strengthen strategic stability by, among other things, limitations on strategic offensive arms most destabilizing to the strategic balance and by measures to reduce and to avert the risk of surprise attack.

Second. Further limitations and reductions of strategic arms must be subject to adequate verification by national technical means, using additionally, as appropriate, cooperative measures contributing to the effectiveness of verification by national technical means. The Parties will seek to strengthen verification and to perfect the operation of the Standing Consultative Commission in order to promote assurance of compliance with the obligations assumed by the Parties.

Third. The Parties shall pursue in the course of these negotiations, taking into consideration factors that determine the strategic situation, the following objectives:

1) significant and substantial reductions in the numbers of strategic offensive arms;
2) qualitative limitations on strategic offensive arms, including restrictions on the development, testing, and deployment of new types of strategic offensive arms and on the modernization of existing strategic offensive arms;
3) resolution of the issues included in the Protocol to the Treaty Between the United States of America and the Union of Soviet Socialist Republics on the Limitation of Strategic Offensive Arms in the context of the negotiations relating to the implementation of the principles and objectives set out herein.

Fourth. The Parties will consider other steps to ensure and enhance strategic stability, to ensure the equality and equal security of the Parties, and to implement the above principles and objectives. Each Party will be free to raise any issue relative to the further limitation of strategic arms. The Parties will also consider further joint measures, as appropriate, to strengthen international peace and security and to reduce the risk of outbreak of nuclear war.

Vienna, June 18, 1979

FOR THE UNITED STATES
OF AMERICA

PRESIDENT OF THE UNITED
STATES OF AMERICA

FOR THE UNION OF SOVIET
SOCIALIST REPUBLICS

GENERAL SECRETARY OF
THE CPSU, CHAIRMAN OF
THE PRESIDIUM OF THE
SUPREME SOVIET OF THE
USSR

60. U.S. Catholic Bishops' Pastoral Letter on War and Peace, May 3, 1983[1]

The Second Vatican Council opened its evaluation on modern warfare with the statement: "The whole human race faces a moment of supreme crisis in its advance toward maturity." We agree with the council's assessment; the crisis of the moment is embodied in the threat which nuclear weapons pose for the world and much that we hold dear in the world. We have seen and felt the effects of the crisis of the nuclear age in the lives of people we serve. Nuclear weaponry has drastically changed the nature of warfare and the arms race poses a threat to human life and human civilization which is without precedent.

We write this letter from the perspective of Catholic faith. Faith does not insulate us from the daily challenges of life, but intensifies our desire to address them precisely in light of the gospel which has come to us in the person of the Risen Christ. Through the resources of faith and reason we desire in this letter to provide hope for people in our day and direction toward a world freed of the nuclear threat.

As Catholic bishops we write this letter as an exercise of our teaching ministry. The Catholic tradition on war and peace is a long and complex one; it stretches from the Sermon on the Mount to the statements of Pope John Paul II. We wish to explore and explain the resources of the moral-religious teaching and to apply it to specific questions of our day. In doing this we realize and we want readers of this letter to recognize that not all statements in this letter have the same moral authority. At times we state universally binding moral principles found in the teaching of the church; at other times the pastoral letter makes specific applications, observations and recommendations which allow for diversity of opinion on the part of those who assess the factual data of situations differently. However, we expect Catholics to give our moral judgments serious consideration when they are forming their own views on specific problems.

Summary of National Conference of Catholic Bishops, "The Challenge of Peace: God's Promise and Our Response; A Pastoral Letter on War and Peace," May 3, 1983. Copyright © 1983 by the United States Catholic Conference, Inc., reprinted by permission.
    1. William P. Clark based his essay (Document No. 61) on an earlier draft.

The experience of preparing this letter has manifested to us the range of strongly held opinion in the Catholic community on questions of both fact and judgment concerning issues of war and peace. We urge mutual respect among individuals and groups in the church as this letter is analyzed and discussed. Obviously, as bishops, we believe that such differences should be expressed within the framework of Catholic moral teaching. We need in the church not only conviction and commitment, but also civility and charity.

While this letter is addressed principally to the Catholic community, we want it to make a contribution to the wider public debate in our country on the dangers and dilemmas of the nuclear age. Our contribution will not be primarily technical or political, but we are convinced that there is no satisfactory answer to the human problems of the nuclear age which fails to consider the moral and religious dimensions of the questions we face.

Although we speak in our own name as Catholic bishops of the church in the United States, we have been conscious in the preparation of this letter of the consequences our teaching will have not only for the United States, but for other nations as well. One important expression of this awareness has been the consultation we have had, by correspondence and in an important meeting held at the Vatican (Jan. 18–19, 1983), with representatives of European bishops' conferences. This consultation with bishops of other countries and, of course, with the Holy See has been very helpful to us.

Catholic teaching has always understood peace in positive terms. In the words of Pope John Paul II: "Peace is not just the absence of war. . . . Like a cathedral, peace must be constructed patiently and with unshakable faith" (Coventry, England, 1982). Peace is the fruit of order. Order in human society must be shaped on the basis of respect for the transcendence of God and the unique dignity of each person, understood in terms of freedom, justice, truth and love. To avoid war in our day we must be intent on building peace in an increasingly interdependent world. In Part III of this letter we set forth a positive vision of peace and the demands such a vision makes on diplomacy, national policy, and personal choices.

While pursuing peace incessantly, it is also necessary to limit the use of force in a world comprised of nation states, faced with common problems but devoid of an adequate international political authority. Keeping the peace in the nuclear age is a moral and political imperative. In Parts I and II of this letter we set forth both the principles of Catholic teaching on war and a series of judgments, based on these principles, about concrete policies. In making these judgments we speak as moral teachers, not as technical experts.

I. Some Principles, Norms and Premises of Catholic Teaching

A. *On War:*
1. Catholic teaching begins in every case with a presumption against war and for peaceful settlement of disputes. In exceptional cases, determined by the moral principles of the Just War tradition, some uses of force are permitted.
2. Every nation has a right and duty to defend itself against unjust aggression.
3. Offensive war of any kind is not morally justifiable.
4. It is never permitted to direct nuclear or conventional weapons to "the indiscriminate destruction of whole cities or vast areas with their populations . . ." ("The Pastoral Constitution on the Church in the Modern World," No. 80). The intentional killing of innocent civilians or noncombatants is always wrong.
5. Even defensive response to unjust attack can cause destruction which violates the principle of proportionality, going far beyond the limits of legitimate defense. This judgment is particularly important when assessing planned use of nuclear weapons. No defensive strategy, nuclear or conventional which exceeds the limits of proportionality is morally permissible.

B. *On Deterrence:*
1. "In current conditions 'deterrence' based on balance, certainly not as an end in itself but as a step on the way toward a progressive disarmament, may still be judged morally acceptable. Nonetheless, in order to ensure peace, it is indispensable not to be satisfied with this minimum which is always susceptible to the real danger of explosion." (Pope John Paul II, Message to U.N. Special Session on Disarmament, No. 8; June 1982.)
2. No use of nuclear weapons which would violate the principles of discrimination or proportionality may be intended in a strategy of deterrence. The moral demands of Catholic teaching require absolute willingness not to intend or to do moral evil even to save our own lives or the lives of those we love.
3. Deterrence is not an adequate strategy as a long-term basis for peace; it is a transitional strategy justifiable only in conjunction with resolute determination to pursue arms control and disarmament. We are convinced that "the fundamental principle on which our present peace depends must be replaced by another, which declares that the true and solid peace of nations consists not in equality of arms but in mutual trust alone." (Pope John XXIII, "Peace On Earth," No. 113.)

C. *The Arms Race and Disarmament:*

1. The arms race is one of the greatest curses on the human race; it is to be condemned as a danger, an act of aggression against the poor, and a folly which does not provide the security it promises. (Cf: "The Pastoral Constitution . . . ," No. 81; Statement of the Holy See to the United Nations, 1976.)

2. Negotiations must be pursued in every reasonable form possible; they should be governed by the "demand that the arms race should cease; that the stockpiles which exist in various countries should be reduced equally and simultaneously by the parties concerned; that nuclear weapons should be banned; and that a general agreement should eventually be reached about progressive disarmament and an effective method of control." (Pope John XXIII, "Peace On Earth," No. 112.)

D. *On Personal Conscience:*

1. Military Service: "All those who enter the military service in loyalty to their country should look upon themselves as the custodians of the security and freedom of their fellow countrymen; and when they carry out their duty properly, they are contributing to the maintenance of peace." ("The Pastoral Constitution . . . ," No. 79.)

2. Conscientious Objection: "Moreover, it seems just that laws should make humane provision for the case of conscientious objectors who refuse to carry arms, provided they accept some other form of community service." ("The Pastoral Constitution . . . ," No. 79.)

3. Non-violence: "In this same spirit we cannot but express our admiration for all who forgo the use of violence to vindicate their rights and resort to other means of defense which are available to weaker parties, provided it can be done without harm to the rights and duties of others and of the community." ("The Pastoral Constitution . . . ," No. 78.)

4. Citizens and Conscience: "Once again we deem it opportune to remind our children of their duty to take an active part in public life, and to contribute towards the attainment of the common good of the entire human family as well as to that of their own political community. . . . In other words, it is necessary that human beings, in the intimacy of their own consciences, should so live and act in their temporal lives as to create a synthesis between scientific, technical and professional elements on the one hand, and spiritual values on the other." (Pope John XXIII, "Peace On Earth," No. 146; 150.)

II. Moral Principles and Policy Choices:

As bishops in the United States, assessing the concrete circumstances of our society, we have made a number of observations and recommendations in the process of applying moral principles to specific policy choices.

A. *On the Use of Nuclear Weapons:*
1. Counter-Population Use: Under no circumstances may nuclear weapons or other instruments of mass slaughter be used for the purpose of destroying population centers or other predominantly civilian targets. Retaliatory action which would indiscriminately and disproportionately take many wholly innocent lives, lives of people who are in no way responsible for reckless actions of their government, must also be condemned.
2. The Initiation of Nuclear War: We do not perceive any situation in which the deliberate initiation of nuclear war, on however restricted a scale, can be morally justified. Non-nuclear attacks by another state must be resisted by other than nuclear means. Therefore, a serious obligation exists to develop morally acceptable non-nuclear defensive strategies as rapidly as possible. In this letter we urge NATO to move rapidly toward the adoption of a "no first use" policy but we recognize this will take time to implement and will require the development of an adequate alternative defense posture.
3. Limited Nuclear War: Our examination of the various arguments on this question makes us highly skeptical about the real meaning of "limited." One of the criteria of the just war teaching is that there must be a reasonable hope of success in bringing about justice and peace. We must ask whether such a reasonable hope can exist once nuclear weapons have been exchanged. The burden of proof remains on those who assert that meaningful limitation is possible. In our view the first imperative is to prevent any use of nuclear weapons and we hope that leaders will resist the notion that nuclear conflict can be limited, contained or won in any traditional sense.

B. *On Deterrence:*
In concert with the evaluation provided by Pope John Paul II, we have arrived at a strictly conditioned moral acceptance of deterrence. In this letter we have outlined criteria and recommendations which indicate the meaning of conditional acceptance of deterrence policy. We cannot consider such a policy adequate as a long-term basis for peace.

C. On Promoting Peace:

1. We support immediate, bilateral, verifiable agreements to halt the testing, production and deployment of nuclear weapons systems. This recommendation is not to be identified with any specific political initiative.

2. We support efforts to achieve deep cuts in the arsenals of both superpowers; efforts should concentrate first on systems which threaten the retaliatory forces of either major power.

3. We support early and successful conclusion of negotiations of a comprehensive test ban treaty.

4. We urge new efforts to prevent the spread of nuclear weapons in the world, and to control the conventional arms race, particularly the conventional arms trade.

5. We support, in an increasingly interdependent world, political and economic policies designed to protect human dignity and to promote the human rights of every person, especially the least among us. In this regard, we call for the establishment of some form of global authority adequate to the needs of the international common good.

This letter includes many judgments from the perspective of ethics, politics, and strategy needed to speak concretely and correctly to the "moment of supreme crisis" identified by Vatican II. We stress again that readers should be aware, as we have been, of the distinction between our statement of moral principles and of official Church teaching and our application of these to concrete issues. We urge that special care be taken not to use passages out of context; neither should brief portions of this document be cited to support positions it does not intend to convey or which are not truly in accord with the spirit of its teaching.

In concluding this summary we respond to two key questions often asked about this pastoral letter:

Why do we address these matters fraught with such complexity, controversy and passion? We speak as pastors, not politicians. We are teachers, not technicians. We cannot avoid our responsibility to lift up the moral dimensions of the choices before our world and nation. The nuclear age is an era of moral as well as physical danger. We are the first generation since Genesis with the power to threaten the created order. We cannot remain silent in the face of such danger. Why do we address these issues? We are simply trying to live up to the call of Jesus to be peacemakers in our own time and situation.

What are we saying? Fundamentally, we are saying that the decisions about nuclear weapons are among the most pressing moral questions of our age. While these decisions have obvious military and political aspects, they involve fundamental moral choices. In simple terms, we are saying that good ends (defending one's country, protecting freedom, etc.) cannot justify immoral means (the use of weapons which kill indiscriminately and threaten whole societies). We fear that our world and nation are headed in the wrong direction. More weapons with greater destructive potential are produced every day. More and more nations are seeking to become nuclear powers. In our quest for more and more security we fear we are actually becoming less and less secure.

In the words of our Holy Father, we need a "moral about-face." The whole world must summon the moral courage and technical means to say no to nuclear conflict; no to weapons of mass destruction; no to an arms race which robs the poor and the vulnerable; and no to the moral danger of a nuclear age which places before humankind indefensible choices of constant terror or surrender. Peacemaking is not an optional commitment. It is a requirement of our faith. We are called to be peacemakers, not by some movement of the moment but by our Lord Jesus. The content and context of our peacemaking is set not by some political agenda or ideological program, but by the teaching of his Church.

Ultimately, this letter is intended as an expression of Christian faith, affirming the confidence we have that the Risen Lord remains with us precisely in moments of crisis. It is our belief in his presence and power among us which sustains us in confronting the awesome challenge of the nuclear age. We speak from faith to provide hope for all who recognize the challenge and are working to confront it with the resources of faith and reason.

To approach the nuclear issue in faith is to recognize our absolute need for prayer: We urge and invite all to unceasing prayer and works of penance for peace with justice for all people. In a spirit of prayerful hope we present this message of peace.

61. William P. Clark to Archbishop Joseph L. Bernadin,[1] November 16, 1982

I would like to take this opportunity to respond, on behalf of President Reagan, Secretary Shultz, Secretary Weinberger, Director Rostow, and other Administration officials, to the request for our views on the second draft of the pastoral letter recently prepared by the Ad Hoc Committee on War and Peace for review by the National Conference of Catholic Bishops. Let me assure you that we have read these drafts with great interest and care.

I believe we can agree that the purpose of any moral theory of defense is "not, in the first place, to legitimize war, but to prevent it," and this, of course, is what American deterrence policy is designed to achieve. I believe we can also agree that any proposed change in strategic systems or doctrines, as well as any recommendation, whether proposed by the U.S. Government or by your committee, should be judged "in light of whether it will render steps toward arms control and disarmament more, or less, likely." We believe that our weapons systems (which are not designed to be "first-strike" systems), our deterrence posture (which is defensive), and our arms control initiatives (which call for deep and verifiable reductions) do conform to these objectives.

As with the committee's first draft, I am especially troubled in reading the second draft of the pastoral letter to find none of the serious U.S. arms control efforts, including major initiatives and ongoing U.S.-Soviet negotiations, described or even noted in the text. Ours are not proposals for freezes on current high ceilings. Such freezes would remove incentives for achieving reductions and would, in any case, require extensive prior negotiations to reach agreement on what numbers and systems to freeze, and on how such freezes might be effectively verified. Ours are initiatives for reduction, or even elimination, of the most destabilizing systems. They involve new verification and confidence-building measures designed both to build trust and to assure compliance.

Because these important initiatives and negotiations have again been ignored in the draft pastoral letter, although they so clearly conform to the

This essay appears in the *New York Times*, November 17, 1982, p. 11. Copyright © 1982 by the New York Times Company. Reprinted by permission.

1. William P. Clark was Chairman of the National Security Council. Archbishop Joseph L. Bernadin was Chairman of the National Conference of Catholic Bishops' Committee on War and Peace.

hopes of all concerned with reducing the arsenals and the risks of war and promoting the path of peace, I would like to summarize them for you again. I do so with a renewed hope that the comments your committee receives from U.S. Government officials in response to the committee's requests will be carefully considered, just as your committee asks that its draft letter "receive a respectful consideration" from others.

This Administration's arms control efforts include the following major initiatives:

In the U.S.-Soviet negotiations on strategic arms (START), which began on June 30, 1982, we are proposing to being with a one-third reduction in the number of warheads on the land and sea-based ballistic missiles and a reduction in the most destabilizing systems of all, the land-based ballistic missiles, to about one-half of the current U.S. levels. In a second phase, we propose to reduce the destructive potential of the remaining missiles to equal levels, lower than we now have, and we could include other strategic systems as well.

In the U.S.-Soviet negotiations on intermediate-range nuclear forces (INF), which began on Nov. 30, 1981, we have proposed to begin with the total elimination of the forces considered the most destabilizing and threatening by both sides, the land-based missile systems. We and our NATO allies have offered to cancel plans for the deployment of U.S. Pershing and ground-launched cruise missiles in exchange for the corresponding destruction of Soviet SS-20, SS-4 and SS-5 missiles. Other elements of the balance could be limited subsequently.

In the multilateral negotiations on mutual and balanced force reductions (MBFR), the U.S. and its NATO allies are proposing to the Warsaw Pact nations major initial reductions in military personnel to common ceilings and a wide range of new verification measures.

In the areas of limiting nuclear testing and chemical and biological weapons, the U.S. is actively participating in discussions in the Committee on Disarmament in Geneva to develop the verification and compliance procedures that would make such limitations truly effective. We are, of course, particularly distressed by the extensive and inhuman use by the Soviet Union and its allies of toxins and chemicals against the defenseless populations of Afghanistan, Laos and Cambodia.

In all of our ongoing arms control negotiations and discussions, we are emphasizing the importance of substantial early reductions and of effective verification and confidence-building measures. Your committee will surely recognize the Administration's nuclear reductions proposals clearly conform to the pastoral letter's recommendations for cuts in nuclear arsenals, and that the other multilateral efforts in which we are currently engaged conform closely with the letter's call for efforts "aimed at reducing and limiting conventional forces and at building confidence between possible adversaries, especially in regions of major military confrontation, as well as those addressed to outlawing effectively the use of chemical and biological weapons."

I continue to believe that as the Bishops Conference reviews new drafts of the pastoral letter, a clear presentation of these initiatives should lead to the Bishops Conference's strong support for them. As I noted in my comments on the first draft, such support would prove enormously helpful in making clear to the world America's seriousness in our efforts and would, in particular, add to Soviet incentives to agree to the reductions and verifiable limitations that we are seeking.

---

62. Treaty Between the United States of America and the Union of Soviet Socialist Republics on the Elimination of Their Intermediate-Range and Shorter-Range Missiles, December 8, 1987

The United States of America and the Union of Soviet Socialist Republics, hereinafter referred to as the Parties,

Conscious that nuclear war would have devastating consequences for all mankind,

Guided by the objective of strengthening strategic stability,

Convinced that the measures set forth in this Treaty will help to reduce the risk of outbreak of war and strengthen international peace and security, and

Mindful of their obligations under Article VI of the Treaty on the Non-Proliferation of Nuclear Weapons,

Have agreed as follows:

Article I

In accordance with the provisions of this Treaty which includes the Memorandum of Understanding and Protocols which form an integral part

This document appears in *Department of State Bulletin* 88, 2131 (February 1988): 22, 24–30.

thereof, each Party shall eliminate its intermediate-range and shorter-range missiles, not have such systems thereafter, and carry out the other obligations set forth in this Treaty.

Article II

For the purposes of this Treaty:

1. The term "ballistic missile" means a missile that has a ballistic trajectory over most of its flight path. The term "ground-launched ballistic missile (GLBM)" means a ground-launched ballistic missile that is a weapon-delivery vehicle.

2. The term "cruise missile" means an unmanned, self-propelled vehicle that sustains flight through the use of aerodynamic lift over most of its flight path. The term "ground-launched cruise missile (GLCM)" means a ground-launched cruise missile that is a weapon-delivery vehicle.

3. The term "GLBM launcher" means a fixed launcher or a mobile land-based transporter-erector-launcher mechanism for launching a GLBM.

4. The term "GLCM launcher" means a fixed launcher or a mobile land-based transporter-erector-launcher mechanism for launching a GLCM.

5. The term "intermediate-range missile" means a GLBM or a GLCM having a range capability in excess of 1000 kilometers but not in excess of 5500 kilometers.

6. The term "shorter-range missile" means a GLBM or a GLCM having a range capability equal to or in excess of 500 kilometers but not in excess of 1000 kilometers.

7. The term "deployment area" means a designated area within which intermediate-range missiles and launchers of such missiles may operate and within which one or more missile operating bases are located.

8. The term "missile operating base" means:

(a) in the case of intermediate-range missiles, a complex of facilities located within a deployment area at which intermediate-range missiles and launchers of such missiles normally operate, in which support structures associated with such missiles and launchers are also located and in which support equipment associated with such missiles and launchers is normally located; and

(b) in the case of shorter-range missiles, a complex of facilities located any place at which shorter-range missiles and launchers of such missiles normally operate and in which support equipment associated with such missiles and launchers is normally located.

9. The term "missile support facility" as regards intermediate-range or

shorter-range missiles and launchers of such missiles, means a missile production facility or a launcher production facility, a missile repair facility or a launcher repair facility, a training facility, a missile storage facility or a launcher storage facility, a test range, or an elimination facility as those terms are defined in the Memorandum of Understanding.

10. The term "transit" means movement, notified in accordance with paragraph 5(f) of Article IX of this Treaty, of an intermediate-range missile or a launcher of such a missile between missile support facilities, between such a facility and a deployment area or between deployment areas, or of a shorter-range missile or a launcher of such a missile from a missile support facility or missile operating base to an elimination facility.

11. The term "deployed missile" means an intermediate-range missile located within a deployment area or a shorter-range missile located at a missile operating base.

12. The term "non-deployed missile" means an intermediate-range missile located outside a deployment area or a shorter-range missile located outside a missile operating base.

13. The term "deployed launcher" means a launcher of an intermediate-range missile located within a deployment area or a launcher of a shorter-range missile located at a missile operating base.

14. The term "non-deployed launcher" means a launcher of an intermediate-range missile located outside a deployment area or a launcher of a shorter-range missile located outside a missile operating base.

15. The term "basing country" means a country other than the United States of America or the Union of Soviet Socialist Republics on whose territory intermediate-range or shorter-range missiles of the Parties, launchers of such missiles or support structures associated with such missiles and launchers were located at any time after November 1, 1987. Missiles or launchers in transit are not considered to be "located."

Article III

1. For the purposes of this Treaty, existing types of intermediate-range missiles are:

(a) for the United States of America, missiles of the types designated by the United States of America as the Pershing II and the BGM-109G, which are known to the Union of Soviet Socialist Republics by the same designations; and

(b) for the Union of Soviet Socialist Republics, missiles of the types designated by the Union of Soviet Socialist Republics as the RSD-10, the

RD-12 and the R-14, which are known to the United States of America as the SS-20, the SS-4 and the SS-5, respectively.

2. For the purposes of this Treaty, existing types of shorter-range missiles are:

(a) for the United States of America, missiles of the type designated by the United States of America as the Pershing IA, which is known to the Union of Soviet Socialist Republics by the same designation; and

(b) for the Union of Soviet Socialist Republics, missiles of the types designated by the Union of Soviet Socialist Republics as the OTR-22 and the OTR-23, which are known to the United States of America as the SS-12 and the SS-23, respectively.

Article IV

1. Each Party shall eliminate all its intermediate-range missiles and launchers of such missiles, and all support structures and support equipment of the categories listed in the Memorandum of Understanding associated with such missiles and launchers so that no later than three years after entry into force of this Treaty and thereafter no such missiles, launchers, support structures or support equipment shall be possessed by either Party.

2. To implement paragraph 1 of this Article, upon entry into force of this Treaty, both Parties shall begin and continue throughout the duration of each phase, the reduction of all types of their deployed and non-deployed intermediate-range missiles and deployed and non-deployed launchers of such missiles and support structures and support equipment associated with such missiles and launchers in accordance with the provisions of this Treaty. These reductions shall be implemented in two phases so that:

(a) by the end of the first phase, that is, no later than 29 months after entry into force of this Treaty:

(i) the number of deployed launchers of intermediate-range missiles for each Party shall not exceed the number of launchers that are capable of carrying or containing at one time missiles considered by the Parties to carry 171 warheads;

(ii) the number of deployed and non-deployed intermediate-range missiles for each Party shall not exceed the number of such missiles considered by the Parties to carry 180 warheads;

(iii) the aggregate number of deployed and non-deployed launchers of intermediate-range missiles for each Party shall not exceed the number of launchers that are capable of carrying or containing at one time missiles considered by the Parties to carry 200 warheads;

(iv) the aggregate number of deployed and non-deployed intermediate-range missiles for each Party shall not exceed the number of such missiles considered by the Parties to carry 200 warheads; and

(v) the ratio of the aggregate number of deployed and non-deployed intermediate-range GLBMs of existing types for each Party to the aggregate number of deployed and non-deployed intermediate-range missiles of existing types possessed by that Party shall not exceed the ratio of such intermediate-range GLBMs to such intermediate-range missiles for that Party as of November 1, 1987, as set forth in the Memorandum of Understanding; and

(b) by the end of the second phase, that is, no later than three years after entry into force of this Treaty, all intermediate-range missiles of each Party, launchers of such missiles and all support structures and support equipment of the categories listed in the Memorandum of Understanding associated with such missiles and launchers, shall be eliminated.

Article V

1. Each Party shall eliminate all its shorter-range missiles and launchers of such missiles, and all support equipment of the categories listed in the Memorandum of Understanding associated with such missiles and launchers, so that no later than 18 months after entry into force of this Treaty and thereafter no such missiles, launchers or support equipment shall be possessed by either Party.

2. No later than 90 days after entry into force of this Treaty, each Party shall complete the removal of all its deployed shorter-range missiles and deployed and non-deployed launchers of such missiles to elimination facilities and shall retain them at those locations until they are eliminated in accordance with the procedures set forth in the Protocol on Elimination. No later than 12 months after entry into force of this Treaty, each Party shall complete the removal of all its non-deployed shorter-range missiles to elimination facilities and shall retain them at those locations until they are eliminated in accordance with the procedures set forth in the Protocol on Elimination.

3. Shorter-range missiles and launchers of such missiles shall not be located at the same elimination facility. Such facilities shall be separated by no less than 1000 kilometers.

Article VI

1. Upon entry into force of this Treaty and thereafter, neither Party shall:

(a) produce or flight-test any intermediate-range missiles or produce any stages of such missiles or any launchers of such missiles; or

(b) produce, flight-test or launch any shorter-range missiles or produce any stages of such missiles or any launchers of such missiles.

2. Notwithstanding paragraph 1 of this Article, each Party shall have the right to produce a type of GLBM not limited by this Treaty which uses a stage which is outwardly similar to, but not interchangeable with, a stage of an existing type of intermediate-range GLBM having more than one stage, providing that that Party shall not produce any other stage which is outwardly similar to, but not interchangeable with, any other stage of an existing type of intermediate-range GLBM.

Article VII

For the purposes of this Treaty:

1. If a ballistic missile or a cruise missile has been flight-tested or deployed for weapon delivery, all missiles of that type shall be considered to be weapon-delivery vehicles.

2. If a GLBM or GLCM is an intermediate-range missile, all GLBMs or GLCMs of that type shall be considered to be intermediate-range missiles. If a GLBM or GLCM is a shorter-range missile, all GLBMs or GLCMs of that type shall be considered to be shorter-range missiles.

3. If a GLBM is of a type developed and tested solely to intercept and counter objects not located on the surface of the earth, it shall not be considered to be a missile to which the limitations of this Treaty apply.

4. The range capability of a GLBM not listed in Article III of this Treaty shall be considered to be the maximum range to which it has been tested. The range capability of a GLCM not listed in Article III of this Treaty shall be considered to be the maximum distance which can be covered by the missile in its standard design mode flying until fuel exhaustion, determined by projecting its flight path onto the earth's sphere from the point of launch to the point of impact GLBMs or GLCMs that have a range capability equal to or in excess of 500 kilometers but not in excess of 1000 kilometers shall be considered to be shorter-range missiles. GLBMs or GLCMs that

have a range capability in excess of 1000 kilometers but not in excess of 5500 kilometers shall be considered to be intermediate-range missiles.

5. The maximum number of warheads an existing type of intermediate-range missile or shorter-range missile carries shall be considered to be the number listed for missiles of that type in the Memorandum of Understanding.

6. Each GLBM or GLCM shall be considered to carry the maximum number of warheads listed for a GLBM or GLCM of that type in the Memorandum of Understanding.

7. If a launcher has been tested for launching a GLBM or a GLCM, all launchers of that type shall be considered to have been tested for launching GLBMs or GLCMs.

8. If a launcher has contained or launched a particular type of GLBM or GLCM, all launchers of that type shall be considered to be launchers of that type of GLBM or GLCM.

9. The number of missiles each launcher of an existing type of intermediate-range missile or shorter-range missile shall be considered to be capable of carrying or containing at one time is the number listed for launchers of missiles of that type in the Memorandum of Understanding.

10. Except in the case of elimination in accordance with the procedures set forth in the Protocol on Elimination, the following shall apply:

(a) for GLBMs which are stored or moved in separate stages, the longest stage of an intermediate-range or shorter-range GLBM shall be counted as a complete missile;

(b) for GLBMs which are not stored or moved in separate stages, a canister of the type used in the launch of an intermediate-range GLBM, unless a Party proves to the satisfaction of the other Party that it does not contain such a missile, or an assembled intermediate-range or shorter-range GLBM, shall be counted as a complete missile; and

(c) for GLCMs, the airframe of an intermediate-range or shorter-range GLCM shall be counted as a complete missile.

11. A ballistic missile which is not a missile to be used in a ground-based mode shall not be considered to be a GLBM if it is test-launched at a test site from a fixed land-based launcher which is used solely for test purposes and which is distinguishable from GLBM launchers. A cruise missile which is not a missile to be used in a ground-based mode shall not be considered to be a GLCM if it is test-launched at a test site from a fixed land-based launcher which is used solely for test purposes and which is distinguishable from GLCM launchers.

12. Each Party shall have the right to produce and use for booster systems, which might otherwise be considered to be intermediate-range or

shorter-range missiles, only existing types of booster stages for such booster systems. Launches of such booster systems shall not be considered to be flight-testing of intermediate-range or shorter-range missiles provided that:

(a) stages used in such booster systems are different from stages used in those missiles listed as existing types of intermediate-range or shorter-range missiles in Article III of this Treaty;

(b) such booster systems are used only for research and development purposes to test objects other than the booster systems themselves;

(c) the aggregate number of launchers for such booster systems shall not exceed 35 for each Party at any one time; and

(d) the launchers for such booster systems are fixed, emplaced above ground and located only at research and development launch sites which are specified in the Memorandum of Understanding.

Research and development launch sites shall not be subject to inspection pursuant to Article XI of this Treaty.

Article VIII

1. All intermediate-range missiles and launchers of such missiles shall be located in deployment areas, at missile support facilities or shall be in transit. Intermediate-range missiles or launchers of such missiles shall not be located elsewhere.

2. Stages of intermediate-range missiles shall be located in deployment areas, at missile support facilities or moving between deployment areas, between missile support facilities or between missile support facilities and deployment areas.

3. Until their removal to elimination facilities as required by paragraph 2 of Article V of this Treaty, all shorter-range missiles and launchers of such missiles shall be located at missile operating bases, at missile support facilities or shall be in transit. Shorter-range missiles or launchers of such missiles shall not be located elsewhere.

4. Transit of a missile or launcher subject to the provisions of this Treaty shall be completed within 25 days.

5. All deployment areas, missile operating bases and missile support facilities are specified in the Memorandum of Understanding or in subsequent updates of data pursuant to paragraphs 3, 5(a) or 5(b) of Article IX of this Treaty. Neither Party shall increase the number of, or charge the location or boundaries of, deployment areas, missile operating bases or missile support facilities, except for elimination facilities, from those set forth in the

Memorandum of Understanding. A missile support facility shall not be considered to be part of a deployment area even though it may be located within the geographic boundaries of a deployment area.

6. Beginning 30 days after entry into force of this Treaty neither Party shall locate intermediate-range or shorter-range missiles, including stages of such missiles, or launchers of such missiles at missile production facilities, launcher production facilities or test ranges listed in the Memorandum of Understanding.

7. Neither Party shall locate any intermediate-range or shorter-range missiles at training facilities.

8. A non-deployed intermediate-range or shorter-range missile shall not be carried on or contained within a launcher of such a type of missile, except as required for maintenance conducted at repair facilities or for elimination by means of launching conducted at elimination facilities.

9. Training missiles and training launchers for intermediate-range or shorter-range missiles shall be subject to the same locational restrictions as are set forth for intermediate-range and shorter-range missiles and launchers of such missiles in paragraphs 1 and 3 of this Article.

Article IX

1. The Memorandum of Understanding contains categories of data relevant to obligations undertaken with regard to this Treaty and lists all intermediate-range and shorter-range missiles, launchers of such missiles, and support structures and support equipment associated with such missiles and launchers, possessed by the Parties as of November 1, 1987. Updates of that data and notifications required by this Article shall be provided according to the categories of data contained in the Memorandum of Understanding.

2. The Parties shall update that data and provide the notifications required by this Treaty through the Nuclear Risk Reduction Centers, established pursuant to the Agreement Between the United States of America and the Union of Soviet Socialist Republics on the Establishment of Nuclear Risk Reduction Centers of September 15, 1987.

3. No later than 30 days after entry into force of this Treaty, each Party shall provide the other Party with updated data, as of the date of entry into force of this Treaty, for all categories of data contained in the Memorandum of Understanding.

4. No later than 30 days after the end of each six-month interval following the entry into force of this Treaty, each Party shall provide updated data

for all categories of data contained in the Memorandum of Understanding by informing the other Party of all changes, completed and in process, in that data, which have occurred during the six-month interval since the preceding data exchange, and the net effect of those changes.

5. Upon entry into force of this Treaty and thereafter, each Party shall provide the following notifications to the other Party:

(a) notification, no less than 30 days in advance, of the scheduled date of the elimination of a specific deployment area, missile operating base or missile support facility;

(b) notification, no less than 30 days in advance, of changes in the number or location of elimination facilities, including the location and scheduled date of a change;

(c) notification, except with respect to launches of intermediate-range missiles for the purpose of their elimination, no less than 30 days in advance, of the scheduled date of the initiation of the elimination of intermediate-range and shorter-range missiles, and stages of such missiles, and launchers of such missiles and support structures and support equipment associated with such missiles and launchers, including:

(i) the number and type of items of missile systems to be eliminated;

(ii) the elimination site;

(iii) for intermediate-range missiles, the location from which such missiles, launchers of such missiles and support equipment associated with such missiles and launchers are moved to the elimination facility; and

(iv) except in the case of support structures, the point of entry to be used by an inspection team conducting an inspection pursuant to paragraph 7 of Article XI of this Treaty and the estimated time of departure of an inspection team from the point of entry to the elimination facility;

(d) notification, no less than ten days in advance, of the scheduled date of the launch, or the scheduled date of the initiation of a series of launches, of intermediate-range missiles for the purpose of their elimination, including:

(i) the type of missiles to be eliminated;

(ii) location of the launch, or, if elimination is by a series of launches, the location of such launches and number of launches in the series;

(iii) the point of entry to be used by an inspection team conducting an inspection pursuant to paragraph 7 of Article XI of this Treaty; and

(iv) the estimated time of departure of an inspection team from the point of entry to the elimination facility;

(e) notification, no later than 48 hours after they occur, of changes in the number of intermediate-range and shorter-range missiles, launchers of such missiles and support structures and support equipment associated with such missiles and launchers resulting from elimination as described in the Protocol on Elimination, including:

(i) the number and type of items of a missile system which were eliminated; and

(ii) the date and location of such elimination; and

(f) notification of transit of intermediate-range or shorter-range missiles or launchers of such missiles, or the movement of training missiles or training launchers for such intermediate-range and shorter-range missiles, no later than 48 hours after it has been completed, including:

(i) the number of missiles or launchers;

(ii) the points, dates and times of departure and arrival;

(iii) the mode of transport; and

(iv) the location and time at that location at least once every four days during the period of transit.

6. Upon entry into force of this Treaty and thereafter, each Party shall notify the other Party, no less than ten days in advance, of the scheduled date and location of the launch of a research and development booster system as described in paragraph 12 of Article VII of this Treaty.

Article X

1. Each Party shall eliminate its intermediate-range and shorter-range missiles and launchers of such missiles and support structures and support equipment associated with such missiles and launchers in accordance with the procedures set forth in the Protocol on Elimination.

2. Verification by on-site inspection of the elimination of items of missile systems specified in the Protocol on Elimination shall be carried out in accordance with Article XI of this Treaty, the Protocol on Elimination and the Protocol on Inspection.

3. When a Party removes its intermediate-range missiles, launchers of such missiles and support equipment associated with such missiles and launchers from deployment areas to elimination facilities for the purpose of their elimination, it shall do so in complete deployed organizational units. For the United States of America, these units shall be Pershing II batteries and BGM-109G flights. For the Union of Soviet Socialist Republics, these units shall be SS-20 regiments composed of two or three battalions.

4. Elimination of intermediate-range and shorter-range missiles and launchers of such missiles and support equipment associated with such missiles and launchers shall be carried out at the facilities that are specified in the Memorandum of Understanding or notified in accordance with

paragraph 5(b) of Article IX of this Treaty, unless eliminated in accordance with Sections IV or V of the Protocol on Elimination. Support structures, associated with the missiles and launchers subject to this Treaty, that are subject to elimination shall be eliminated *in situ*.

5. Each Party shall have the right, during the first six months after entry into force of this Treaty, to eliminate by means of launching no more than 100 of its intermediate-range missiles.

6. Intermediate-range and shorter-range missiles which have been tested prior to entry into force of this Treaty, but never deployed, and which are not existing types of intermediate-range or shorter-range missiles listed in Article III of this Treaty, and launchers of such missiles, shall be eliminated within six months after entry into force of this Treaty in accordance with the procedures set forth in the Protocol on Elimination. Such missiles are:

(a) for the United States of America, missiles of the type designated by the United States of America as the Pershing IB, which is known to the Union of Soviet Socialist Republics by the same designation; and

(b) for the Union of Soviet Socialist Republics, missiles of the type designated by the Union of Soviet Socialist Republics as the RK-55, which is known to the United States of America as the SSC-X-4.

7. Intermediate-range and shorter-range missiles and launchers of such missiles and support structures and support equipment associated with such missiles and launchers shall be considered to be eliminated after completion of the procedures set forth in the Protocol on Elimination and upon the notification provided for in paragraph 5(e) of Article IX of this Treaty.

8. Each Party shall eliminate its deployment areas, missile operating bases and missile support facilities. A Party shall notify the other Party pursuant to paragraph 5(a) of Article IX of this Treaty once the conditions set forth below are fulfilled:

(a) all intermediate-range and shorter-range missiles, launchers of such missiles and support equipment associated with such missiles and launchers located there have been removed;

(b) all support structures associated with such missiles and launchers located there have been eliminated; and

(c) all activity related to production, flight-testing, training, repair, storage or deployment of such missiles and launchers has ceased there.

Such deployment areas, missile operating bases and missile support facilities shall be considered to be eliminated either when they have been inspected pursuant to paragraph 4 of Article XI of this Treaty or when 60 days have elapsed since the date of the scheduled elimination which was notified pursuant to paragraph 5(a) of Article IX of this Treaty. A deployment area, missile operating base or missile support facility listed in the

Memorandum of Understanding that met the above conditions prior to entry into force of this Treaty, and is not included in the initial data exchange pursuant to paragraph 3 of Article IX of this Treaty, shall be considered to be eliminated.

9. If a Party intends to convert a missile operating base listed in the Memorandum of Understanding for use as a base associated with GLBM or GLCM systems not subject to this Treaty, then that Party shall notify the other Party, no less than 30 days in advance of the scheduled date of the initiation of the conversion, of the scheduled date and the purpose for which the base will be converted.

Article XI

1. For the purpose of ensuring verification of compliance with the provisions of this Treaty, each Party shall have the right to conduct on-site inspections. The Parties shall implement on-site inspections in accordance with this Article, the Protocol on Inspection and the Protocol on Elimination.

2. Each Party shall have the right to conduct inspections provided for by this Article both within the territory of the other Party and within the territories of basing countries.

3. Beginning 30 days after entry into force of this Treaty, each Party shall have the right to conduct inspections at all missile operating bases and missile support facilities specified in the Memorandum of Understanding other than missile production facilities, and at all elimination facilities included in the initial data update required by paragraph 3 of Article IX of this Treaty. These inspections shall be completed no later than 90 days after entry into force of this Treaty. The purpose of these inspections shall be to verify the number of missiles, launchers, support structures and support equipment and other data, as of the date of entry into force of this Treaty, provided pursuant to paragraph 3 of Article IX of this Treaty.

4. Each Party shall have the right to conduct inspections to verify the elimination, notified pursuant paragraph 5(a) of Article IX of this Treaty, of missile operating bases and missile support facilities other than missile production facilities, which are thus no longer subject to inspections pursuant to paragraph 5(a) of this Article. Such an inspection shall be carried out within 60 days after the scheduled date of the elimination of that facility. If a Party conducts an inspection at a particular facility pursuant to paragraph 3 of this Article after the scheduled date of the elimination of that facility,

then no additional inspection of that facility pursuant to this paragraph shall be permitted.

5. Each Party shall have the right to conduct inspections pursuant to this paragraph for 13 years after entry into force of this Treaty. Each Party shall have the right to conduct 20 such inspections per calendar year during the first three years after entry into force of this Treaty, 15 such inspections per calendar year during the subsequent five years, and ten such inspections per calendar year during the last five years. Neither Party shall use more than half of its total number of these inspections per calendar year within the territory of any one basing country. Each Party shall have the right to conduct:

(a) inspections, beginning 90 days after entry into force of this Treaty, of missile operating bases, and missile support facilities other than elimination facilities and missile production facilities, to ascertain, according to the categories of data specified in the Memorandum of Understanding, the numbers of missiles, launchers, support structures and support equipment located at each missile operating base or missile support facility at the time of the inspection; and

(b) inspections of former missile operating bases and former missile support facilities eliminated pursuant to paragraph 8 of Article X of this Treaty other than former missile production facilities.

6. Beginning 30 days after entry into force of this Treaty, each Party shall have the right, for 13 years after entry into force of this Treaty, to inspect by means of continuous monitoring:

(a) the portals of any facility of the other Party at which the final assembly of a GLBM using stages, any of which is outwardly similar to a stage of a solid-propellant GLBM listed in Article III of this Treaty, is accomplished; or

(b) if a Party has no such facility, the portals of an agreed former missile production facility at which existing types of intermediate-range or shorter-range GLBMs were produced.

The Party whose facility is to be inspected pursuant to this paragraph shall ensure that the other Party is able to establish a permanent continuous monitoring system at that facility within six months after entry into force of this Treaty or within six months of initiation of the process of final assembly described in subparagraph (a). If, after the end of the second year after entry into force of this Treaty, neither Party conducts the process of final assembly described in subparagraph (a) for a period of 12 consecutive months, then neither Party shall have the right to inspect by means of continuous monitoring any missile production facility of the other Party unless the process of final assembly as described in subparagraph (a) is initiated again. Upon entry into force of this Treaty, the facilities to be inspected by continuous monitoring shall be: in accordance with subparagraph (b), for the United States of America, Hercules Plant Number 1, at Magna, Utah; in accordance with subparagraph (a), for the Union of Soviet Socialist Republics, the Votkinsk Machine Building Plant, Udmurt Autonomous Soviet Socialist Republic, Russian Soviet Federative Socialist Republic.

7. Each Party shall conduct inspections of the process of elimination, including elimination of intermediate-range missiles by means of launching, of intermediate-range and shorter-range missiles and launchers of such missiles and support equipment associated with such missiles and launchers carried out at elimination facilities in accordance with Article X of this Treaty and the Protocol on Elimination. Inspectors conducting inspections provided for in this paragraph shall determine that the processes specified for the elimination of the missiles, launchers and support equipment have been completed.

8. Each Party shall have the right to conduct inspections to confirm the completion of the process of elimination of intermediate-range and shorter-range missiles and launchers of such missiles and support equipment associated with such missiles and launchers eliminated pursuant to Section V of the Protocol on Elimination, and of training missiles, training missile stages, training launch canisters and training launchers eliminated pursuant to Sections II, IV and V of the Protocol on Elimination.

Article XII

1. For the purpose of ensuring verification of compliance with the provisions of this Treaty, each Party shall use national technical means of verification at its disposal in a manner consistent with generally recognized principles of international law.

2. Neither Party shall:

(a) interfere with national technical means of verification of the other Party operating in accordance with paragraph 1 of this Article; or

(b) use concealment measures which impede verification of compliance with the provisions of this Treaty by national technical means of verification carried out in accordance with paragraph 1 of this Article. This obligation does not apply to cover or concealment practices, within a deployment area, associated with normal training, maintenance and operations, including the use of environmental shelters to protect missiles and launchers.

3. To enhance observation by national technical means of verification, each Party shall have the right until a treaty between the Parties reducing

and limiting strategic offensive arms enters into force, but in any event for no more than three years after entry into force of this Treaty, to request the implementation of cooperative measures at deployment bases for road-mobile GLBMs with a range capability in excess of 5500 kilometers, which are not former missile operating bases eliminated pursuant to paragraph 8 of Article X of this Treaty. The Party making such a request shall inform the other Party of the deployment base at which cooperative measures shall be implemented. The Party whose base is to be observed shall carry out the following cooperative measures:

(a) No later than six hours after such a request, the Party shall have opened the roofs of all fixed structures for launchers located at the base, removed completely all missiles on launchers from such fixed structures for launchers and displayed such missiles on launchers in the open without using concealment measures; and

(b) The Party shall leave the roofs open and the missiles on launchers in place until twelve hours have elapsed from the time of the receipt of a request for such an observation.

Each Party shall have the right to make six such requests per calendar year. Only one deployment base shall be subject to these cooperative measures at any one time.

Article XIII

1. To promote the objectives and implementation of the provisions of this Treaty, the Parties hereby establish the Special Verification Commission. The Parties agree that, if either Party so requests, they shall meet within the framework of the Special Verification Commission to:

(a) resolve questions relating to compliance with the obligations assumed; and

(b) agree upon such measures as may be necessary to improve the viability and effectiveness of this Treaty.

2. The Parties shall use the Nuclear Risk Reduction Centers, which provide for continuous communication between the Parties, to:

(a) exchange data and provide notifications as required by paragraphs 3, 4, 5 and 6 of Article IX of this Treaty and the Protocol on Elimination;

(b) provide and receive the information required by paragraph 9 of Article X of this Treaty;

(c) provide and receive notifications of inspections as required by Article XI of this Treaty and the Protocol on Inspection; and

(d) provide and receive requests for cooperative measures as provided for in paragraph 3 of Article XII of this Treaty.

Article XIV

The Parties shall comply with this Treaty and shall not assume any international obligations or undertakings which would conflict with its provisions.

Article XV

1. This Treaty shall be of unlimited duration.

2. Each Party shall, in exercising its national sovereignty, have the right to withdraw from this Treaty if it decides that extraordinary events related to the subject matter of this Treaty have jeopardized its supreme interests. It shall give notice of its decision to withdraw to the other Party six months prior to withdrawal from this Treaty. Such notice shall include a statement of the extraordinary events the notifying Party regards as having jeopardized its supreme interests.

Article XVI

Each Party may propose amendments to this Treaty. Agreed amendments shall enter into force in accordance with the procedures set forth in Article XVII governing the entry into force of this Treaty.

Article XVII

1. This Treaty, including the Memorandum of Understanding and Protocols, which form an integral part thereof, shall be subject to ratification, in accordance with the constitutional procedures of each Party. This Treaty shall enter into force on the date of the exchange of instruments of ratification.

2. This Treaty shall be registered pursuant to Article 102 of the Charter of the United Nations.

DONE at Washington on December 8, 1987, in two copies, each in the English and Russian languages, both texts being equally authentic,

FOR THE UNITED STATES
OF AMERICA:

FOR THE UNION OF SOVIET
SOCIALIST REPUBLICS:

RONALD REAGAN

M. GORBACHEV

President of the United States
of America

General Secretary of the Central
Committee of the CPSU

63. Agreement Between the United States of America and the Union of Soviet Socialist Republics on the Conduct of a Joint Verification Experiment, May 31, 1988

The United States of America and the Union of Soviet Socialist Republics, hereinafter referred to as the Parties,

Reaffirming the statement of the Secretary of State of the United States and the Foreign Minister of the Union of Soviet Socialist Republics of December 9, 1987,

Proceeding from the agreement to conduct a Joint Verification Experiment, hereinafter referred to as JVE, for the purpose of the elaboration of effective verification measures for the Treaty Between the United States of America and the Union of Soviet Socialist Republics on the Limitation of Underground Nuclear Weapon Tests, hereinafter referred to as the 1974 Treaty on the Limitation of Underground Nuclear Weapon Tests,

Taking into account the agreements reached by the U.S. and Soviet delegations at the negotiations in Geneva on specific JVE technical procedures and organizational plans in full conformity with the December 9, 1987, ministerial statement,

Have agreed as follows:

1. For purposes of the JVE, there shall be two nuclear explosions, one at the U.S. Nevada Test Site and one at the USSR Semipalatinsk Test Site, each hereinafter being referred to as a JVE explosion.

2. The planned yield of the JVE explosion at each test site shall be not less than 100 kilotons and shall approach 150 kilotons.

This text appears in *Department of State Bulletin* 88, 2137 (August 1988): 67–68.

3. Each Party shall have the opportunity to measure, on the basis of reciprocity, the yield of the JVE explosion conducted at the other Party's test site using teleseismic methods and, at the other's test site, using hydrodynamic yield measurement methods.

4. Each Party shall also perform teleseismic measurements with its national seismic station network for both JVE explosions. To assist in teleseismic measurement, the Parties shall exchange data on five nuclear explosions conducted after January 1, 1978 but before January 1, 1988 to include yield, date and time, geographic coordinates, depth of burial, and associated geological and geophysical data. For each of these historical explosions, the Parties shall exchange teleseismic recordings taken at five designated stations on each side including station corrections and the best network seismic magnitude.

5. Each Party shall perform hydrodynamic yield measurements within the satellite hole provided for that purpose of the JVE explosions at both Parties' test sites using the methods it has identified in this Agreement.

6. As a yield standard, the experiment will include yield measurement within the emplacement hole of the JVE explosions at both Parties' test sites using the hydrodynamic methods each Party has identified in this Agreement. Each Party shall report to the other Party the yield values of each of the JVE explosions that are derived by each Party on the basis of hydrodynamic yield measurements undertaken within the satellite hole and within the emplacement hole. Each Party shall undertake for the purpose of the JVE to ensure at its test site a test configuration that will allow each Party to obtain an accurate yield standard of the JVE explosion. The use of hydrodynamic yield measurement methods within the emplacement hole by the visiting Party is being undertaken only in the JVE, and such measurement methods within the emplacement hole shall not be proposed by either Party for verification of the 1974 Treaty on the Limitation of Underground Nuclear Weapon Tests.

7. In the course of the JVE, each Party shall carry out teleseismic measurements of both JVE explosions at its five seismic stations for which historical data were exchanged. The Parties shall exchange the seismic data obtained in the JVE in corresponding detail to that exchanged for the historical explosions.

8. The JVE will provide information on the basis of which each Party can demonstrate the effectiveness of its hydrodynamic yield measurement methods at the test site of the other Party. Because the JVE is not designed to produce statistically significant results, it cannot by itself establish statistical proof of the accuracy of any particular yield measurement method.

9. The JVE conducted at both test sites will provide sufficient information

to resolve all concerns, except those of a statistical nature, that have been identified by either Party regarding methods proposed by the other Party for verification of the 1974 Treaty on the Limitation of Underground Nuclear Weapon Tests by providing an example of the effectiveness of the verification methods used in the JVE and by demonstrating their practicability and non-intrusiveness.

10. Specific design procedures of the JVE configuration within the emplacement hole that may have been necessary to accommodate technical objectives of the JVE shall not provide a basis for objections by either Party regarding the use of hydrodynamic yield measurements within the satellite hole for future nuclear tests. Such design procedures of the JVE configuration shall not establish a precedent for requiring similar design procedures in the two Parties' future tests as a condition for agreement on measures permitting effective verification of the 1974 Treaty on the Limitation of Underground Nuclear Weapon Tests.

11. The JVE will assist the Parties in: finalizing operational procedures for the conduct of hydrodynamic yield measurements within the satellite hole and teleseismic yield measurements for verification of future nuclear tests; establishing procedures for gathering the geological and geophysical data that is to be exchanged in accordance with any future yield measurement method proposed by either Party; determining procedures for exchange of data by the Parties on shock-wave properties of rock; comparing procedures to be used by the Parties for analyzing results of either hydrodynamic or teleseismic yield measurement methods proposed by either Party; and considering improved measures for reducing any intrusiveness associated with the verification methods proposed by each Party.

12. The Parties will use their best efforts to conduct the JVE explosions in accordance with the schedule specified in the Annex.

13. The exchange of the data obtained in the preparation for and conduct of the JVE and of the results of the analysis by each Party will be done in accordance with the schedule specified in the Annex with a view toward agreement on measures providing for effective verification of the 1974 Treaty on the Limitation of Underground Nuclear Weapon Tests.

14. Upon request by either Party, the Parties shall meet promptly to discuss any question or concern that may arise concerning the provisions of this Agreement.

15. Each Party shall treat with due respect the personnel of the other Party in its territory in connection with the preparatory work for, and execution of, the JVE and shall take all appropriate steps to prevent any attack on the person, freedom and dignity of such personnel.

16. To ensure the effective implementation of the foregoing provisions,

the Parties have reached the agreements set forth in the Annex, which form an integral part of this Agreement.

This Agreement, including the Annex hereto, shall enter into force upon signature.

DONE at Moscow on May 31, 1988, in two copies, each in the English and Russian languages, both texts being equally authentic.

FOR THE UNITED STATES OF AMERICA
GEORGE P. SHULTZ

FOR THE UNION OF SOVIET SOCIALIST REPUBLICS
E. SHEVARDNADZE

---

64. Frank Carlucci on Nuclear Deterrence and Strategic Defenses, January 17, 1989

a. Flexible Response: Foundation of U.S. Nuclear Deterrence[1]

For the past 40 years, U.S. nuclear doctrine has been characterized by remarkable consistency. Since 1945, there has been only one major change in our nuclear doctrine—the shift, during the Kennedy Administration, from massive retaliation to flexible response. Despite this continuity, three secretaries of defense since then have had to respond to charges that U.S. strategic nuclear doctrine had changed during their tenure. This section states clearly what our nuclear strategy is—and what it is not.

Whereas massive retaliation sought to deter any form of Soviet aggression through the threat of immediate, large-scale, nuclear attacks against military, leadership, and urban industrial targets in the Soviet Union, the key to flexible response is explicit in its name. Massive retaliation provided only two options to a president in the event of Soviet aggression—do nothing, or launch a massive attack against the Soviet Union. As the Soviets

This text appears in *Report of the Secretary of Defense to Congress*, January 17, 1989, pp. 34–37.

1. Parts b, c not reproduced.

acquired a nuclear capability, including the ability to strike targets in the United States, the credibility of a deterrent based solely upon this threat declined. The new flexible response doctrine increased the number of options available to the president, and provided the capability either to respond to Soviet aggression at the level at which it was initiated, or to escalate the conflict to a higher level.

Flexible response confronts Soviet attack planners with the possibility that we may respond to a conventional attack with conventional forces, or, if these fail to defeat the aggression, with land- and/or sea-based nonstrategic nuclear weapons, or with limited or massive use of U.S. strategic nuclear weapons against targets in the Soviet homeland. Flexible response has enhanced deterrence, multiplying the uncertainties confronting the Soviet leadership, and confronting them with the threat of costs that would far outweigh any gains that might be achieved through aggression.

Nuclear weapons are incorporated into our flexible response doctrine at two levels. On one level, U.S. non-strategic weapons—both land- and sea-based—are incorporated into U.S. and NATO planning. These weapons could be employed to degrade Soviet military operations in a particular theater, and to induce the Soviet leadership to cease its aggression through the threat of further escalation. Strategic nuclear weapon systems are also included in planning for limited strikes to provide a capability to retaliate against military installations deeper in Eastern Europe or the Soviet homeland. The incorporation of U.S. nonstrategic and strategic systems in these options provides a president with greater flexibility.

On a second level, strategic nuclear systems are incorporated into U.S. nuclear war planning to provide the president with a series of large-scale alternative responses to a massive Soviet nuclear attack. These systems also provide the backbone for our alliance commitments. Since the inception of flexible response, planning for large-scale retaliatory options has emphasized the capability to strike at Soviet military targets separately, or in combination with attacks on Soviet leadership installations and/or the industrial base. The intent of these attacks is to deny the Soviet Union the ability to achieve its war aims. By providing credible responses to the various potential levels of a major Soviet attack, these options fortify deterrence. In this context, our ability to withhold attacks against particular targets—including installations in a subset of cities particularly valuable to the Soviet leadership—is intended both to influence the Soviet attack planners' prewar planning, and—in the event of war—to provide to the Soviet leadership an incentive to terminate their attacks short of an all-out exchange. Secretary McNamara's 1963 description of the rationale behind these options, during testimony before the House Armed Services Committee, remains valid today: "In talking about global nuclear war, the Soviet leaders always say that they would strike at the entire complex of our military power including government and production centers, meaning our cities. If they were to do so, we would, of course, have no alternative but to retaliate in kind. But we have no way of knowing whether they would actually do so. It would certainly be in their interest as well as ours to try to limit the terrible consequences of a nuclear exchange. By building into our forces a flexible capability, we at least eliminate the prospect that we could strike back in only one way, namely, against the entire Soviet target system including their cities. Such a prospect would give the Soviet Union no incentive to withhold attacks against our cities in a first strike. We want to give them a better alternative."

There certainly have been evolutionary adjustments to U.S. nuclear planning since 1963. For example, the massive buildup of Soviet strategic nuclear forces, changes to the Soviet target base, and a better understanding of Soviet strategy and war aims led to shifts in the targeting of U.S. nuclear weapons systems. Deployment of more accurate weapon systems; improvements to the capability, survivability, and endurance of our command, control, and communications systems; and upgrades to our nuclear planning system also have facilitated the construction of more selective and limited options. All of these modifications, however, have taken place in an evolutionary manner, within the framework of our flexible response doctrine, not as a series of different strategies imposed by each administration. In returning to the original term—flexible response—our intent has been to emphasize the continuity of our approach to this element of our defense strategy. Yet after more than 25 years of continuity, several myths have developed regarding U.S. nuclear policy. The following discussion is intended to dispel these myths and clarify our nuclear policy aims.

Myth 1: U.S. Nuclear Strategy is Based on Mutual Assured Destruction.
Many critics have alleged that flexible response is simply massive retaliation by another name. In their view, the United States would respond to any Soviet nuclear attack with an immediate, massive strike against the Soviet homeland, including its cities. Some even believe that the U.S. response should be directed solely against Soviet cities and population, and that this was at one time U.S. policy. But this mutual assured destruction philosophy has *never* been U.S. policy. As noted, for over a generation we have looked for ways to develop multiple options as a means of enhancing deterrence, increasing flexibility, and controlling escalation. As early as 1963 Secretary McNamara emphasized the importance of multiple options in U.S. nuclear planning. He noted that "we have to build and maintain a second strike force that has sufficient flexibility to permit a choice of strategies. . . ."

Secretary James Schlesinger, in his FY 1975 Report to the Congress, reaffirmed the importance of strategic force flexibility, noting that "If anything, the need for options other than suicide or surrender, and other than escalation to all-out nuclear war, is more important for us today than it was in 1960. . . . The Soviet Union now has the capability in its missile forces to undertake selective attacks against targets other than cities. This poses for us an obligation, if we are to ensure the credibility of our strategic deterrent, to be certain that we have a comparable capability in our strategic systems and in our targeting doctrine, and to be certain that the U.S.S.R. has no misunderstanding on this point. . . ."

In his FY 1982 Report to the Congress, Secretary Harold Brown again reaffirmed the importance of selective and limited options, observing that "Our planning must provide a continuum of options, ranging from small numbers of strategic and/or theater nuclear weapons aimed at narrowly defined targets, to employment of large portions of our nuclear forces against a broad spectrum of targets."

The capability to respond in an across-the-board manner has always been *one* of the components of U.S. nuclear strategy planned under flexible response. Indeed, that capability—to inflict unacceptable damage on the Soviets' military, leadership, and industrial infrastructure—may be the key deterrent to a massive Soviet attack. Deterrence, however, may fail on *less* than a massive scale. The importance of this fact was noted by Secretary Weinberger when he discussed what would happen if deterrence failed: "If that were to occur we cannot predict the nature of a Soviet nuclear strike, nor assure with any certainty that what may have started out as a limited Soviet attack would remain confined at that level. Nevertheless, we must plan for flexibility in our forces and in our response options so that there is a possibility of reestablishing deterrence at the lowest possible level of violence, and avoiding further escalation."

The declining credibility of a single massive response as the sole deterrent to less than all-out aggression was recognized even in 1961, when we still had significant nuclear superiority. In fact, that recognition played a significant part in the shift to flexible response. Indeed, the key element which has, from the outset, differentiated flexible response from massive retaliation is the provision for options apart from an all-out response.

Myth 2: U.S. Strategy is Based on "Nuclear War Fighting."

Many of those who believe mistakenly that U.S. nuclear strategy was once based on MAD have also criticized the U.S. government for "shifting" from this strategy. They contend that we have adopted a nuclear warfighting strategy. These critics seem to believe that our mere possession of nuclear weapons is sufficient to deter Soviet aggression. In their view, if deterrence ever fails, the inevitable outcome will be a spasm nuclear war immediately involving massive attacks on cities. According to this philosophy, developing plans and acquiring capabilities for more selective employment options undermines stability and deterrence, and suggests our intention to fight a "limited" and/or "protracted" nuclear war.

If a limited nuclear warfighting capability is one in which a single or small number of nuclear weapons are used in an attempt to end a major conventional war before it escalates to all-out nuclear war, then, in fact, we do possess such a capability. If a protracted nuclear warfighting capability is one in which nuclear forces and their supporting command and control structure might be available and effectively employed for more than 30 minutes following the onset of a Soviet nuclear attack, then we also possess this capability. The critical question is: Do these capabilities strengthen our ability to deter? The answer is "yes."

It is not our intention to fight a nuclear war of any description: "limited" or "massive," "prompt" or "protracted." It is our policy to *prevent* nuclear war. In doing so, we must determine what would deter the *Soviet* leadership from considering aggression—*not* what would deter us. In that regard, we have watched the steady buildup of Soviet strategic nuclear forces for over two decades, and the Soviet leadership's preparations for nuclear war, along with evidence that reflects their belief that such a war may, under certain circumstances, be fought and won. That evidence includes:

- The Soviets' capability to reload many of their ICBM silos after launch of the first ICBM; a capability supported by spare ICBMs and reloading exercises.
- Their continued expansion of a nationwide network of over 1,500 buried command bunkers for the Communist Party and military leadership, plus an extensive mobile command system—both supported by an extensive communications network.
- Increasing Soviet deployments of mobile ICBMs—the SS-24 and SS-25— which, with their greater survivability, could be employed over an extended period.

The Soviets clearly can conduct both limited and protracted nuclear attacks. We must deter them from these types of aggression. Indeed, we must make a Soviet victory, as seen through Soviet eyes and measured by Soviet standards, impossible across the broad range of scenarios the Kremlin leadership might consider. We may not agree with the assumptions upon which Soviet strategy appears to be founded, but we must design a deterrent strategy that takes these factors into account to remove any temptation for the Soviet leadership to believe they could fight and win a nuclear war. Our forces and our flexible response doctrine are designed to maximize the uncertainties that Soviet leaders would face, and confront them with an unfavorable outcome in *any* contingency in which they may contemplate aggression.

Myth 3: As Part of its Nuclear Strategy, the U.S. Relies on
a Launch-Under-Attack Policy.

Over the past decade, as Soviet ICBM counter-silo capabilities improved, some have questioned the continued survivability of the ICBM leg of the Triad. Rather than abandon one leg of the Triad, however, successive administrations chose to modernize the ICBM force by deploying the Peacekeeper ICBM in a survivable basing mode. In 1986, we decided to deploy Peacekeeper in a highly survivable rail-based system. Predictably, many of the critics who question the continued value of the ICBM force began to assert that no truly survivable basing mode could be established. They contend, therefore, that the United States has shifted to a launch-under-attack posture, since our ICBMs would be destroyed unless launched prior to the impact of the incoming Soviet attack.

As noted, successive administrations have devoted considerable effort and resources to increasing the flexibility and the number of choices available to the president should deterrence fail and the use of nuclear weapons become necessary. Asserting that the United States maintains a launch-under-attack policy ignores these efforts, and the deterrent provided by the Triad. We have not spent billions of dollars to modernize and increase the capabilities of the bomber and sea-based legs of the Triad only to leave the president with a single effective option with which to respond to a massive Soviet attack. We do not, however, intend to reduce the uncertainties facing Soviet attack planners—or the Soviet leadership. In order to increase the uncertainties in the minds of Soviet planners, it is not our policy to explain in detail how we would respond to a Soviet missile attack. However, the United States does not rely on its capability for launch-on-warning or launch-under-attack to ensure the credibility of its deterrent. At the same time, our ability to carry out such options complicates Soviet assessments of war outcomes and enhances deterrence.

IX

Nuclear Power

*E*vents between 1951 and 1954 accelerated the drive for commercial nuclear power. In the winter of 1951, a small government reactor at the Idaho National Engineering Laboratory produced electricity for the first time. Eisenhower's "Atoms for Peace" speech in December 1953 emphasized an international program of civilian uses of atomic energy, especially power reactors for export to fuel-hungry Europe.

That same December, the Atomic Energy Commission selected the Duquesne Light and Power Company, a Pittsburgh utility, to join a three-way partnership in the construction and operation of the first civilian nuclear power plant. Duquesne would provide the land and about $300 million and would operate and maintain the plant. Duquesne's partners would be the federal government and the Westinghouse Corporation, which would manufacture the reactor. "Off the shelf" technology would be used. The reactor was a civilian version of a pressurized water reactor Westinghouse was building for Hyman G. Rickover's Navy program. Admiral Lewis L. Strauss, a science adviser to Eisenhower who would later chair the Atomic Energy Commission, recommended adapting the Navy's pressurized water reactor to civilian purposes, otherwise "much time and momentum would be lost by turning over this project to a new contractor. . . ." To act quickly was paramount.

Congress encouraged further private reactor development when it amended the Atomic Energy Act of 1946 in the summer of 1954. Under the 1954 Act, free enterprise would provide the impetus for civilian nuclear power. The new legislation permitted the government to license private corporations to build and operate nuclear-fueled power plants. The Atomic Energy Commission could continue to fund research and development, but the act prohibited the Commission from constructing power "reactors for the purpose of selling or distributing electricity." Industry, so Congress and *the Eisenhower administration believed, could do it better. "The goal of atomic power at competitive prices will be reached more quickly if private enterprise, using private funds, is . . . encouraged to play a far larger role in the development of atomic power. . . ." That role, as the congressmen perceived it, would be plant construction, while the Atomic Energy Commission would continue to subsidize research and development projects shunned by industry.*

Nuclear power advocates exuded optimism in 1954. The U.S.S. Nautilus, the world's first nuclear-powered submarine, was launched that year. The Atomic Energy Act had been amended to provide more government assistance for private development of nuclear energy. President Eisenhower continued to stress the peaceful uses of the atom. The announcement of the construction of a nuclear power plant at Shippingport, Pennsylvania, by the Atomic Energy Commission and Duquesne Electric Company and a five-year commercial reactor demonstration program highlighted this emphasis. Ike predicted the day when the atom "will be wholly devoted to the constructive purposes of man." Congressmen spoke about "cheap power" within a decade. Utilities, experts argued, would save billions of dollars. The president of General Electric estimated that 50 percent of the new electric power plants would be atomic by 1976. And Lewis Strauss predicted that "our children will enjoy in their homes electrical energy too cheap to meter."

By the winter of 1954–55, the Atomic Energy Commission and several utility company consortiums were ready to demonstrate the feasibility of nuclear power. In December the Atomic Energy Commission announced the Power Demonstration Reactor Program and asked for proposals from the power companies by April 1, 1955. The Atomic Energy Commission hoped to encourage industry activity with government support limited to research and development. At the same time, the commission sought to extend technical knowledge about a wide range of reactor types and, in the end, make nuclear power economically feasible. Four groups replied. Detroit Edison proposed a fast-breeder reactor; Yankee Atomic, a consortium of thirteen New England utilities, suggested a pressurized water reactor on the Shippingport model; the Nuclear Power Group, headed by Commonwealth Edison in Chicago, proposed a boiling water reactor; and the Consumers Public Power District in Nebraska submitted plans for a sodium-graphite reactor.

In addition, another utility offered to build its own nuclear plant without any outside assistance either from the government or in consort with other companies. The Consolidated Edison Company of New York, managed by men steeped in the tradition of private enterprise, submitted an application to construct a nuclear plant at Indian Point, some thirty-five miles up the Hudson River from New York City. By May 1956 the Atomic Energy

Commission had granted Con Edison's plans for a pressurized water reactor, designed by Babcock & Wilcox, the first public nuclear power construction permit.

In encouraging private development of nuclear power, the Atomic Energy Commission had created a new area of regulatory concerns. Government reactors had always been located in remote areas. But to give nuclear energy a chance to be economically feasible, power companies sought to locate plants near their customers and existing power grids. Therefore, the commission had to draw up new siting plans, and radiation standards had to ensure the public's health and safety. A dilemma emerged.

The Atomic Energy Act of 1954 gave the Atomic Energy Commission responsibility for regulating and licensing power reactors and, at the same time, the task of promoting nuclear power. Both the regulatory and non-regulatory functions were held by the Division of Civilian Application (the Division of Military Application was responsible for weapons) until 1957 when a separate regulatory staff was established. In addition, an Advisory Committee on Reactor Safeguards, a group of fifteen experts, provided advice to the Commission on the safety and licensing of nuclear reactors. The Advisory Committee, nicknamed the "Brake Department," spoke out at times against the commission's policy on safety, especially in its concern about Detroit Edison's Fermi breeder reactor. In 1974, Congress emphasized this functional split when it abolished the Atomic Energy Commission and established the Nuclear Regulatory Commission to regulate nuclear power and the Energy Research and Development Administration (later the Department of Energy) to promote its use.

Shippingport began producing electricity for commercial use in 1957. Commonwealth Edison's Dresden plant near Chicago began operating in 1960, the Yankee Atomic plant near Rowe, Massachusetts, went on line in 1961, and Con Edison's Indian Point plant started up in 1962. By late 1970, the nuclear power industry had eighteen operating nuclear plants and had applications pending for twenty more. By 1975, fifty-five plants were in operation and one hundred seventy-eight more were planned or under construction.

By the mid-1960s, experimentation over power reactor types had ceased. The pressurized water reactor, built by Westinghouse and other corporations, and the boiling water reactor built by General Electric became the industry standards, and plant orders increased. Industry spokesmen believed over 50 percent of the nation's electric power would be nuclear by the year 2000.

The glowing promise of nuclear energy, however, was never fulfilled. Opposition to a proposed nuclear plant in New York City in early 1963 marked the beginnings of public concern. Widespread focus on the danger of fallout, the moratorium on atmospheric testing of nuclear weapons, and President John F. Kennedy's initiatives toward a test-ban treaty contributed to linking nuclear weapons with nuclear power plants in the public perception.

In addition, the Union of Concerned Scientists and other groups feared the increased size of the nuclear reactors and the ability of the industry to make them safe enough to contain a serious accident. Consequently, opponents to nuclear power led a growing number of attempts to block the construction or licensing of new atomic power stations. At public hearings, these groups intervened to delay or halt plant operation for reasons other than radiation safety. A proposed plant in Easton, New York, for example, failed to gain approval because it was located too close to the Saratoga National Historical Park and received adverse comments from the Advisory Council on Historic Preservation. A planned reactor in Oregon failed to win voter approval in a special referendum that restricted the use of its public funding.

The National Environmental Policy Act of 1969 became the chief weapon in changing the Atomic Energy Commission's licensing procedures. In the Calvert Cliffs decision in 1971, the federal courts ruled that the commission had to consider the total impact a nuclear plant would have on the environment, including water quality, and perform a cost-benefit analysis on the economic, technical, and other benefits against alternatives, including not building the plant. The decision restricted the commission's licensing program and slowed down the planning and construction of nuclear power plants.

Even as the cost of oil soared after the 1973 boycott, the cost of nuclear plants also rose. Many in the planning and construction stage could no longer be competitive as building costs doubled, then tripled, as legal delays and inflation pushed costs higher. By 1976, utility order for new nuclear plants had already begun to decline, several years before the accident at Three Mile Island seemed to doom nuclear power in the public's perception.

The debate over nuclear power intensified. The 1979 Three Mile Island accident provided new ammunition for opponents who sought to use the accident to close down all atomic power plants. Industry proponents argued that the containment system at Three Mile Island worked, in spite of human errors that compounded the problem. Moreover, they said, the steady depletion of fossil fuels would make nuclear power inevitable. By the end of the 1980s, controversies over the effect of acid rain, climate change, and energy imports gave new life to the nuclear power industry, which had experienced a steadily shrinking infrastructure since 1973 when utilities began *to cancel all orders for new plants. To regain public confidence in nuclear power, engineers spoke of "inherently safe" or "passively safe" reactors. Construction costs would be trimmed by instituting standardized designs and simplified licensing procedures. Nonetheless, scientists had yet to find an acceptable solution to the nuclear waste disposal problem. The peaceful atom continued to stir as much controversy as the bomb.*

The best short history of the development of the nuclear power industry is by Irwin Bupp and J. C. Derian, Light Water: How the Nuclear Dream Dissolved *(New York: Basic Books, 1978). A more technical study is Frank G. Dawson,* Nuclear Power: Development and Management of a Technology *(Seattle and London: University of Washington Press, 1976). On the politics of the antinuclear movement in Europe, see Dorothy Nelkin and Michael Pollak,* The Atom Besieged *(Cambridge, Mass., and London: MIT Press, 1981).*

On the Chernobyl nuclear accident of April 1986, see Grigori Medvedev, The Truth About Chernobyl *(New York: Basic Books, 1991). On Three Mile Island, see Philip L. Cantelon and Robert C. Williams,* Crisis Contained: The Department of Energy at Three Mile Island *(Carbondale: Southern Illinois University Press, 1982).*

The history of regulating the peaceful atom will be written in a series of official histories of the Nuclear Regulatory Commission. The first volume, Controlling the Atom: The Beginnings of Nuclear Regulation, 1946–1962 *(Berkeley: University of California Press, 1984) by George T. Mazuzan and J. Samuel Walker is a well-written, highly regarded study of the regulation and licensing of nuclear power by the Atomic Energy Commission and Congress. John W. Johnson,* Insuring Against Disaster: The Nuclear Industry on Trial *(Macon, Ga.: Mercer University Press, 1986) is a thorough history of the Price-Anderson Act, which provides government insurance for nuclear accidents. Richard L. Meehan,* The Atom and the Fault: Experts, Earthquakes, and Nuclear Power *(Cambridge, Mass.: MIT Press, 1986) is critical of some of the Commission's decisions about safety. The most recent published scholarship on the evolving history of nuclear regulations can be found in J. Samuel Walker's "Reactor at the Fault: The Bodega Bay Nuclear Plant Controversy, 1958–1964—A Case Study in the Politics of Technology, 1958–1964,"* Pacific Historical Review *59 (August 1990): 323–48; George T. Mazuzan, "Very Risky Business: A Power Reactor for New York City,"* Technology and Culture *27 (April 1986): 262–84; and J. Samuel Walker, "Nuclear Power and the Environment: The Atomic Energy Commission and Thermal Pollution, 1965–1971,"* Technology and Culture *30 (October 1989): 964–92.*

65. Edward Teller to Sterling Cole, July 23, 1953

The Honorable Sterling Cole
Chairman
Joint Committee on Atomic Energy
The Congress of the United States
Washington, D.C.

Dear Sir:

In response to your invitation to make a statement in connection with the development of atomic energy by private enterprise, I should like to discuss two topics concerning which I have some specific experience. These are: the safety of nuclear reactors and the connection between power production and military application.

Briefly, my opinion can be stated as follows. First, nuclear power-producing units will be dangerous instruments and careful thought will have to be given to their safe construction and operation and, second, there is a great and increasing need for fissionable materials in the military field.

I should like to recommend

First, that an advisory committee should be set up to review planned reactors and supervise functioning reactors under the control of private enterprise. Instead of setting up a new committee, the present Advisory Committee on Reactor Safeguards of the Atomic Energy Commission might serve this purpose, and

Second, that the Government stimulate power production by private enterprise by guaranteeing to buy militarily useful by-products at a pre-determined price and in limited but large quantities for a period of five or ten years.

Safety of Nuclear Reactors

For the past six years I have served as the Chairman of the Reactor Safeguard Committee. Recently, this committee and the Industrial Committee on Reactor Location problems have been merged into the Advisory

This document appears in National Archives, RG 128, Box 397, "LRL NARA LAB."

Committee on Reactor Safeguards, and I am participating in the work of this new committee.

Up to the present time we have been extremely fortunate in that accidents in nuclear reactors have not caused any fatalities. With expanding applications of nuclear reactions and nuclear power, it can not be expected that this unbroken record will be maintained. It must be realized that this good record was achieved to a considerable extent because of safety measures which have necessarily retarded development.

The main factors which influence reactor safety are, in my opinion, reasonably well understood. There have been in the past years a few minor incidents, all of which have been caused by neglect of clearly formulated safety rules. Such occasional accidents can not be avoided. It is rather remarkable that they have occurred in such a small number of instances. I want to emphasize in particular that the operation of nuclear reactors is not mysterious and that the irregularities are no more unexpected than accidents which happen on account of disregard of traffic regulations.

In the popular opinion, the main danger of a nuclear pile is due to the possibility that it may explode. It should be pointed out, however, that such an explosion, although possible, is likely to be harmful only in the immediate surroundings and will probably be limited in its destructive effects to the operators. A much greater public hazard is due to the fact that nuclear plants contain radioactive poisons. In a nuclear accident, these poisons may be liberated into the atmosphere or into the water supply. In fact, the radioactive poisons produced in a powerful nuclear reactor will retain a dangerous concentration even after they have been carried downwind to a distance of ten miles. Some danger might possibly persist to distances as great as 100 miles. It would seem appropriate that Federal regulations should apply to a hazard which is not confined by state boundaries. The various committees dealing with reactor safety have come to the conclusion that none of the powerful reactors built or suggested up to the present time are absolutely safe. Though the possibility of an accident seems small, a release of the active products in a city or densely populated area would lead to disastrous results. It has been therefore the practice of these committees to recommend the observance of exclusion distances, that is, to exclude the public from areas around reactors, the size of the area varying in appropriate manner with the amount of radioactive poison that the reactor might release. Rigid enforcement of such exclusion distances might hamper future development of reactors to an unreasonable extent. In particular, the danger that a reactor might malfunction and release its radioactive poison differs for different kinds of reactors. It is my opinion that reactors of sufficiently safe types might be developed in the near future. Apart from the basic construction of the reactor, underground location or particularly thought-fully constructed safety devices might be considered.

It is clear that no legislation will be able to stop future accidents and avoid completely occasional loss of life. It is my opinion that the unavoidable danger which will remain after all reasonable controls have been employed must not stand in the way of rapid development of nuclear power. It also would seem that proper legislation at the present time might make provisions for safe construction and safe operation of nuclear reactors. In case an accident should occur which involved the lives of many people, pressure for such legislation would become overwhelming. Proper steps taken at the present time could reasonably prepare for accidents and minimize the suffering that is caused, when and if they should occur.

It would seem reasonable to extend the Atomic Energy Commission procedures on reviewing planned reactors and supervising functioning reactors to nuclear plants under the control of private enterprise. To what extent these functions should be advisory or regulatory is a difficult question. I feel that ultimate responsibility for safe operation will have to be placed on the shoulders of the men and the organization most closely connected with the construction and the operation of the reactor.

Power Production and Military Application

The first and best known military application of atomic energy was connected with strategic bombing. In the popular mind, such strategic bombing has been identified with the destruction of cities. The belief is widely held that a relatively limited number of atomic bombs can not only cause terrifying destruction but would produce saturation, that is, only a limited number of atomic bombs would be needed. It is my conviction that this opinion is based on a misconception and that indeed a great stockpile of fissionable material could be usefully applied in warfare. Furthermore, it seems to me that a more general use of fission weapons will not result necessarily in a more thorough destruction of cities but might rather be used against military targets of the more conventional type. It seems to me therefore that a less expensive source of fissionable materials would be desirable. Such a less expensive source could be obtained if atomic reactors were constructed for the dual purpose of providing power and producing fissionable materials.

Strategic targets include industrial plants and military installations far behind the enemy's lines. Depending on the vulnerability of these targets and on their contribution to the enemy's war effort, one may well be justified in using atomic bombs against these targets. The size of the target need not be decisive and the number of such targets may be quite appreciable.

The possible tactical targets are even more numerous. Any concentration of fighting forces or of material near the fighting lines constitutes tactical targets. Strongly defended positions might be attacked by atomic bombs. Atomic weapons could be used against beachheads or against enemy forces attempting to cross a natural obstacle. Conversely, atomic weapons could be employed to prepare a landing on a beachhead or the attack of a parachute force. The vulnerability of naval vessels to atomic bombs has been demonstrated in the Bikini tests. Vehicles less expensive than naval units may present atomic bomb targets, particularly if the cost of the bomb is lower than the cost of the vehicle which one attempts to destroy. An enemy bomber or even an enemy fighter plane might be considered as a possible target for an atomic bomb.

It might seem extravagant to use atom bombs for all these different types of targets. The question of extravagance or of sound economy must be considered, however, in connection with the ease of delivery, with the expense of delivery and with the expense of the fissionable materials. I can think of no exception to the rule that the cost of delivery will be less if one produces a certain damage by atomic weapons rather than by more conventional means. It is therefore the cost of fissionable materials which will decide how extensively one can use atomic weapons in warfare. The more the cost of atomic weapons can be reduced, the greater will be the number of applications where relatively cheap delivery systems can replace the much more expensive conventional methods. Increase in our stockpile of fissionable materials may therefore reduce the military expenditure without reducing military potential.

It seems to be doubtful whether, on the basis of present technology, atomic energy can produce power in an economically profitable manner. Power production can, however, be conducted in such a manner as to produce militarily useful materials. It would seem to me reasonable to stimulate the construction of power-producing reactors by guaranteeing a price at which the Government will buy the militarily useful by-products. This price should of course be set lower than the price at which the Atomic Energy Commission is producing fissionable materials at the present time. It probably will be necessary to set a limit to the amount of fissionable material which the Government is prepared to purchase and also to set a limit to the time during which such purchases will be made at the fixed price. Nevertheless, it seems probable that if a fair price is guaranteed for a period like five or ten years, this will be an effective stimulant to the nation's atomic power industry. This industry is likely to become a factor in national defense which may not be second even to the steel or aircraft industries.

The above contains the substance of the testimony which I have prepared for the Joint Congressional Committee. I should like to express my very great regret that at the date set for the hearing it was completely impossible for me to leave Livermore. It would be a great pleasure to appear before the Joint Congressional Committee at any time to amplify the above statements or else to help in any other way that you can think of.

Yours very truly,
Edward Teller

66. United States Atomic Energy Commission Report on the Need for Nuclear Power, 1962

Our technological society requires ample sources of energy. Although large, the supplies of fossil fuels are not unlimited and, furthermore, these materials are especially valuable for many specific purposes such as transportation, small isolated heat and power installations, and as sources of industrial chemicals. Reasonable amounts should be preserved for future generations.

Comparison of estimates of fossil fuel resources with projections of the rapidly increasing rate of energy consumption predicts that, if no additional forms of energy were utilized, we would exhaust our readily available, low-cost fossil fuels in a century or less and our presently visualized total supplies in about another century. In actual fact, long before they become exhausted we will be obliged to taper off their rate of use by supplementing them increasingly from other sources.

In contrast, our supplies of uranium and thorium contain almost unlimited amounts of latent energy that can be tapped provided "breeder" reactors are developed to convert the fertile materials, uranium-238 and

This text appears in U.S. Atomic Energy Commission, *Annual Report to Congress of the Atomic Energy Commission for 1962* (Washington, D.C.: U.S. Government Printing Office, 1963), pp. 109–16.

thorium-232, to fissionable plutonium-239 and uranium-233, respectively.*
Successfully done, this will render relatively unimportant the cost of nuclear
raw materials so that even very low-grade sources will become economically
acceptable.

The use of nuclear energy for electric power and, less immediately, for
industrial process heat and other purposes is technically feasible and economically reasonable. In addition to its ultimate importance as a means of
exploiting a large new energy resource, nuclear electric power holds important near-term possibilities: as a means of significantly reducing power generation costs, especially in areas where fossil fuel costs are high; as an
important contributor to new industrial technology and to our technological
world leadership; as a significant positive element in our foreign trade; and,
potentially, as a means of strengthening our national defense.

In view of the above we have concluded that: *Nuclear energy can and
should make an important and, ultimately, a vital contribution toward meeting our long-term energy requirements,* and, in particular, that: *The development and exploitation of nuclear electric power is clearly in the near- and
long-term national interest and should be vigorously pursued.*

The Role of the Federal Government

The technological development of nuclear power is expensive. The reactors are complex, and operating units, even of a scaled-down test variety,
must of necessity be large and costly. Furthermore, nuclear power does not
meet a hitherto unfilled need but must depend for marketability on purely
economic advantages that will return the development investment slowly.
Hence, the equipment industry could not have afforded to undertake the
program by itself. The Government must clearly play a role.

An early objective should be to reach the point where, with appropriate encouragement and support, industry can provide nuclear power installations of economic attractiveness sufficient to induce utilities to install them
at their own expense. Once this is achieved the Government should devote
itself to advanced developments designed to meet long-range objectives,
leaving to industry responsibility for nearer-term improvements. Gradually,

---

*The readily fissionable material found in nature is confined to uranium-235 which consti-
tutes only 0.7% of normal uranium. The energy contained in this isotope in uranium mineable
at near present costs is only a small fraction of that contained in our fossil fuel reserves. Fortunately, the so-called "fertile" isotopes, uranium-238, constituting the remainder of normal
uranium, and thorium-232 constituting practically all normal thorium can be converted to
fissionable plutonium-239 and uranium-233 by absorption of neutrons in a nuclear reactor.

as technological maturity is reached, the transition to industry should become complete.

Thus, *the proper role of Government is to take the lead in developing and demonstrating the technology in such ways that economic factors will promote industrial applications in the public interest and lead to a self-sustaining and growing nuclear power industry.*

The Present Situation

Accordingly, in keeping with national policy, and with the responsibilities assigned to it by the Atomic Energy Act, the Atomic Energy Commission has conducted and encouraged a vigorous program directed toward the development and extensive exploitation of nuclear energy for civilian purposes, with emphasis on nuclear electric power. About $1.275 billion has been expended by the AEC to date* on the civilian power program. This program has included both research and development and a "power demonstration" program, involving aid in the construction and operation of practical reactors on utility grids. Several reactor types are under development. Most highly developed are "converter" reactors that produce less fissionable material than they consume; much less far along are "breeder" reactors that produce more than they consume.

In one segment of the power demonstration program, Commission-built and -owned "prototype" reactors are operated by utilities that buy the steam; in another segment, utilities are given research and development assistance in designing and constructing their own reactors and, for a few years no charge is made for the lease of Government-owned nuclear fuel. Six sizeable reactors of the more highly developed types are in successful operation on utility grids (the two largest without AEC assistance); seven more will be completed by the end of 1963; a few others are under construction or nearly so.

Experience has shown that nuclear electric power is readily achieved technically but difficulties have been met in developing a technology that is economically competitive with conventional power generation methods. Happily, in recent years these difficulties have been progressively overcome.

Certain classes of power reactors, notably water-cooled converters producing saturated steam are now on the threshold of economic competitive-

---

*We estimate that industry has expended approximately $0.5 billion of its own funds, mostly for plant and equipment.

ness with conventional power in large installations in high fossil fuel cost areas of the country. Foreseeable improvements will substantially increase the areas of competitiveness.

Technical Considerations

Saturated steam reactors, however, have certain inherent limitations. They produce relatively low temperature saturated steam which limits their efficiencies and requires the use of large, expensive turbines; they are only moderately effective converters.* Consequently, converter concepts utilizing other moderators and coolants and promising improved economics and fuel utilization are being actively pursued with encouraging results; early competitiveness seems assured for some of them. All of these are "thermal"† reactors. They include the "spectral shift" reactor, the high temperature gas-cooled reactor, and the sodium-graphite reactor. All have relatively high efficiencies and excellent economic promise. The first two will have excellent conversion ratios; indeed they may eventually be made to breed in the thorium-uranium cycle.§ The sodium-graphite reactor can achieve quite high temperatures, has good safety features and helps develop the liquid sodium technology necessary for fast breeders. The heavy water moderated reactor also shows promise of high conversion ratios but present designs are not so attractive economically as other types in the United States. The organic-cooled and -moderated reactor may have application for process heat. Some of these should be carried to the stage of operating prototypes during the next several years, and some will reach the full-scale operational phase by the early 1970's. Operating reactors of these types will help accelerate the industry, will increase operating experience and will help provide plutonium needed for the breeder program.

Although much technical progress has been made, breeder reactors have not yet reached an economically useful stage of development. Even when they do, they will not, initially at least, make new material fast enough to

---

*They convert 0.5 to 0.7 as much material as they consume. Compounded, this results in doubling to tripling the energy finally made available.

† In a "thermal" reactor most of the fission neutrons are slowed-down (moderated) before interacting with the nuclear materials; this is accomplished through many collisions with light nuclei such as hydrogen (in water or organic compounds), carbon (in graphite) or beryllium. In a "fast" reactor, little or no moderation is used, so that most of the neutrons retain the high energies and velocities with which they were emitted in the fission process. "Intermediate" reactors lie between.

§ See footnote, page 7 [p. 313 in this volume].

provide the fuel for new plants at the rate required if nuclear power is to increase its proportional share of the national electric power load. Hence, even after breeders become available, it will be necessary to fuel some portion of the installations with uranium-235 until such time as improved breeding gains and reductions in the relative rate of growth in power consumption enable the breeders to be self-sufficient. For the thermal reactors used to make U-233 from thorium, this need can be met by substituting U-235 for U-233 in some of them, at a sacrifice in fuel produced. A similar procedure would, however, be uneconomic in the "fast" reactors required to breed plutonium. Hence, in the transition stage, which will last for many decades, fast breeders that burn as well as make plutonium will probably be augmented by thermal converters both for economic reasons and because it is important that the combination of breeders and converters reaches an overall net breeding capability, or very nearly so, while relatively cheap fuel supplies are still available.

In our opinion, economic nuclear power is so near at hand that only a modest additional incentive is required to initiate its appreciable early use by the utilities. Should this occur the normal economic processes would, we feel, result in expansion at a rapid rate. The Government's investment would be augmented manyfold by industry. Equipment manufacturers could finance major technical developments, thus reducing the future need for Government participation.

Continuation of the Commission's present effort, with some augmentation in support for the power demonstration program, and with program adjustments to give added emphasis to breeders, would, we believe, provide industry with the needed stimulus to build a significant number of large reactors in the near future, would bring nuclear power to a competitive status with conventional power throughout most of the country during the 1970's, and would make breeder reactors economically attractive by the 1980's.

Under these conditions, we estimate that by the end of the century nuclear power would be assuming the total increase in national electric energy requirements and would be providing half the energy generated.* This rate of progress, projected into the next century, would be an important step in conservation of the fossil fuels and, unless breeders lagged the converters much more than we predict, would raise no problems in nuclear fuel supplies.

---

*Since, by Federal Power Commission estimates, the total use of electric energy will grow tenfold in the same period, fossil fuel consumption for this purpose would still increase by a factor of from four to five.

Under conservative cost assumptions, it is estimated that by the end of the century the above projected use of nuclear power would result in cumulative savings in generation costs of about $30 billion.* The annual saving would be between $4 and $5 billion. High cost power areas would no longer exist, since, in the absence of significant fuel transportation expenses, the cost of nuclear power is essentially the same everywhere. This would be an economic boon to areas of high cost fossil fuels and, by enabling them to compete better, should increase the industrial potential of the entire country.

More generally, the introduction of nuclear power technology on a significant scale would add to the health and vigor of our industry and general economy. Technical progress would assist the space and military programs and have other ancillary benefits. Our international leadership in the field would be maintained, with benefit to our prestige and our foreign trade. Nuclear power could also improve our defense posture; it would not burden the transportation system during national emergencies; furthermore, the "containment" required for safety reasons could, if desired, be achieved at little, if any, extra cost by underground installations, thus "hardening" the plants against nuclear attack.

A substantially lesser program would sharply reduce these benefits. Too great a slowdown could result in losing significant portions of industry's present nuclear capability thereby seriously delaying the time at which it would assume a major share of the development costs.

On the other hand we do not believe that a major step-up in the whole Commission program is appropriate. Taken as a whole, support of the scientists and engineers engaged in developmental work is about adequate and, in view of the country's other needs, it would seem unwarranted to increase appreciably such manpower in this field.

To summarize we have concluded that *the nuclear power program should continue on an expeditious basis. Commission support should continue with added emphasis on stimulating industrial participation. The program should include: (1) early construction of plants of the presently most competitive reactor types; (2) development, construction and demonstration of advanced converters to improve the economics and the use of nuclear fuels; (3) intensive development and, later, demonstration of breeder reactors to fill the long-range needs of utilizing fertile as well as fissile fuels.*

An important corollary area is the development of economical chemical reprocessing methods whereby useful fissile and fertile materials are

---

*At 5% interest these cumulative savings would have a discounted value of about $10 billion in 1970.

recaptured from used fuel assemblies and the fission products are removed. Another important line of work concerns the ultimate storage or disposal of the large amounts of radioactive fission products that will be generated when a major power industry comes into being.

An overriding consideration is that of safety. Not only must inherent safety be assured in fact but its existence must be conclusively demonstrated to the public. With adequate technical improvements and the accumulation of satisfactory experience, it should be possible gradually to remove many of the siting restrictions in force today, thus permitting plant locations closer to the large load centers.

Possible Construction Program

A composite construction program for the next dozen years might entail the following: (1) the construction and placing into operation of seven or eight power-producing prototype reactors, approximately half of which would be advanced converters and the rest breeders; most of their cost would probably be borne by the AEC; (2) assistance, as necessary, to industry in the construction of 10–12 full-scale power plants of improving design as time goes on; hopefully, industry will concurrently bear full costs of many more of well proven design.

This construction would, of course, be backed by specific development programs directed at the more advanced reactor types, especially breeders, and by research and development related to the underlying technology.

Legal, Financial and Administrative Matters

Careful attention must be paid to several legal, financial and administrative questions, among them (1) private ownership of special nuclear materials and related policies on fuel pricing and "toll enrichment"; (2) policies relating to the raw material and other supporting industries; (3) licensing and regulation, including reactor siting criteria.

The commission has recommended that private ownership of special nuclear materials be authorized at an early date, thus permitting the free play of normal economic forces and minimizing economic distortions of the technology. To prevent sudden dislocations such ownership should not be made mandatory for a decade or so.

The Commission further believes that a policy of "toll enrichment" or equivalent should be adopted. Industry could then buy its raw materials on the open market, use privately owned plants to prepare them for enrichment, and depend upon the Government only for the actual enrichment in the diffusion plants. This service should also be extended to our friends abroad, subject to proper safeguards against diversion for military use.

Before and during the period of transition to private ownership the value set by the Commission on enriched uranium for lease or sale should, as at present, be determined by the actual cost, with appropriate allowances for depreciation and other indirect expenses. The Commission has recommended that prices for the purchase of plutonium be in accordance with its "near-term" value as a reactor fuel. We believe that consideration should be given to scaling the price in accordance with the content of fissionable isotopes. The same pricing policies should apply to purchases abroad of plutonium made from uranium enriched in the U.S.

The Commission's contracts with uranium miners and processors expire at the end of 1966. Since it seems probable that the requirements for new uranium for weapons, the dominating use to date, will decrease in the next decade, careful planning is necessary to so guide further procurement that the uranium industry will be kept viable during any slack period before civilian power creates another large demand. With this in mind the Commission is planning to offer the industry a "stretchout" program under which an AEC commitment to purchase additional material after January 1, 1967 would be used as an incentive to induce industry to delay until after that date delivery of part of the uranium presently under contract. If successful, this program would result in a leveling-off process that should carry through the period of slack use without injuring the industry substantially or resulting in an unreasonably large surplus.

The Commission intends to continue and extend encouragement to the industrial activities ancillary to the major equipment industry. Many that could start on a small scale are already well underway. There are, however, a few activities, such as the chemical separation of used fuels, that are attractive to industry only on a fairly substantial scale and for which there will be little private business until civilian reactors have operated for an appreciable period. Strong encouragement is being given to private industry to embark in these fields with some prospect of success. As rapidly as a private capability comes into being the Commission should withdraw from all such work deriving from industry and should utilize private plants to fill its own requirements except, perhaps, for those related to materials for weapons.

Recognizing that simplifying and streamlining licensing and regulatory procedures can be a major help in encouraging the utility industry to adopt nuclear power, the Congress and the AEC have been taking steps in this direction. A major step is the recent enactment of laws that will reduce greatly the number of mandatory public hearings for reactor licensing. The Commission is studying means of simplifying its own licensing procedures by reducing the volume and complexity of administrative processes. Further operating experience should reduce the time and effort required for technical analysis and review.

Objectives for the Future

Clearly: *The overall objective of the Commission's nuclear power program should be to foster and support the growing use of nuclear energy and, importantly, to guide the program in such directions as to make possible the exploitation of the vast energy resources latent in the fertile materials, uranium-238 and thorium.*

More specific objectives may be summarized as follows:

1. *The demonstration of economic nuclear power by assuring the construction of plants incorporating the presently most competitive reactor types;*
2. *The early establishment of a self-sufficient and growing nuclear power industry that will assume an increasing share of the development costs;*
3. *The development of improved converter and, later, breeder reactors to convert the fertile isotopes to fissionable ones, thus making available the full potential of the nuclear fuels.*
4. *The maintenance of U.S. technological leadership in the world by means of a vigorous domestic nuclear power program and appropriate cooperation with, and assistance to, our friends abroad.*

The role of the Commission in achieving these objectives must be one of positive and vigorous leadership, both to achieve the technical goals and to assure growing participation by the equipment and utility industry as nuclear power becomes economic in increasing areas of this country and the world at large.

67. Proposed AEC Licensing Procedure and Related Legislation, June–July 1971

CONCLUSIONS OF REPORT OF ADVISORY TASK FORCE ON POWER REACTOR
EMERGENCY COOLING (1967)[1]

CONCLUSION 1. PHENOMENA ASSOCIATED WITH LOSS-OF-COOLANT

Current technology is sufficient to enable predicting with reasonable assurance the key phenomena associated with the loss-of-coolant; for quantitative understanding of the accident, the analysis of such an event requires that the core be maintained in place and essentially intact to preserve the heat-transfer area and coolant-flow geometry. Without preservation of heat-transfer area and coolant-flow geometry, fuel-element melting and core disassembly would be expected. With the start of core disassembly there would be a major increase in the uncertainty of prediction of core behavior, and degeneration of the core to a meltdown situation could not be ruled out.

Although basic analytical techniques are available to adequately predict the complex behavior characteristics of a loss-of-coolant, further assurance of the understanding of the event would result from additional experimental and analytical information. Hence, experiments in geometries representative of reactor coolant systems should be conducted, and more precise, analytical representations should also be developed.

CONCLUSION 2. STRUCTURAL RESPONSE REQUIREMENTS TO BLOWDOWN

The mechanical or structural response to blowdown of key primary-system components must be such that the extent of deformations which could occur do not interfere with effective cooling of the core, do not preclude reactor shutdown, and do not cause further consequential primary system damage. The structural integrity of emergency core-cooling sys-

---

This text appears in U.S. Congress, Joint Committee on Atomic Energy, AEC Licensing Procedure and Related Legislation: Hearings Before the Subcommittee on Legislation, Part I, 92nd Congress, 1st session, June–July 1971, pp. 538–43.

1. Notes in original not reproduced.

tems themselves must also be such that emergency core cooling can be accomplished.

As discussed in Conclusion 1, it is within the state of technology to predict, within conservative bounds, the hydraulic forces associated with blowdown. Methods are available for predicting the structural response to these forces, including prediction in the region of limited plastic deformation. The magnitude of these forces is within the range that can be handled with practicable engineering designs. Designs involving more extensive plastic deformation should not *a priori* be excluded, but the extent of deformation is currently difficult to predict.

CONCLUSION 3. REQUIREMENTS OF EMERGENCY CORE-COOLING SYSTEM

The design requirements for the emergency core-cooling system must be:

a. First, to terminate in a loss-of-coolant accident core-temperature transients which could otherwise result in the loss of a definable core heat-transfer and coolant-flow geometry;

b. Then, to reduce the core to emergency core-coolant temperatures; and,

c. Finally, to maintain the core in this condition until full recovery from the loss-of-coolant accident is achieved.

It is important to recognize that fulfillment of the first requirement necessitates the prevention of bulk melting of the clad. At the present time and in the context of present peaking factors, a conservative interpretation of this requirement would be that the emergency core-cooling system be designed to prevent clad melt. Currently, the accepted procedure for fulfillment of the above requirement is to analytically demonstrate by means of a conservatively bounded evaluation that the core cladding in its normal geometry does not melt. This procedure is considered to be sufficient. However, it must be emphasized that this interpretation of "no clad melt" is not a requirement in itself since it may be possible to demonstrate that temperature transients can still be terminated in the presence of some clad melting; and that therefore, the overall objective for emergency core cooling would be satisfied.

CONCLUSION 4. TECHNOLOGY OF EMERGENCY CORE COOLING

Sufficient test data are available to indicate that the phenomena of spray cooling and flooding represent satisfactory approaches to emergency core cooling. The implementation of these phenomena as cooling techniques is amenable to experimental verification. While there has been considerable effort expended in such experimental verification of core-cooling techniques, further testing at higher temperatures and degenerated conditions as well as general evaluations should be conducted.

CONCLUSION 5. PRACTICABILITY OF EMERGENCY CORE-COOLING SYSTEMS

The requirements for emergency core cooling are such that it is practical to design adequate emergency core-cooling systems within the current engineering technology.

The determination that the emergency core-cooling system used on a particular plant will be adequate requires detailed systems engineering evaluation. It is suggested that the elements of this evaluation be developed into a standardized procedure to insure that the evaluation is complete in all cases.

CONCLUSION 6. RELIABILITY ANALYSIS

The concept of reliability analysis has proven a useful and effective tool for systems evaluation in other industries. It is concluded that this concept can likewise be used to similar advantage in the assessment of emergency core-cooling systems. It would be of particular use in the relative comparison of systems and would also serve to aid in the identification of areas within a system network which are critical to its reliability. It is, therefore, recommended that the necessary reliability discipline and techniques be established within the nuclear industry and that this be placed on a formal basis to facilitate its implementation.

CONCLUSION 7. PRIMARY SYSTEM INTEGRITY

A main line of defense against the possibility of a core meltdown is the integrity of the primary system boundary. Much has been done already to assure an acceptable level of integrity; however, the large number of plants now being constructed and planned for the future makes it prudent that even greater assurance be provided henceforth. Accordingly, we recommend that improvements, of the types suggested below, be made both from a short-range and long-range standpoint.

Short Range

a. As a minimum, those parts of the primary system whose failure could lead to large breaks should be designed, manufactured, and inspected to the high degree of reliability comparable to that presently used for reactor vessels, and to the additional requirements enumerated below. The present efforts on preparation of nuclear piping and nuclear value and pump codes should be expedited and these codes put into effect without delay to reflect these high standards. These standards should also be applied to those components critical to emergency core cooling. Thorough reviews of the design of each component and subsystem making up the entire primary coolant system should be made by a qualified group separate from the one that has responsibility for the design. This separate group could be within or without the same organization. These design reviews should also include systems and components other than the primary system which are critical to the problem of core cooling.

b. Adequate allowance should be made in the design and operation of components and systems for the effects on materials resulting from neutron irradiation, such as the shifts which occur in the nilductility transition temperature. In addition, reactor vessel material, weldment, and heat-affected zone samples, should be included in the reactor vessel for periodically monitoring changes in reactor-vessel-material and weldment properties during the life of the vessel. These considerations should be included in an appropriate standard or code. It should be noted that safety limits and conditions to assure that a plant is operated within approved design limits have to be specified in Plant Technical Specifications as required for obtaining AEC operating licenses.

c. Further emphasis should be placed on using overlapping inspection techniques, on greater quality control, and on the training of inspectors, and test personnel. Areas suggested for consideration include:

(1) Apply more than one nondestructive-test method in order to increase the assurance of flaw detection where special considerations such as geometry, accessibility, or variation in test technique warrant. This overlapping in inspection could include, for example, the ultrasonic testing of weld joints as well as their radiography. In this connection it is urged that standards and procedures be established to further the use of ultrasonic testing in the inspection of primary components.

(2) Establish qualification standards for all nondestructive-test inspectors and test personnel. (It is understood that the ASME Boiler and Pressure Vessel Committee is presently working on establishing such standards.) Such personnel should be required to formally pass these

standards before they can be used to inspect any primary coolant component or system. Further, the personnel should be re-examined periodically (every two years) to assure that they are fully knowledgeable and up-to-date with all latest testing techniques and requirements.

(3) Have a formal quality-assurance plan, prepared by the primary-component manufacturer and approved by the organization responsible for the plant design, which delineates the quality control that will be used in the manufacture of the component.

(4) Establish a separate monitoring system to assure that all phases of the quality-assurance program for the manufacture of each component are fully implemented.

d. Review and upgrading of Section III of the ASME Code, other appropriate codes, and inspection standards should be performed frequently to keep pace with improvements in technology, design techniques, inspection methods, and test equipment. Require that such codes and standards be used by all fabricators of primary coolant components and systems. (Ultrasonic testing of plates and forging is an example where the development of tighter inspection standards is underway.)

e. Prepare and keep on file accurate manufacturing and inspection records of primary system components signed by responsible company representatives.

f. Require a leak detection system (such as air-activity detectors) external to the primary system and not connected to it so as to provide early warning if a leak develops in the primary system. (Experience as summarized in Appendix 3[2] indicates that leaks occurring in the primary systems are small and any propagation would be very gradual.)

Long Range

In addition to the relatively short-range action outlined above, effort should proceed toward the development of reliable and repeatable in-service techniques and associated standards for detecting flaws in primary system components, especially reactor vessels, during plant shutdowns. It should be noted that effective utilization of such shutdown inspections will require a reference inspection before the component is placed in service. The purpose of the periodic inspections is to determine whether any change has occurred since the previous inspection. It is understood that a program on this subject is being initiated by the Pressure Vessel Research Committee together with fundamental work on pressure vessel materials.

2. Appendixes not reproduced.

CONCLUSION 8. BREAK SIZE FOR EMERGENCY CORE-COOLING DESIGN

a. We consider it unnecessary to assume that large and rapid failures will occur in any component or system which is designed, manufactured, inspected, protected against missiles, and operated in accordance with the requirements given in Conclusion 7 or their equivalent.

b. Because the record of conventional as well as nuclear plant performance to date clearly indicates that small leaks from a pressurized system can occur, we consider it necessary that back-up means be provided for introducing water into the primary system to assure continued core cooling.

c. In addition to a. and b., the emergency core-cooling system should also be capable of handling a large and rapid failure of those components and systems which are not designed, fabricated, inspected, protected against missiles, and operated in accordance with Conclusion 7 or its equivalent.

d. We expect that, as recommended herein, more and more elements of the primary system will be designed, manufactured, and inspected to the same degree of high standards as required by Section III of the ASME Code, its revisions in process, and additional requirements such as those recommended in this report, to give the same reliability as reactor vessels. This evolution, which will further assure primary system integrity, should make it possible to design emergency core-cooling systems for reduced break sizes, because large and rapid failure of components meeting the recommended standards will not have to be considered. Eventually, a minimum in the reduced break size would still have to be specified as an acceptable basis for designing emergency core-cooling systems. In establishing such a minimum, a prudent safety factor based on engineering experience and judgment should be used. We consider that even with this safety factor the minimum acceptable break size eventually will be considerably smaller than the current design basis.

CONCLUSION 9. SAFEGUARDS ROLE OF CONTAINMENT

The present concepts of containment, with their cooling systems, can provide an adequate barrier to the release of fission products to the environs when emergency core-cooling systems fulfill their design objectives. Both energy release and fission product release can be effectively contained.

Since the performance of the containment as a safeguard system is related to the performance of the other safeguard systems, we recommend that its design basis be chosen accordingly. Containment design should be based upon the energy released by the coolant, decay heat, and metal-water reactions consistent with functioning of emergency core-cooling systems and a prudent safety margin.

CONCLUSION 10. CORE MELTDOWN

If emergency core-cooling systems do not function and meltdown of a substantial part of an irradiated core occurs, the current state of knowledge regarding the sequence of events and the consequences of the meltdown is insufficient to conclude with certainty that integrity of containments of present designs, with their cooling systems, will be maintained.

CONCLUSION 11. COUNTERMEASURES PRIOR TO LOSS OF CONTAINMENT INTEGRITY

Although containment integrity cannot be assured in the event of a postulated core meltdown, a significant period of time may elapse before breachment of the containment occurs. It may be possible to develop preventive measures which are effective during this period and which could reduce the hazards resulting from subsequent failure of the containment. The desirability of utilizing such systems and the merits of requiring containments to be designed to assure such time availability should be evaluated after the effectiveness of these systems has been established through necessary development work. The use of such safeguards will depend on weighing their merits with those of other safety features to obtain the desired objectives in overall reactor safety.

CONCLUSION 12. HANDLING OF LARGE MOLTEN MASSES

Reliable and practical methods of handling large molten masses of fuel for long periods of time do not exist today. The desirability of seeking such methods in order to improve the independence of the containment as an engineered safeguard should be considered in the light of primary system integrity and emergency core cooling effectiveness. It should be recognized that effective means of holding the molten core are not in themselves adequate to prevent containment violation from overpressure.

ADVISORY COMMITTEE ON REACTOR SAFEGUARDS,
U.S. ATOMIC ENERGY COMMISSION,
Washington, D.C., February 26, 1968.

Subject. Report of Advisory Task Force on Power Reactor Emergency
Cooling.
HON. GLENN T. SEABORG,
Chairman, U.S. Atomic Energy Commission,
Washington, D.C.

DEAR DR. SEABORG: The Advisory Committee on Reactor Safeguards
offers the following comments on the recently issued Report of the Advisory
Task Force on Power Reactor Emergency Cooling.

The Committee believes that the Task Force has performed a valuable
service by assembling in a single document discussions covering many of the
problems associated with postulated loss-of-coolant accidents and the phenomena important to proper functioning of emergency core cooling systems.
Also, the Task Force has reviewed in a useful manner the many phenomena
involved in the course of a postulated large-scale core meltdown.

Certain of the report's conclusions and recommendations appear to constitute expressions of judgment as to the adequacy or sufficiency of particular
reactor safety provisions in respect to their capability for providing assurance
against undue risk to the health and safety of the public. No attempt is made
to comment on these. There are, however, a number of other conclusions in
the report concerning which the Committee wishes to recommend emphasis, supplementation, or a differing viewpoint. Comments on these are given
below.

In Conclusion 1, the report states in connection with the loss-of-coolant
accident: " . . . for quantitative understanding of the accident, the analysis
of such an event requires that the core be maintained in place and essentially
intact to preserve the heat-transfer area and coolant-flow geometry. Without
preservation of heat-transfer area and coolant-flow geometry, fuel-element
melting and core disassembly would be expected. With the start of core
disassembly there would be a major increase in the uncertainty of prediction
of core behavior, and degeneration of the core to a meltdown situation
could not be ruled out." The ACRS is in substantial agreement with this
observation.

With respect to assuring that the core remains essentially intact during a
loss-of-coolant accident, the report emphasizes the importance of properly
assessing and designing for the hydraulic effects incurred, and lists several
important specific aspects of the problem that must be recognized and dealt

with in designing to cope with such effects. The ACRS agrees with this emphasis.

The possibility of fuel element failure from high internal pressure and high clad temperature during a loss-of-coolant accident is mentioned. In this connection, the Committee notes that present license applications show that a large fraction of fuel rods may fail in such accidents even though the emergency core cooling system works as designed. The Committee believes that, in addition to the work proposed by the Task Force, further research is needed to ascertain the modes of fuel rod failure and to determine that failures will not propagate or tend to block coolant flow excessively.

Conclusion 2 discusses further the importance of controlled and acceptable structural deformation during reactor blowdown in a loss-of-coolant accident. The ACRS agrees with this and calls attention to the need for considering deterioration during the life of the reactor and the role that periodic inspection could play in alleviating this potential difficulty. Also, more conservatism in design and fabrication may be needed where structural member response to accident-induced hydraulic forces is not testable. Further, the Committee continues to be concerned with the possibility of thermal shock effects on the pressure vessel, or other parts of the primary system, as a consequence of the rapid introduction of emergency cooling water in the unlikely event of a large primary system rupture.

The Committee endorses Conclusion 4 which recommends further testing of emergency core cooling at higher temperatures and for degenerated conditions such as core distortion.

The systematized approach to the design and evaluation of emergency core cooling systems described in Conclusion 5 appears potentially useful. However, deliberate allowance should be made for the possibility of aggravated accident conditions introduced by possible design errors, by weaknesses common to redundant components, or by other unexpected conditions, and full attention should be given to the potential advantage of diverse approaches to the design of emergency core cooling subsystems. It should be recognized, also, that new design features may introduce new potential safety issues in specific reactor designs.

The Committee endorses the recommendation of Conclusion 7 for improvements in primary system integrity to reduce still further the already low probability of primary system boundary failure.

The ACRS agrees with the statement of Conclusion 10 that: "If emergency core-cooling systems do not function . . . the current state of knowledge regarding the sequence of events and the consequences of the meltdown is insufficient to conclude with certainty that integrity of containments of present designs, with their cooling systems, will be maintained."

Recognizing that absolute certainty cannot exist concerning any facet of safety, the Committee strongly recommends that a positive approach be adopted toward studying the workability of protective measures to cope with core meltdown. Basic safety research experiments would provide valuable insight and, possibly, direct attention to potentially profitable avenues of design which eventually could lead to substantial additional protection in this area. The proposal in Conclusion 11 for study of preventive measures to be made effective prior to loss of containment integrity to minimize the ultimate hazard is a helpful step in this direction.

In summary, the Task Force Report presents considerable information of interest on primary system integrity, key phenomena effective during loss of coolant and core heatup, functional considerations for emergency core cooling systems, and phenomena and effects associated with large-scale core meltdown. However, there is provided little examination or discussion of the degree to which the efficacy of core cooling systems might be augmented by way of design modifications or new design concepts. Similarly, the report does not provide recommendations on design approaches to cope with large molten masses of fuel, or on the particular research and development problems related to these approaches.

The ACRS endorses the Task Force recommendations for improvement in primary system boundary integrity and for additional research and development work on emergency core cooling systems. The Committee further recommends, as it did in its 1966 report on safety research, that a vigorous program be aimed at gaining better understanding of the phenomena and mechanisms important to the course of large-scale core meltdown. It also recommends that additional design and development effort be aimed at means for providing protection against the extremely low probability type of loss-of-coolant accident in which emergency core cooling systems of current design may not be effective. The ACRS urges that these matters be pursued vigorously by manufacturers of nuclear equipment, the electric utilities, and the AEC, as appropriate.

Sincerely yours,

CARROLL W. ZABLE, *Chairman.*

68. Atomic Energy Act of 1954

"Sec. 103. Commercial Licenses. —

"a. Subsequent to a finding by the Commission as required in section 102, the Commission may issue licenses to transfer or receive in interstate commerce, manufacture, produce, transfer, acquire, possess, import, or export under the terms of an agreement for cooperation arranged pursuant to section 123, such type of utilization or production facility. Such licenses shall be issued in accordance with the provisions of chapter 16 and subject to such conditions as the Commission may by rule or regulation establish to effectuate the purposes and provisions of this Act.

"b. The Commission shall issue such licenses on a nonexclusive basis to persons applying therefor (1) whose proposed activities will serve a useful purpose proportionate to the quantities of special nuclear material or source material to be utilized; (2) who are equipped to observe and who agree to observe such safety standards to protect health and to minimize danger to life or property as the Commission may by rule establish; and (3) who agree to make available to the Commission such technical information and data concerning activities under such licenses as the Commission may determine necessary to promote the common defense and security and to protect the health and safety of the public. All such information may be used by the Commission only for the purposes of the common defense and security and to protect the health and safety of the public.

"c. Each such license shall be issued for a specified period, as determined by the Commission, depending on the type of activity to be licensed, but not exceeding forty years, and may be renewed upon the expiration of such period.

"d. No license under this section may be given to any person for activities which are not under or within the jurisdiction of the United States, except for the export of production or utilization facilities under terms of an agreement for cooperation arranged pursuant to section 123, or except under the provisions of section 109. No license may be issued to any corporation or other entity if the Commission knows or has reason to believe it is owned, controlled, or dominated by an alien, a foreign corporation, or a foreign government. In any event, no license may be issued to any person within

These selections from the full document appear in *Laws of 83rd Congress—2nd Session*, August 30, 1954, pp. 1098, 1118–21.

the United States if in the opinion of the Commission, the issuance of a license to such person would be inimical to the common defense and security or to the health and safety of the public.

"Chapter 16. Judicial Review and Administrative Procedure

"Sec. 181. General.—The provisions of the Administrative Procedure Act (Public Law 404, Seventy-ninth Congress, approved June 11, 1946) shall apply to all agency action taken under this Act, and the terms 'agency' and 'agency action' shall have the meaning specified in the Administrative Procedure Act: Provided, however, that in the case of agency proceedings or actions which involve Restricted Data or defense information, the Commission shall provide by regulation for such parallel procedures as will effectively safeguard and prevent disclosure of Restricted Data or defense information to unauthorized persons with minimum impairment of the procedural rights which would be available if Restricted Data or defense information were not involved.

"Sec. 182. License Applications.—

"a. Each application for a license hereunder shall be in writing and shall specifically state such information as the Commission, by rule or regulation, may determine to be necessary to decide such of the technical and financial qualifications of the applicant, the character of the applicant, the citizenship of the applicant, or any other qualifications of the applicant as the Commission may deem appropriate for the license. In connection with applications for licenses to operate production or utilization facilities, the applicant shall state such technical specifications, including information of the amount, kind, and source of special nuclear material required, the place of the use, the specific characteristics of the facility, and such other information as the Commission may, by rule or regulation, deem necessary in order to enable it to find that the utilization, or production of special nuclear material will be in accord with the common defense and security and will provide adequate protection to the health and safety of the public. Such technical specifications shall be a part of an license issued. The Commission may at any time after the filing of the original application, and before the expiration of the license, require further written statements in order to enable the Commission to determine whether the application should be granted or denied or whether a license should be modified or revoked. All applications and statements shall be signed by the applicant or licensee under oath or affirmation.

"b. The Commission shall not issue any license for a utilization or production facility for the generation of commercial power under section 103, until it has given notice in writing to such regulatory agency as may have jurisdiction over the rates and services of the proposed activity, to municipalities, private utilities, public bodies, and cooperatives within transmission distance authorized to engage in the distribution of electric energy and until it has published notice of such application once each week for four consecutive weeks in the Federal Register, and until four weeks after the last notice.

"c. The Commission, in issuing any license for utilization or production facility for the generation of commercial power under section 103, shall give preferred consideration to applications for such facilities which will be located in high cost power areas in the United States if there are conflicting applications resulting from limited opportunity for such license. Where such conflicting applications resulting from limited opportunity for such license include those submitted by public or cooperative bodies such applications shall be given preferred consideration.

"Sec. 183. Terms of Licenses.—Each license shall be in such form and contain such terms and conditions as the Commission may, by rule or regulation, prescribe to effectuate the provisions of this Act, including the following provisions:

"a. Title to all special nuclear material utilized or produced by facilities pursuant to the license, shall at all times be in the United States.

"b. No right to the special nuclear material shall be conferred by the license except as defined by the license.

"c. Neither the license nor any right under the license shall be assigned or otherwise transferred in violation of the provisions of this Act.

"d. Every license issued under this Act shall be subject to the right of recapture or control reserved by section 108, and to all of the other provisions of this Act, now or hereafter in effect and to all valid rules and regulations of the Commission.

"Sec. 184. Inalienability of Licenses.—No license granted hereunder and no right to utilize or produce special nuclear material granted hereby shall be transferred, assigned or in any manner disposed of, either voluntarily or involuntarily, directly or indirectly, through transfer of control of any license to any person, unless the Commission shall, after securing full information, find that the transfer is in accordance with the provisions of this Act, and shall give its consent in writing. The Commission may give such consent to the creation of a mortgage, pledge, or other lien upon any facility owned or thereafter acquired by a licensee, or upon any leasehold or other interest in such property, and the rights of the creditors so secured may thereafter be enforced by any court subject to rules and regulations established by the Commission to protect public health and safety and promote the common defense and security.

"Sec. 185. Construction Permits.—All applicants for licenses to construct or modify production or utilization facilities shall, if the application is otherwise acceptable to the Commission, be initially granted a construction permit. The construction permit shall state the earliest and latest dates for the completion of the construction or modification. Unless the construction or modification of the facility is completed by the completion date, the construction permit shall expire, and all rights thereunder be forfeited, unless upon good cause shown, the Commission extends the completion date. Upon the completion of the construction or modification of the facility, upon the filing of any additional information needed to bring the original application up to date, and upon finding that the facility authorized has been constructed and will operate in conformity with the application as amended and in conformity with the provisions of this Act and of the rules and regulations of the Commission, and in the absence of any good cause being shown to the Commission why the granting of a license would not be in accordance with the provisions of this Act, the Commission shall thereupon issue a license to the applicant. For all other purposes of this Act, a construction permit is deemed to be a 'license.'

"Sec. 186. Revocation.—

"a. Any license may be revoked for any material false statement in the application or any statement of fact required under section 182, or because of conditions revealed by such application or statement of fact or any report, record, or inspection or other means which would warrant the Commission to refuse to grant a license on an original application, or for failure to construct or operate a facility in accordance with the terms of the construction permit or license or the technical specifications in the application, or for violation of, or failure to observe any of the terms and provisions of this Act or of any regulation of the Commission.

"b. The Commission shall follow the provisions of section 9(b) of the Administrative Procedure Act in revoking any license.

"c. Upon revocation of the license, the Commission may immediately retake possession of all special nuclear material held by the licensee. In cases found by the Commission to be of extreme importance to the national defense and security or to the health and safety of the public, the Commission may recapture any special nuclear material held by the licensee or may enter upon and operate the facility prior to any of the procedures provided under the Administrative Procedure Act. Just compensation shall be paid for the use of the facility.

"Sec. 187. Modification of License.—The terms and conditions of all licenses shall be subject to amendment, revision, or modification, by reason of amendments of this Act or by reason of rules and regulations issued in accordance with the terms of this Act.

"Sec. 188. Continued Operation of Facilities.—Whenever the Commission finds that the public convenience and necessity or the production program of the Commission requires continued operation of a production facility or utilization facility the license for which has been revoked pursuant to section 186, the Commission may, after consultation with the appropriate regulatory agency, State or Federal, having jurisdiction, order that possession be taken of and such facility be operated for such period of time as the public convenience and necessity or the production program of the Commission may, in the judgment of the Commission, require, or until a license for the operation of the facility shall become effective. Just compensation shall be paid for the use of the facility.

"Sec. 189. Hearings and Judicial Review.—

"a. In any proceeding under this Act, for the granting, suspending, revoking, or amending of any license or construction permit, or application to transfer control, and in any proceeding for the issuance or modification of rules and regulations dealing with the activities of licensees, and in any proceeding for the payment of compensation, an award or royalties under sections 153, 157, 186 c., or 188, the Commission shall grant a hearing upon the request of any person whose interest may be affected by the proceeding, and shall admit any such person as a party to such proceeding.

"b. Any final order entered in any proceeding of the kind specified in subsection a. above shall be subject to judicial review in the manner prescribed in the Act of December 29, 1950, as amended (ch. 1189, 64 Stat. 1129), and to the provisions of section 10 of the Administrative Procedure Act as amended."

69. Calvert Cliffs Coordinating Committee versus U.S. Atomic Energy Commission, 1971

CALVERT CLIFFS COORDINATING
COMMITTEE, INC., et al.,
Petitioners,
v.
UNITED STATES ATOMIC ENERGY
COMMISSION and United States
of America, Respondents.
Nos. 24839, 24871.
District of Columbia Circuit.
Argued April 16, 1971.
Decided July 23, 1971.

Proceeding to review order of Atomic Energy Commission. The Court of Appeals, J. Skelly Wright, Circuit Judge, held that the courts have power to require agencies to comply with procedural directions of National Environmental Policy Act of 1969 and that the Commission's rules precluding review consideration of nonradiological environmental issues unless specifically raised, prohibiting raising such issues in certain pending proceedings or when issues have been passed on by other agencies, and precluding consideration between grant of construction permit and consideration of grant of operating license, did not comply with Act.

Remanded . . .

J. Skelly Wright, Circuit Judge:

These cases are only the beginning of what promises to become a flood of new litigation—litigation seeking judicial assistance in protecting our natural environment. Several recently enacted statutes attest to the commitment of the Government to control, at long last, the destructive engine of material "progress." But it remains to be seen whether the promise of this legislation will become a reality. Therein lies the judicial role. In these cases, we must for the first time interpret the broadest and perhaps most

This document appears in Calvert Cliffs Coordinating Committee, Inc., et al., Petitioners, *v.* USAEC and USA, Respondents, July 23, 1971; *449F. 2d 1109 (1971).* Federal Reporter Series, pp. 1111–31 passim.

important of the recent statutes: the National Environmental Policy Act of 1969 (NEPA). We must assess claims that one of the agencies charged with its administration has failed to live up to the congressional mandate. Our duty, in short, is to see that important legislative purposes, heralded in the hall of Congress, are not lost or misdirected in the vast hallways of the federal bureaucracy.

NEPA, like so much other reform legislation of the last 40 years, is cast in terms of a general mandate and broad delegation of authority to new and old administrative agencies. It takes the major step of requiring all federal agencies to consider values of environmental preservation in their spheres of activity, and it prescribes certain procedural measures to ensure that those values are in fact fully respected. Petitioners argue that rules recently adopted by the Atomic Energy Commission to govern consideration of environmental matters fail to satisfy the rigor demanded by NEPA. The Commission, on the other hand, contends that the vagueness of the NEPA mandate and delegation leaves much room for discretion and that the rule challenged by petitioners falls well within the broad scope of the Act. We find the policies embodied in NEPA to be a good deal clearer and more demanding than does the Commission. We conclude that the Commission's procedural rules do not comply with the congressional policy. Hence we remand these cases for further rule making.

The special importance of the pre-operating license stage is not difficult to fathom. In cases where environmental costs were not considered in granting a construction permit, it is very likely that the planned facility will include some features which do significant damage to the environment and which could not have survived a rigorous balancing of costs and benefits. At the later operating license proceedings, this environmental damage will have to be fully considered. But by that time the situation will have changed radically. Once a facility has been completely constructed, the economic cost of any alteration may be very great. In the language of NEPA, there is likely to be an "irreversible and irretrievable commitment of resources," which will inevitably restrict the Commission's options. Either the licensee will have to undergo a major expense in making alterations in a completed facility or the environmental harm will have to be tolerated. It is all too probable that the latter result would come to pass.

By refusing to consider requirement of alterations until construction is completed, the Commission may effectively foreclose the environmental protection desired by Congress. It may also foreclose rigorous consideration of environmental factors at the eventual operating license proceedings. If "irreversible and irretrievable commitment(s) of resources" have already been made, the license hearing (and any public intervention therein) may become a hollow exercise. This hardly amounts to consideration of environmental values "to the fullest extent possible."

A full NEPA consideration of alterations in the original plans of a facility, then, is both important and appropriate well before the operating license proceedings. It is not duplicative if environmental issues were not considered in granting the construction permit. And it need not be duplicated, absent new information or new developments, at the operating license stage. In order that the pre-operating license review be as effective as possible, the Commission should consider very seriously the requirement of a temporary halt in construction pending its review and the "backfitting" of technological innovation. For no action which might minimize environmental damage may be dismissed out of hand. Of course, final operation of the facility may be delayed thereby. But some delay is inherent whenever the NEPA consideration is conducted—whether before or at the license proceedings. It is far more consistent with the purposes of the Act to delay operating at a stage where real environmental protection may come about than at a stage where corrective action may be so costly as to be impossible.

(12) Thus we conclude that the Commission must go farther than it has in its present rules. It must consider action, as well as file reports and papers, at the pre-operating license stage. As the Commission candidly admits, such consideration does not amount to a retroactive application of NEPA. Although the projects in question may have been commenced and initially approved before January 1, 1970, the Act clearly applies to them since they must still pass muster before going into full operation. All we demand is that the environmental review be as full and fruitful as possible.

70. Three Mile Island, March 1979

A combination of design deficiency, mechanical failure, and human error contributed to the ill-controlled accident that was touched off at about 4 A.M. on Wednesday, March 28, at Unit 2 of the Three Mile Island nuclear power station of Metropolitan Edison Company, a member company of General

This essay appears in *Nuclear News*, Special Report, April 6, 1979. Copyright © 1979 by Nuclear News. Reprinted by permission.

Public Utilities (GPU). The initial event at the unit, located near Harrisburg, Pa., has been characterized as a loss-of-normal-feedwater turbine trip—with complications. As Norman Rasmussen of M.I.T. explained the following Sunday on ABC's television program "Issues and Answers," the event could not be considered, in the parlance of reactor safety studies, a "major event" ("an event of major consequences" to the public health and safety, he quickly went on to clarify). The TMI-2 "event," however, certainly promises to have major consequences for the utilization of nuclear power in America and elsewhere—this in spite of the fact that the accident was contained and the amounts and forms of radioactivity that did escape from the plant, appeared, by most accounts, to have been of no major consequence.

If not calamitous, however, the accident was indeed most serious. Considerable fuel failure occurred, fission products were released into and out of the containment and into the atmosphere, and a tremendous loss, in terms of failed and suspect fuel, the staggering costs of eventual cleanup, and lost confidence, has been visited upon Metropolitan Edison—not to speak of the entire nuclear effort.

Buildup to a Crisis

The TMI-2 incident occurred one year to the day after it had first achieved criticality. It was operating at 98 percent of full power. In a matter of minutes during the early morning hours of March 28, it was off line, the secondary system was out of service, the pressurizer relief valve was stuck open, and primary system water was spilling into the containment structure, subsequently to be pumped out to the reactor auxiliary building, from which radioactive gases would escape to the atmosphere (see "How It Happened").

In spite of the severity of the situation, it was not until almost 7 A.M. that a site emergency was declared. High activity readings in the reactor building were then noted, prompting the call for a general emergency at about 7:30 A.M. The utility also notified the Nuclear Regulatory Commission's Region I office outside Philadelphia, and Civil Defense in nearby Harrisburg. State and federal officials alike have taken Metro Ed to task for not issuing notification until nearly four hours after the initiation of the accident.

The NRC dispatched a team to the site and began taking radiation readings—as did the Pennsylvania Department of Environmental Resources. Ventilation in the auxiliary building continued, radioactivity in the containment remained high, and the "bubble" in the reactor presented a major obstacle to achievement of cold shutdown. By the end of Wednesday, it was generally agreed that this was the most severe commercial nuclear accident to date, and that the worst may not have been over.

As the gravity of the situation became clear, the NRC beefed up its presence at the site. Harold Denton, director of the Office of Nuclear Reactor Regulation, led a contingent from NRC headquarters and became the agency's spokesman at the frequent news conferences held during the next few days. Help was also brought in from other NRC regional offices, and from virtually every national laboratory.

Also sent in were utility and industry teams (including, of course, the B&W personnel), teams from state and federal agencies, and scientists and engineers from outside the United States. Other teams in Washington, D.C., and elsewhere in the United States worked full-time on the problem.

A huge press contingent also gathered in the towns of Middletown and Goldsboro very close to the site, and in Harrisburg. The press generally is credited with doing a good job of reporting on the incident in spite of some confusion during the initial states.

There was often disagreement in the reports from the Metropolitan Edison Company and those of the NRC concerning the status of the plant and the progress in the efforts to cool down the reactor. Some of this was understandable as it was apparently not known at first how serious the problem was and, in some instances, because one source had more recent information than the other. But even when working from similar information, the reports from the company and NRC differed, with the reports from the utility almost always carrying a more optimistic tone than those from the Commission. As the press increasingly noted these differences, and as the situation was becoming acrimonious between the NRC and the utility, the White House ordered on April 1 that all further technical reports on the status of the plant were to come from the Commission.

Radioactivity

By the morning of April 3, the NRC had announced the following cumulative data from environmental monitoring:

- No radioiodine was detected in any of the 130 water samples taken by the NRC, the Department of Energy, and the Commonwealth of Pennsylvania since the accident. From what was calculated as the amount of radioiodine released from the plant by water, the thyroid dose to anyone drinking the water was estimated to be less than 0.2 millirem.

- Eight of the 152 offsite air samples showed radioiodine present, the most activity measured being 2.4 × 10 microcuries per millilitre. Thyroid dose to anyone at the site boundary was estimated to be less than 50 mrem over a five-day period. Radioactive inert gases were detected in the sample, taken within a 65-km radius of the plant, but the maximum activity detected was about one-fourth of the permissible concentration established in 10CFR20.
- Milk samples tested by the state did not yield radioiodine, but the federal Food and Drug Administration found results ranging from 14 to 40 picocuries of I-131 per litre in the nine samples it took. Herds are to be placed on stored feed, according to the Department of Health, Education and Welfare, when the I-131 count reaches 12,000 pCi/l. The thyroid dose for anyone drinking milk is estimated to be less than 0.5 mrem per day.
- No radioiodine was detected in any of 147 soil samples taken by the NRC and the DOE, nor in 171 vegetation samples taken within 3 km of the site by the NRC, the DOE, and the state.

Radioiodine in the plant was determined (activity adjusted to decay as of midnight, April 2–3) to be 5.1 × 10^6 Ci in the core, 3.32 × 10^6 Ci in the coolant, and 0.235 Ci in the industrial waste treatment system; in all cases, I-131 accounts for at least 96 percent of the activity, and I-133 for the remainder. The industrial waste treatment system was continuing to be filled through the turbine building sump, and capacity was limited. Releases from this system to the Susquehanna River since the accident have stirred much public resentment.

Three plant employees were exposed to radiation beyond the 3 rem per quarter level by the end of the quarterly monitoring period on March 31. Dosages of 3.1, 3.4, and 3.8 rem were received when the workers tried to drain and secure the auxiliary building. At least 12 other workers received considerable doses below 3 rem.

Hitting Home

Technical aspects aside, what established this as the most serious nuclear accident was its effect upon the daily lives of the public. For the first time, a problem in a nuclear power plant disrupted the personal and business activities of many thousands of people.

Initially, most of the residents in the area went about their daily routines, but this quickly changed. While some residents did continue to state they felt they were in no great danger, many others expressed feelings ranging from anger at the inconvenience caused by the situation to fear for their health and safety. Widespread distrust of the statements by utility spokesmen was reported. "I just don't believe them" was a typical response. Others expressed concerns for their property values and their businesses.

On Friday, March 30, Gov. Richard Thornburgh ordered several schools closed in the area of the plant and recommended the evacuation of preschool children and pregnant women within five miles of the plant. Also, he urged people living within 10 miles to remain indoors. These actions were taken on the recommendation of NRC Chairman Joseph Hendrie and following an "unplanned and unexpected release" of radioactive gas. The following day, Thornburgh said it was no longer seen as necessary that people living within 10 miles of the plant remain indoors, but he still advised that pregnant women and preschool children move from the area five miles around the plant.

Also, plans were prepared for a general evacuation for everyone out to a distance of 20 miles from the plant. Thornburgh announced that a precautionary evacuation might be ordered should the efforts in bringing the reactor to a cold shutdown state be considered sufficiently risky to require evacuation. Such an evacuation was most discussed on March 30–31 and April 1, when the hydrogen "bubble" in the reactor was complicating efforts to cool down the core.

On Sunday, April 1, President Carter visited the site, with the reason for his visit interpreted widely as that of reassuring residents while preparing them for a general evacuation, if that became necessary. If it did, he said, the evacuation instructions should be followed "calmly and exactly." By April 2, definite progress was noted in removing the hydrogen bubble, and Thornburgh announced that a general evacuation would probably not be needed.

Estimates varied greatly about the number of people who moved out of the area, but civil defense officials said that of the 200,000 people living within 20 miles of the plant, an estimated 80,000 left over the weekend of March 31–April 1; 40,000 were still away on Tuesday, April 3; and 30,000 to 35,000 had not yet returned on April 5. In addition, an undetermined number of businesses were closed.

Helping somewhat to alleviate the disgruntlement of the area people was the presence of American Nuclear Insurers, which set up an emergency claims office in Harrisburg. ANI staff members distributed checks to evacuees to cover their hotel and meal expenses, and by Tuesday, April 3, about $200,000 had been paid out. A spokesman from ANI said that the pool of nuclear insurance firms had "no idea" how much would be paid out in such claims, and that ANI had "no estimates on either side" on property and liability claims. Prior to the TMI-2 situation, ANI had paid a total of only $620,000 in claims since the insurance pools were formed in 1957.

On April 3 and 4, with the situation improving at the plant, and with people returning and some schools and businesses reopening, the area was starting to return to some semblance of normalcy. Still to be determined were just how permanent would be the loss of faith by the area residents in the plant's safety, and any adverse impact on the local economy.

The accident sharply changed the plans of the whole General Public Utilities Corporation group (which also includes Jersey Central Power and Light Company and Pennsylvania Electric Company). Even if the outcome is the most favorable for the utility—i.e., if TMI-2 can be repaired and returned to service at a tolerable cost—the expenditure in money and personnel will be sizable. The situation will not be helped if TMI-1, which was off line for routine maintenance at the time of the TMI-2 accident, is forced to remain out of service very long; TMI-1 has been a dependable, highly productive unit during its four and a half years of operation, and a prolonged outage will add replacement power costs to GPU's other worries.

Aware of its predicament, GPU moved on April 3 to suspend all construction work on new power plants (including the Forked River nuclear unit) and all but the most essential construction on existing facilities. Nonessential maintenance will be reduced or postponed wherever possible, and scheduled outages may be postponed.

Political Reactions

At the political level, reactions came quickly, and, not unexpectedly, there were few to take up any defense of nuclear power. On Thursday, March 29, the morning after the accident, all five NRC commissioners, along with top NRC aides, were called before the House Interior Committee's Subcommittee on Energy and the Environment (Rep. Morris K. Udall [D., Ariz.], chairman) to brief the committee on the accident. Reflecting the obvious concern of their constituents, the Congressmen were at times sharp in their questioning of NRC Chairman Hendrie and his colleagues. Hendrie was asked by Rep. James Weaver (D., Ore.), "How close did we come to the China Syndrome?" Hendrie replied, "We were nowhere near it, nowhere near it." The Congressmen persisted, "Could it happen?" Hendrie answered, "In principle, yes," but he insisted that the chances that it would happen were very small. Weaver indicated he was not assured by the NRC Chairman's responses.

Representative Udall did not hide his skepticism concerning the safety of nuclear power either at the hearing or on the same "Issues and Answers" program alluded to earlier, where he urged that America's reliance on nuclear power for the future be seriously questioned. He said that since the country is now dependent on this source for no more than 10–12 percent of its electricity, it would not be too difficult to curtail future efforts in this direction and to shift to other alternatives. He said the people had been told that this type of accident would not happen, and now it has happened. Rasmussen later corrected this by saying that this type of event was among those covered in the Reactor Safety Study (WASH-1400). Rasmussen agreed with Udall that a thorough review of similar reactors would be necessary to prevent similar occurrences.

The NRC, in the meantime, has directed the operators of all nine Babcock & Wilcox reactors having operating licenses to review their facilities and procedures in light of the TMI-2 incident (see "Other B&W Units Cleared to Operate"). The Commission also, as part of its investigation of the incident, is setting up a special review group similar to the one established in the case of the Brown's Ferry fire.

Over and above these efforts by the NRC, Pennsylvania's Sen. Richard Schweiker (R.) has called for the formation of a Presidential commission to analyze "the full implications of the Three Mile Island accident" and has asked that this analysis include what role nuclear power should play in the future of U.S. electricity generation.

Other legislation doubtless will be generated from the hearings being conducted by various committees on both sides of Congress. Sen. Gary Hart (D., Colo.), chairman of the Subcommittee on Nuclear Regulation, said he plans to propose legislation that would strengthen federal oversight of all nuclear power plants. His legislation would put the government, rather than the utility, clearly in charge if and when a serious accident should occur. Hart suggested that all plants be monitored remotely by the NRC, possibly by satellite communication.

Somewhat along those same lines, President Carter signed almost immediately after the accident an executive order centralizing much of the federal emergency planning in a single agency, the Federal Emergency Management Agency. This agency combines the Defense Civil Preparedness Agency, the Federal Disaster Assistance Administration, the Federal Preparedness Agency, the Federal Insurance Administration, and the National Fire and Control Administration.

Actions were taken, too, at the state level. California Gov. Jerry Brown sent a telegram to the NRC after the accident urging that the Rancho Seco nuclear plant near Sacramento, a B&W unit, be ordered shut down until the

Three Mile Island incident was fully studied and corrective measures could be taken.

A few brave souls in the public eye did make statements in defense of nuclear power—notably, Department of Energy Secretary James Schlesinger, who two days after the accident cited the "historic record" of nuclear power as "excellent" and urged that the accident be viewed in proper perspective. He noted that the radiation exposures experienced were "very limited" and contended that nuclear power must continue to be "an essential element" in the nation's drive for energy independence. Sen. J. Bennett Johnston (D., La.) was quoted soon after the accident as saying that it "could lead to an agonizing reappraisal of our use of nuclear power," but adding that "if you're going to have a viable economy, you've simply got no choice but to have nuclear power." The Senator's ambivalence clearly describes the quandry in which many public officials—and the American people—now find themselves with respect to nuclear power.

A President Treads Lightly

In his energy speech on April 5—one week after the Three Mile Island accident—President Carter had little to say about the future role of nuclear power in achieving the stated goal of energy sufficiency. Just as he wore protective plastic shoe covers in his postaccident visit to TMI-2, the President trod lightly over the subject, saying at the outset of his speech that the accident "demonstrated dramatically that we have *other* energy problems" and adding later that he had established an independent Presidential commission of experts "to investigate the causes of the accident and to make recommendations on how we can improve the safety of nuclear power plants. You deserve a full accounting and you will get it."

Among the avenues toward energy independence, he mentioned solar energy, shale oil, advanced coal technology, improved automobile efficiency, improved mass transit systems, and even wood-burning stoves. He made no mention, however, of nuclear power, whether fission or fusion.

Other B&W Units Cleared to Operate

Directed to the Nuclear Regulatory Commission to review their operations in light of the TMI-2 incident and to report within 10 days on the status of their plants and procedures, the operators of eight other Babcock & Wilcox reactors will be allowed to continue operating, this being judged "without undue risk to the public," as stated by Edson Case, NRC deputy director of nuclear reactor regulation at a public briefing on April 4. Case noted that the NRC would take whatever action might be necessary after the reports are received.

Of the eight plants, roughly the same but differing slightly (some have supplemental safety systems not found at TMI-2), only five are operating currently. These are: Rancho Seco, operated by Sacramento Municipal Utility District; the three Oconee units of Duke Power Company; and Crystal River-3 (Florida Power Corporation), scheduled for an outage later this month. The other three units are presently down for scheduled outages. These are: TMI-1; Unit 1 of Arkansas Nuclear One (Arkansas P&L); and Davis-Besse-1 (Central Area Power Coordination Group).

Addendum . . . How It Happened

The actual sequence of events beginning early in the morning of March 28 is believed to have gone as follows:

- At about 4 A.M., the main feedwater system of Three Mile Island-2 malfunctioned, apparently as a result of failure either of the demineralizer or of the air supply to an air-operated valve.
- With the main feedwater supply system out of operation the auxiliary feedwater system was to have started automatically. It did not, however, because a number of manual valves in the auxiliary system had inadvertently been left closed after a test of the system in the days prior to the accident.
- Without feedwater supply, the steam generators dried out, resulting in a rise in the primary coolant temperature and pressure. The turbine generator tripped. The overpressurization generated a signal that scrammed the reactor.
- Within seconds, primary loop pressure rose to 1.625×10 Pascal (2355 psi), triggering a relief valve at the top of the pressurizer. The primary coolant, thus relieved, was piped to a quench tank, located within the containment. The relief valve failed in the open position, providing a continuous release of coolant from the system. This failure and the failure in the main feedwater system (both of a mechanical nature), and the malfunction of the auxiliary feedwater system were the main initiators of the incident.
- The emergency core-cooling system started automatically at two minutes into the accident sequence, and began to raise the coolant level. Within

a few minutes, the level indicator for the pressurizer is reported to have started giving erroneously high readings and in fact went off scale on the high side—while actually the opposite condition should have been indicated. An operator manually overrode first one high-pressure injection pump (ECCS) and, a few minutes later, the other. It is believed that this was done to prevent the pressurizer from becoming "solid" (filled entirely with water); if the pressurizer were filled, pressure control would be lost.

- Within two minutes of the first ECCS shutoff, it has been reported, primary coolant was flashing to steam in the reactor.

- Also after the first ECCS shutoff, containment basement sump pumps began to operate automatically (containment isolation at TMI-2 is not automatic with ECCS initiation, but only after pressure within the containment reaches four psig). Several minutes later the feedwater flow was restarted, increasing pressure in the steam generators.

- With flow through the pressurizer relief valve unchecked, the quench tank continued to fill, and some fifteen minutes into the accident sequence its rupture disc blew open, releasing ultimately upwards of 40,000 litres of primary coolant onto the containment floor. From here the water was pumped into a storage tank in the auxiliary building, where water spilled again after the tank, already partly filled, exceeded capacity. Radioactive isotopes of krypton, iodine, and xenon began to escape to the environment through the building's ventilation system; vent filters trapped much of the iodine but did not stop the noble gases.

- The primary coolant continued to be released from the primary system through the open pressurizer relief valve, resulting in a decrease in primary pressure. After one hour and fifteen minutes after the initial transient, and at about 1300 psi, two primary coolant pumps on one steam generator loop were manually tripped, apparently to prevent cavitation and resultant damage to the pumps. The other two pumps were tripped, again manually, 25 minutes later. At this point the reactor system was without forced convection and without a heat sink.

- Control of the pressurizer relief system was regained, and the ECCS turned on again. It is believed, however, that coolant level had dropped so low that part of the core had been uncovered resulting in substantial fuel damage. Although temperatures in the core registered quite high, no actual melting is believed to have occurred.

- At the three-hour point it was noted that primary-to-secondary leaks in the "B" steam generator had occurred, providing a pathway for more radioactive material to the environment from the secondary system. The "B" steam generator was subsequently isolated; the "A" steam generator remained functional.

- It is postulated that the substantial core damage, combined with high fuel temperatures, resulted in the formation within the reactor vessel of a gas bubble, consisting of hydrogen (from Zircaloy-water reaction) and gaseous fission products. The bubble presented a problem in cooldown: If primary system pressure were reduced, the bubble would expand and possibly impair the cooling mode (e.g., cause pump cavitation).
- Hydrogen also became a problem in the containment. In the beginning phases of the accident the containment pressure had risen above atmospheric. Later on, hydrogen was detected, which, if allowed to accumulate, had the potential for leading to an explosion.
- By Wednesday evening, cooling of the damaged core was restored by activation of one of the primary coolant pumps on the "A" loop, using the "A" steam generator as a heat sink. The degasifying process involved drawing dissolved gas out of the coolant from the cold leg of the primary system and spraying it through the pressurizer and out into the containment. This eventually reduced the size of the bubble but added to the hydrogen in the containment. By April 3, a hydrogen recombiner was in place outside the containment, and hydrogen concentration dropped gradually (2.3 percent, late P.M., April 1; 2.1 percent, early A.M., April 4). It was expected that the concentration would be brought down to 1 percent by about April 15.
- At press time, some coolant temperatures above the core were still in excess of 200 degrees C, requiring primary pressure to be maintained at about 6. × 10 Pa (1000 psi). The approach to cold shutdown that was preferred as of April 6 was to continue cooling with pumped circulation until about April 11, and then convert to "natural circulation" in the primary system and solid water in the secondary. Use of the residual heat removal system (RHRS) was not actively being considered.

71. Alvin M. Weinberg, "A Nuclear Power Advocate Reflects on Chernobyl"[1]

Fifteen years have passed since I first alluded to nuclear energy as a Faustian bargain: humans, in opting for nuclear energy, must pay the price of extraordinary technical vigilance for the energy they derive from nuclear fission if they are to avoid serious trouble. At the time, I could not say exactly what degree of technical vigilance would be required, nor exactly what the consequences of a lapse in vigilance would be. As long as nuclear energy was small and unimportant, as it was in 1971, the likelihood of having to pay up on our bargain would be small; but as the number of reactors grew, as nuclear energy became large and very important, the likelihood, and therefore the frequency, of accidents would increase. Experience would then enable us to judge, rather than speculate on, the risk fission imposed on society.

Chernobyl has given us a second point on the risk versus benefit curve. The first point was Three Mile Island, where, after about 400 reactor years of operation of pressurized water reactors, about 20 curies of iodine 131 were released in a meltdown. No individual was exposed to more than 100 millirem of radiation, and, very probably, no one suffered physical harm. Nevertheless Three Mile Island profoundly affected the nuclear enterprise throughout the world. Perhaps most important, all members of the nuclear community—engineers, managers, reactor scientists—now realized that accidents described in Rasmussen's famous probabilistic risk assessment could actually occur, and with a probability not grossly different from the theoretical probability he had calculated.[*]

Chernobyl, which occurred after about 250 reactor years of operation with reactors of its type has given us another point though on a somewhat different curve. The Chernobyl reactor, being graphite-moderated and with pressure tubes rather than water-moderated with a single pressure vessel, is not really comparable to the reactors that account for most of the world's nuclear power. No one in the West, as far as I know, has seen a Rasmussen-type probabilistic risk analysis for Chernobyl, nor do we know whether such

This article appears in *Bulletin of the Atomic Scientists* 43, 1 (Aug/Sept 1986): 57–60. Copyright © 1986 by the Educational Foundation for Nuclear Science. Reprinted by permission.

1. Alvin M. Weinberg is the former director of the Institute for Energy Analysis at the Oak Ridge Associated Universities in Oak Ridge, Tennessee.

*N. Rasmussen, *The Three Mile Island Accident: Lessons and Implications* (New York: New York Academy of Sciences, 1981), pp. 20–37.

an a priori analysis of accident in such reactors exists. In any case, had the sequence of events that precipitated Chernobyl been identified in probabilistic risk analysis as having a probability as high as one in 250 reactor years, I should think the designers would have eliminated the sequences that were responsible for such high probability. But this is speculation. We in the West do not really know what happened at Chernobyl. We eagerly await the results of the Soviet investigation of the accident.

We do know that the consequences have been substantial. According to Dr. Robert Gale, the American physician who helped with bone marrow transplants, some 300 people received doses around 200 rads or more; and the roughly 100,000 reported to have been evacuated have received doses "large enough to warrant annual examination."

Perhaps of greater significance than the actual physical harm caused by Chernobyl is the near panic the incident precipitated in much of Europe. When the Rasmussen report was issued in 1981, we certainly did not anticipate that a reactor accident of the magnitude of Chernobyl would provoke the public concern that has occurred. Nor did we recognize the full social impact of land interdiction because of fallout. To be sure, at Missassauga in Ontario, Canada, some 240,000 people were evacuated when a chlorine tank car burst in November 1979; once the cloud passed, however, people returned to their homes. The interdiction of land at Chernobyl is rather more like the aftermath at Seveso in Italy, following the 1976 dioxin contamination, or perhaps the aftermath of a dam failure or even a volcanic eruption. In these cases land is interdicted for a long time. How long the land around Chernobyl will be interdicted remains to be seen.

Chernobyl and Three Mile Island are revealing some of the social costs of nuclear accidents, costs that can hardly be estimated by probabilistic risk assessment, especially since the costs depend very much upon the cultural and political environment of the country in which the accident occurs.

An important social cost of Chernobyl is the possible abandonment of nuclear power in several West European countries. Austria, Denmark, and Norway had already rejected nuclear power even before Chernobyl; Sweden's phasing out of nuclear power by 2010, which had been regarded as increasingly unlikely, is once again taken very seriously. In the United States, an ABC poll suggested that 78 percent of the public is now opposed to nuclear power.* I would not be surprised if the 1988 Democratic Party platform calls for a phase-out of nuclear power, even though 15 percent of U.S. electricity is now generated from fission. At the very least, I would

*Energy Daily, June 22, 1986.

expect the 10-year hiatus in construction of new U.S. nuclear plants to continue.

Of the many arguments against abandoning nuclear power, I shall mention only two:

- The 285 gigawatts of electric capacity now represented by the world's reactors significantly reduces the global requirement for oil. An oil-fired power plant that generates 1,000 megawatts at 60 percent load factor burns about 25,000 barrels of residual oil (that is, oil that remains after refining) each day. Thus were all the nuclear electricity generated instead by oil, the world's demand for oil would be increased by seven million barrels per day, or about 12 percent of today's demand for oil. Of course, many of the world's nuclear reactors displace coal and gas, not oil, but an estimate of several million barrels per day saved seems justified.
- As long as a nuclear plant is operating properly, its impact on the environment is far less than that of fossil burning plants, and even dams. I would insist that we have the strongest incentive to continue to utilize nuclear energy, but we have an equal incentive to reduce both the frequency and consequences of malfunction.

Having stated this general criterion, we face the much more difficult question of how much safer reactors must be in order to restore public confidence in nuclear energy. One view, put forward only partly facetiously, is that some reactor accidents are necessary so that the public can place in proper context the real, as opposed to the imagined, hazards of nuclear energy. Thus as far as mortality and morbidity go, Chernobyl is probably a smaller incident than was Bhopal, or the 1963 Vajont Dam failure in Italy. A case can be made for even the worst nuclear accident being no worse than the worst dam or chemical plant failure; since the latter are now "accepted" by the public, the former should be also—provided we actually experience some reactor disasters.

I do not accept this line of reasoning. Supporters of nuclear power must accept the idea that the nuclear hazard is somehow different, that the notion of interdicting land with an unseen agent is viewed by the public as particularly threatening. In short, Chernobyl demands that we reduce the frequency and consequences of accidents to the point where they do not evoke the near-panic we have recently witnessed.

We must distinguish between reactors already on line and new reactors, possibly of different design, that might be built in the future. A central question is the relevance of the experience at Chernobyl to the Western world's light-water reactors—in particular, whether the massive, and quite leak-tight, containment structures around such Western reactors would have mitigated the consequences of a Chernobyl-like accident. Although important arguments have been made that the Western containments are more substantial than the Chernobyl containment, I am inclined to await a detailed comparison before answering this question. The Western containments certainly appear to be stouter, and a graphite fire is impossible in a light-water reactor. On the other hand, it would be claiming too much to insist an accident in a light-water reactor that breaches a Western containment is impossible. The most that can be said is that the likelihood of such an accident appears to be extremely small, probably much smaller than in a Chernobyl-type reactor.

As for the safety of plants already on line in the United States and many other countries, much has been done in the wake of Three Mile Island: operating procedures have been modified, the Institute for Nuclear Power Operation—an oversight agency operated by the utility industries—is in full gear, and various pieces of hardware have been improved. Minarick and Kukiela have analyzed precursor events in operating reactors that, had they not been aborted by human or mechanical intervention, could have led to a core-melt.[*] Their semi-empirical estimate of the core-melt probability in 1969–1979 was 0.001 per reactor year (in good agreement with the Three Mile Island experience). This probability in 1980–1981 fell to 0.00015 per reactor year. Based on this analysis, the probability of a core-melt between now and 2000 in one of the 100 U.S. light-water reactors would be about 20 percent. This number can be disputed. D. Phung suggests that on average the core-melt probability in U.S. reactors today is close to Rasmussen's 0.00005 per reactor year or three times lower than the aforementioned estimate.[†] Newer modifications of light-water reactors, such as Westinghouse's advanced pressurized-water reactor and General Electric's boiling-water reactor, designed jointly with Japanese reactor vendors, are estimated to be even safer. Their core-melt probabilities are estimated to lie in the range of 0.000001 to 0.0000001 per reactor year.

No one can say whether the safety goals, if achieved, will be sufficient to restore public confidence in nuclear energy. If one accepts Phung's probability estimate, the chance of a meltdown in a U.S. reactor by 2000 is about one in 12. How much this has been lowered by better operating procedures is unknown. Should a light-water reactor suffer a core-melt during the next 15 years, the public's reaction probably will depend upon how much radio-

[*] J. W. Minarick and C. A. Kukiela, "Precursors to Severe Core-Damage Accidents 1969–79," prepared for the Nuclear Regulatory Commission (NUREG-CR-2797); idem. "Precursors to Severe Core-Damage Accidents, 1980–81" (NUREG-CR-3591).

[†] A. N. Weinberg et al., *The Second Nuclear Era* (New York: Praeger, 1985), chap. 13.

activity is released. Following Three Mile Island, where very little radioactive iodine, cesium, and strontium were released, many in the nuclear community were confident that such low release of fission products was a characteristic of water reactors. Chernobyl, with its huge release, raises questions about the validity of this belief. Clearly we cannot draw the full technical implications of Chernobyl until we have a complete investigation of the incident. What we do have is an impression of the social dislocations caused by the accident, and three dislocations are large.

The ultimate question is whether we can accept a technology whose safety is measured probabilistically. Members of the nuclear community had always assumed that if the probability of a severe accident was sufficiently low, nuclear power would be accepted—even if the consequences of unlikely accidents, especially the social dislocations, were very large. This has traditionally been society's attitude toward technological hazards. If a technology confers an important benefit, society has accepted both the technology and the risk, the latter always being probabilistic. We accept the chemical industry despite Bhopal; we accept dams despite Vajont; we accept fossil fuel despite the greenhouse effect. Why should we not accept nuclear energy despite Chernobyl?

In a democracy such as in the United States, the answer to this question can only be given by the public. Although most nuclear technologists can point to the differences between the Chernobyl reactor and contained light-water reactors, and to the low estimated probabilities of serious accidents, one cannot say whether the public will ultimately be persuaded that the nuclear energy is acceptable.

The improvements promised by the advanced U.S.-Japanese light-water reactors are surely impressive. For example, were the 100 reactors now on line in the United States replaced by reactors with core-melt probability of 0.000001 per reactor year, one could expect an accident no more frequently than every 100 years. Moreover, most meltdowns, at least in contained reactors, ought to release little radioactivity to the public. Yet the safety of such reactors is in the final analysis a matter of probability: even though the probability is very low, an accident can happen.

Can reactors be designed for which the probability of a serious accident is zero—that is, a reactor whose safety depends not on active interventions of safety systems (which with some probability can fail), but rather on physical principles that insure the reactor's safety without active intervention? Were such an inherently safe reactor actually demonstrated, could the public be convinced of its safety? And could such a reactor be the basis for a second nuclear era no longer tormented by profound public disaffection?

David Lilienthal, the first chairman of the Atomic Energy Commission, proposed exactly this in the wake of Three Mile Island. He called on nuclear technologists to develop an inherently safe reactor: one whose safety depended not on mechanical or human intervention, but rather upon immutable physical principles that even in an emergency could not be abrogated. Although the nuclear community at first regarded this challenge as impossible, its position has now changed. At least three different inherently safe reactors have received considerable attention: the Swedish Process Inherent Ultimately Safe Reactor (PIUS); the modular high temperature gas-cooled reactor; and the metal-fueled, sodium-cooled breeder. These reactors have been described elsewhere so I shall not repeat their description.* All of them introduce new inherent safety elements not incorporated in the current generation of reactors. The Swedish PIUS, in particular, appears to me to have as close to a zero probability of a disabling accident as I can imagine.

Since no new nuclear power reactor is likely to be built in the United States for, say, at least the next 15 years, it makes sense to use this time to develop a truly inherently safe reactor. Indeed I would go further: development of inherently safe reactors should be made an international project on which the United States and the Soviet Union would collaborate. I am suggesting that the world seize the opportunity offered by Chernobyl to take up David Lilienthal's challenge to develop one or more inherently safe reactors—devices so transparently safe that both friend and foe of nuclear energy will regard them as adequately safe.

The Chernobyl incident has internationalized the peaceful atom in a way that the world had not anticipated. Secretary Gorbachev's suggestion to strengthen the International Atomic Energy Agency's (IAEA) activities in reactor safety, particularly in the exchange of detailed information on near misses, is most welcome. These steps are important in helping assure that existing reactors will operate without incident. Even more striking is his suggestion, in a letter to the U.N. secretary general, that the IAEA sponsor international efforts to develop "a new generation of economical and reliable reactors with enhanced safety of operations compared to the existing reactors." A precedent for U.S.-Soviet cooperation in such an endeavor is the agreement reached at the November 1985 summit to collaborate on fusion research. Indeed, development of inherently safe reactors could be an agenda item at a future summit meeting.

For those who regard nuclear energy as an abomination that must be extirpated, such a course is unappealing. But to those who insist that the proper response to nuclear energy's deficiencies is to fix the deficiencies, I

*Ibid.

should think serious development of inherently safe reactors is a sensible and perhaps overdue goal.

Chernobyl has once more demonstrated, as did Three Mile Island, that a nuclear accident anywhere is a nuclear accident everywhere. We have yet to see the full impact of Chernobyl on the IAEA, which will almost surely strengthen its programs on nuclear safety. Up to now, the IAEA incident reporting system has been less thorough than has the system sponsored by the Organization for Economic Cooperation and Development, or by the U.S. Nuclear Regulatory Commission. Gorbachev's appeal to the United Nations to strengthen the international reactor safety regime, and particularly to confront the problem of terrorist threats against nuclear power plants, is surely a welcome move.

Bennett Ramberg has summarized the issue in several publications, most recently in the March 1986 *Bulletin*. He points out that, to some extent, the presence of nuclear reactors in belligerent countries has the potential of converting nonradiological war into radiological war. Whether this would serve to deter war (in a miniature replaying of nuclear deterrence) or would simply exacerbate the horrors of a war that such miniature deterrence had not prevented no one can say. On the other hand, these issues would become irrelevant if we gradually switched to inherently safe reactors. PIUS—contained in a 25-foot-thick prestressed concrete structure not unlike an ICBM silo—is all but immune from damage by conventional weapons. Its designers claim it to be invulnerable to all but a direct nuclear hit.

Can we escape the fate of our Faustian bargain? Can the world's nuclear technologists design reactors whose safety depends not upon mechanical interventions, but upon inherent, immutable laws of nature? And would the development of such reactors restore public confidence in nuclear energy?

When a few technologists first took up David Lilienthal's challenge, most of the nuclear community was skeptical about whether inherently safe reactors were feasible, whether their deployment would lead to demands to shut down existing reactors, and whether a technical fix would allay public opposition to nuclear energy. Today, the technical, if not the economic, feasibility of inherently safe reactors is conceded. As for the coexistence of conventional and inherently safe reactors, one can only point to the many propeller-driven DC-3s that are still flying, and are accepted, alongside the much safer jet-propelled commercial aircraft. Whether inherently safe reactors will restore confidence in nuclear energy is hard to say.

That inherently safe reactors are possible is a powerful revelation. Should nuclear energy founder in the wake of Chernobyl—as it may in several countries—I cannot believe that the world will forever foreswear fission.

The possibility of an inherently safe reactor, having been demonstrated theoretically, will tantalize future nuclear technologists. Eventually, some inherently safe reactors will be built and their inherent safety demonstrated. At that time we shall better know whether such devices can produce electricity economically, and whether their use will forever exorcise from the minds of a skeptical public the fear of having to pay for its Faustian bargain. In view of the uncertainties and risks associated with every source of energy, I would hope that this generation takes up David Lilienthal's challenge and devotes the necessary resources to develop inherently safe nuclear reactors. Nuclear power could then be part of the solution to the problems of acid rain and the atmospheric accumulation of carbon dioxide rather than a festering source of political conflict.

X

Afterword

We live in a world where nuclear weapons can literally mean the end of history. History is not so much what has happened as what is remembered, and the destruction of our species would thus eliminate the past in entirety. In these pages we have considered the history of the American atom and thereby the process whereby we have achieved the terrible power that was only imagined in the apocalyptic premonitions of our ancestors: the power of self-destruction and the ability to end history.

In the meantime we muddle through. Living with nuclear energy involves both technology and human choice, necessity and freedom. The necessity involves three fairly intractable realities with which we must live: first, that nuclear weapons exist in the tens of thousands and nuclear reactors in the hundreds worldwide; second, that the human community continues to be organized in units of nation-states according to the principle of national sovereignty; third, that Soviet-American relations are founded on a doctrine of mutual deterrence aimed at avoiding nuclear war.

From this perspective, we are unlikely to eliminate either the nuclear technology we have inherited or the human political framework and superpower relationship that currently exist. We are likely to have to live with and understand these realities to control them. Some choices have been made for us; others still remain.

The American atom is history, an inherited dilemma—how to reconcile national sovereignty and nuclear technology in a manner that allows us to control the dangerous machines that we have historically considered to be essential to our national security and productive well being. Nuclear energy is not the product of a conspiracy, despite its legacy of secrecy. In 1945 an overwhelming majority of Americans supported dropping the first nuclear weapons on Japan; in the 1950s and 1960s a majority supported the development of nuclear power to satisfy the national appetite for electrical energy and its attendant home conveniences.

Thus we may object to the American atom, but we cannot deny our historical choice to benefit from its power in time of both war and peace. The purpose of this volume has been to delineate how that choice was made and what its consequences have been, so that we may choose wisely in a technological future of both risks and benefits. Without question, the American atom is a vital part of our history. It is also the greatest challenge to our future survival as a species on this planet. We must master its past if we are to control its future.

Index

This book has been set in Linotron Caledonia and ITC Avant Garde Gothic.

Caledonia was designed by W. A. Dwiggins in 1938 for the Merganthaler Linotype Company. Dwiggins designed Caledonia based upon the strengths he admired in the Scotch roman typefaces.

ITC Avant Garde Gothic was designed by Lubalin and Carnese in the 1970's for the International Typeface Corporation.

Printed on acid-free paper.